Springer Proceedings in Mathematics & Statistics

Volume 376

This book series features volumes composed of selected contributions from workshops and conferences in all areas of current research in mathematics and statistics, including operation research and optimization. In addition to an overall evaluation of the interest, scientific quality, and timeliness of each proposal at the hands of the publisher, individual contributions are all refereed to the high quality standards of leading journals in the field. Thus, this series provides the research community with well-edited, authoritative reports on developments in the most exciting areas of mathematical and statistical research today.

More information about this series at https://link.springer.com/bookseries/10533

P. Balasubramaniam · Kuru Ratnavelu ·
Grienggrai Rajchakit · G. Nagamani
Editors

Mathematical Modelling and Computational Intelligence Techniques

ICMMCIT-2021, Gandhigram, India
February 10–12

Springer

Editors
P. Balasubramaniam
Department of Mathematics
The Gandhigram Rural Institute (Deemed to be University)
Dindigul, Tamil Nadu, India

Kuru Ratnavelu
Institute of Computer Science and Digital Innovation
UCSI University
Kuala Lumpur, Malaysia

Grienggrai Rajchakit
Department of Mathematics
Faculty of Science
Maejo University
Chiang Mai, Thailand

G. Nagamani
Department of Mathematics
The Gandhigram Rural Institute (Deemed to be University)
Dindigul, Tamil Nadu, India

ISSN 2194-1009 ISSN 2194-1017 (electronic)
Springer Proceedings in Mathematics & Statistics
ISBN 978-981-16-6017-7 ISBN 978-981-16-6018-4 (eBook)
https://doi.org/10.1007/978-981-16-6018-4

Mathematics Subject Classification: 97M10, 03E72, 05C50, 34K20, 34K35, 47H10, 65K10, 68U10, 92B20

This Springer imprint is published by the registered company Springer Nature Singapore Pte Ltd.
The registered company address is: 152 Beach Road, #21-01/04 Gateway East, Singapore 189721, Singapore

Preface

Mathematical modelling is an activity by which a problem involving the real world is translated into mathematics to form a model, which can then be used to provide information about the original real problem. Computational science, also known as scientific computing or scientific computation, is a rapidly growing field that uses advanced computing capabilities to understand and solve complex problems. It is an area of science that spans many disciplines, but at its core, it involves the development of models and simulations to understand natural systems.

Mathematical modelling and computational intelligence techniques are promising, hot areas of current research and development, which can provide significant advantages to users. It also plays a vital role in mathematics as far as applications are concerned. Almost all mathematicians, industrialists, scientists, engineers, and researchers in science disciplines apply mathematical modelling and computing techniques. Many new concepts, methods, and algorithms have emerged frequently with many real-life applications in the past few decades.

Owing to these facts, the Department of Mathematics of The Gandhigram Rural Institute (Deemed to be University), Gandhigram, Dindigul, Tamil Nadu, India, has organized an International Conference on Mathematical Modelling and Computational Intelligence Techniques (ICMMCIT 2021) from 10–12 February 2021. ICMMCIT 2021 is intended to provide a common forum for researchers, scientists, engineers, and practitioners throughout the world to share their ideas, latest research findings, developments, and applications, including their links to mathematical modelling, computational sciences, information sciences, and so forth. This conference is a refereed conference emphasizing different mathematical modelling, computational intelligence techniques, and their science and engineering applications. It will focus on developing mathematical modelling, analysis, and applications from theoretical and numerical perspectives involving different applied sciences and engineering.

Based on the scientific committee's reviews that composed of 104 field experts from all over the world, we accepted only 97 papers for presentation at ICMMCIT 2021 out of 143 research papers submitted. Among these 97 papers, 43 papers were selected for the peer review process. Out of 30% of selected and presented papers, only 15% of the research contributions were finally shortlisted for publication. This volume comprises five parts consisting of 21 accepted papers after peer review. Mathematical modelling-related papers are presented in Part One. Part Two consists of papers related to image processing. Control theory papers are arranged in Part Three. Part Four consists of papers in graphs and networks. Finally, papers on inventory control are presented in Part Five.

Dindigul, India — P. Balasubramaniam
Kuala Lumpur, Malaysia — Kuru Ratnavelu
Chiang Mai, Thailand — Grienggrai Rajchakit
Dindigul, India — G. Nagamani

Acknowledgements

As the Conference Chairs of ICMMCIT 2021, we thank all the funding agencies for their grant support for completing the international conference successfully. ICMMCIT 2021 was supported in part by the following funding agencies.

- Council of Scientific and Industrial Research (CSIR), New Delhi, India.
- Tamil Nadu State Council for Science and Technology, Tamil Nadu, India.

We would like to express our sincere thanks to the Vice-Chancellor and the Registrar, The Gandhigram Rural Institute (Deemed to be University) (GRI-DTBU), Gandhigram, Tamil Nadu, India, for their motivation and support. We also extend our profound thanks to all faculty members and research scholars of the Department of Mathematics and the administrative staff members of GRI—DTBU, Gandhigram.

We especially thank the Honorary Chairs, Co-chairs, Technical Programme Chairs, Organizing Chair, and all the committee members who worked as a team by investing their time to make ICMMCIT 2021 a great success.

We are grateful to the keynote speakers who kindly accepted our invitation. Especially, we would like to thank Dr. Ong Seng Huat, Dr. P. Raveendran (UCSI University, Malaysia), Dr. Chee Pin Tan (Monash University, Malaysia), Dr. Er. Meng Joo (Dalian Maritime University, China), Dr. Mohammad Sajid (Qassim University, Saudi Arabia), Prof. Dr. S. Sanjeewa Nishantha Perera (University of Colombo, Sri Lanka), Dr. Raju K. George (Indian Institute of Space Science and Technology (IIST), India), Dr. K. Balachandran (Bharathiar University, India), Dr. R. Roopkumar (Central University of Tamil Nadu, India), Dr. K. Somasundaram (GRI-DTBU), Dr. Rajeswari Seshadri (Pondicherry University, India), Dr. R. Sakthivel (Bharathiar University, India), and Dr. S. Marshal Anthoni (Anna University Regional Campus, India) for presenting plenary talks and making ICMMCIT 2021 a grand event.

A total of 120 experts on various topics from all around the world reviewed the paper submissions. We express our greatest gratitude for spending their valuable time to review and sort out the papers for presentation at ICMMCIT 2021. We thank Springer for providing an excellent tool called online conference systems (OCS) for managing the papers.

Finally, we would like to thank Dr. Shamim Ahmad, Senior Editor and all other Editors, Mathematical Sciences, Springer India, and members of publishing team for processing our proposal to publish the papers in the Springer series.

July 2021

P. Balasubramaniam
Kuru Ratnavelu
Grienggrai Rajchakit
G. Nagamani

About This Book

This book is a collection of papers presented at the International Conference on Mathematical Modelling and Computational Intelligence Techniques (ICMMCIT 2021), held at the Department of Mathematics, The Gandhigram Rural Institute (Deemed to be University), Gandhigram, Tamil Nadu, India, from 10 to 12 February 2021. It was a prestigious event organized to provide an excellent international platform for the leading academicians, researchers, industrial participants, and budding students worldwide to discuss their research findings with global experts. Significant contributions from researchers worldwide in all major fields of applied analysis, mathematical modelling, and computing techniques have been received for this conference. The accepted papers are organized in topical sections as mathematical modelling, image processing, control theory, graphs and networks, and inventory control.

Contents

Image Processing

Control Theory and Its Applications

Graphs and Networks

About the Editors

P. Balasubramaniam is Professor at the Department of Mathematics, The Gandhigram Rural Institute, Gandhigram, Tamil Nadu, India, since 2006, where he joined as Lecturer in Mathematics in 1997. Earlier, he worked with several engineering colleges from 1994–1997. He earned his Ph.D. in Mathematics from the Department of Mathematics, Bharathiar University, Coimbatore, Tamil Nadu, India, in 1994, with specialization in control theory. He has worked as Visiting Research Professor during 2001 and 2005–2006 for promoting research in the field of control theory and neural networks at Pusan National University, Pusan, South Korea. He also has worked as Visiting Professor at the Institute of Mathematical Sciences, University of Malaya, Malaysia, from September 2011 to March 2012.

He has published more than 260 research papers in various international journals. He is a reviewer of many international journals and on the editorial board of some other journals. He is Editor-in-Chief of the journal *Modern Instrumentation* and Associate Editor of *Neural Processing Letters*. He has received the Tamil Nadu Scientist Award (TANSA) and Tamil Nadu Senior Scientist Award in Mathematical Sciences instituted by the Tamil Nadu State Council for Science and Technology in 2005 and 2018. Recently, he received the Mid-Career Award by the University Grants Commission, the Government of India, New Delhi, in 2019. He is recognized among the top 2% scientist in India, in the field of artificial intelligence and image processing by Stanford University, USA, in 2020. His research interest includes the areas of control theory, fractional differential equations, stochastic differential equations, soft computing, stability analysis, cryptography, neural networks, image processing, and reaction–diffusion equations.

Kuru Ratnavelu is Director of the Institute of Computer Science and Digital Innovations, UCSI University, Kuala Lumpur, Malaysia. Previously, he has served as Deputy Dean of Science and Deputy Vice-Chancellor (Development) and Head of UM Strategic Planning Unit under Vice-Chancellor at the University of Malaya. He played a key role in the High Impact Research Program at the University of Malaya. He obtained his first-class honours in B.Sc. in 1982 and his M.Sc. by research in 1985, at University Malaya. He earned his Ph.D. in atomic physics from Flinders University with a Flinders Research Scholarship in 1990. His recent international recognitions are Council Member of the Association of Asia-Pacific Physical Societies (2016–2019 and 2020–2023).

With more than 30 years of academic and research experience in Malaysia, his outstanding research and academic achievements was recognized, and then, he was awarded the Malaysian Toray Science Foundation Science and Technology Award in 2004 as well as the Young Scientist Award (Strategic Sector) in 1996 by MOSTE. He was elected as Fellow of the Academy of Sciences Malaysia in 2005, Fellow of Institut Fizik Malaysia in 2000. In 2018, he was awarded the 19th Fellow of Persatuan Sains Matematik Malaysia. Internationally, he was awarded the Inaugural Distinguished Alumni Award by Flinders University for his contributions to atomic physics as well as education in Malaysia in 2006. In 2000–2001, his sabbatical at the University of Oklahoma was supported by the US NRC Grant. His professional contribution to Malaysia has been the development of physics and mathematics in Malaysia since the 1990s in his capacity as the Honorary Secretary of the Institut Fizik Malaysia (1996 until now) and Malaysian Mathematical Society (PERSAMA), respectively. His primary interest has been in theoretical atomic collision processes. He has focused on complex and social network analysis and optimal theory.

Grienggrai Rajchakit is Lecturer at the Department of Mathematics, Faculty of Science, Maejo University, Chiangmai, Thailand. He received his B.S. (Mathematics) degree from Thammasat University, Bangkok, Thailand, and his M.S. (Applied Mathematics) degree from Chiangmai University, Chiangmai, Thailand, respectively in 2003 and 2005. He earned his Ph.D. (Applied Mathematics) degree from the King Mongkut's University of Technology Thonburi, Bangkok, Thailand, in the area of stability and control of neural networks. He is the recipient of the Thailand Frontier Author Award by Thomson Reuters Web of Science in 2016 and TRF-OHEC-Scopus Researcher Awards by The Thailand Research Fund (TRF), Office of the Higher Education Commission (OHEC) and Scopus in 2016. His research interests include complex-valued neural networks, complex dynamical networks, control theory, stability analysis, sampled data control, multi-agent systems, T-S fuzzy theory, and cryptography. He is a reviewer for various reputed international journals. He has authored and co-authored more than 133 research articles for various international journals.

G. Nagamani is Assistant Professor at the Department of Mathematics, The Gandhigram Rural University (Deemed to be University), Gandhigram, Tamil Nadu, India, since November 2011. She served as Lecturer in Mathematics at Mahendra Arts and Science College, Namakkal, Tamil Nadu, India, from 2001–2008. She has post-graduated from the Department of Mathematics of Sri Sarada College for women, Salem, Tamil Nadu, India, in 1997, and earned her M.Phil. from the Department of Mathematics, Bharathiar University, Coimbatore, Tamil Nadu, India, in 1999, and did her Ph.D. in Mathematical Sciences with a specialized area of passivity analysis of neural networks with time-varying delays from the Department of Mathematics, The Gandhigram Rural University (Deemed to be University), Gandhigram, Tamil

Nadu, India, in 2011. She has received the Tamil Nadu Young Women Scientist Award (TYWSA) from Tamil Nadu State Council for Science and Technology, in 2012, for her remarkable research contribution in the field of mathematical sciences. Her research interest includes modelling of stochastic differential equations, neural networks, dissipativity and passivity analysis, and control theory.

Mathematical Modelling

Application of Optimal Controls on Dengue Dynamics—A Mathematical Study

Ananya Dwivedi, Vinod Baniya, and Ram Keval

Abstract In the present article, we proposed an epidemiological SIR model of dengue fever with two types of controls and investigated their nature with human and mosquitoes. The stability of equilibrium points has been discussed using the Routh–Hurwitz criterion and the Lyapunov function. We use sensitivity analysis to quantify how variation in our model parameters affects our result and the total number of infective populations. With the help of Pontryagin's minimum principle's, we formulated the optimal control problem, and the existence of optimal solutions is shown graphically. Also, the method of controls is consolidated to detract infected individuals. Therefore, the entire calculation is to minimize the spread of disease via optimization technique as a guideline.

Keywords Dengue fever · Epidemiological model · Optimal control · Sensitivity analysis · Fourth-Order Runge–Kutta method

1 Introduction

Dengue fever is a mosquito-borne disease spread by infected female Aedes aegypti or Aedes albopictus mosquitos, largely found in tropical and subtropical climates, and is caused by four closely related dengue serotypes (DENV 1-4). Dengue fever can affect people of all age groups. Its symptoms appear 3–14 days after bitten by an infected mosquito. While recovery with one dengue serotype, one can become completely resistant to that serotype, but the other three serotypes can still affect you. Dengue control policy in India is difficult due to the country complicated healthcare system. In the last few years, an increase in dengue infections in India has necessitated dengue control policies. It is important to provide prior knowledge of the effects of control measures to implement them in the future. Before 1970, only nine countries confirmed the virus dengue. More than 100 countries have survived this infection in their tropical

A. Dwivedi · V. Baniya · R. Keval (✉)
Department of Mathematics and Scientific Computing, Madan Mohan Malaviya University of technology, Gorakhpur, U.P., India

P. Balasubramaniam et al. (eds.), *Mathematical Modelling and Computational Intelligence Techniques*, Springer Proceedings in Mathematics & Statistics 376,
https://doi.org/10.1007/978-981-16-6018-4_1

and subtropical regions of the world. According to an observation, every year, 50–100 million people infected with dengue virus. Also, from World Health Organization (WHO), dengue fever has been declared as the fastest communicable mosquito-borne disease in terms of human morbidity and mortality. So, timely treatment of infected persons is a crucial decision to control the eradication of dengue fever. Early detection and adequately taking the treatment can reduce the risk of uncertain death and medical complications. In India, this disease is endemic, and it has increase's rapidly in many states of India [1].

Several mathematical models have been proposed and analyzed to study control of dengue disease dynamics. In [2], authors developed single and two-serotype dengue mathematical models that were validated against dengue data from Kupang, Indonesia, to investigate the effect of vaccination on dengue transmission dynamics. In addition, the effects of vaccination on seronegative and seropositive individuals were investigated, as well as a global sensitivity analysis to assess the model's most influential parameters. Iboi et al. [3] designed a mathematical model and used it to analyzed the impact of Dengvaxia vaccine on the transmission dynamics. Kar et al. [4–6] analyzed a vector-borne disease model and applied three various control parameters: vaccination, treatment, and insecticide. Also, we used a different possible combination of controls, and their results are collated with numerical simulation. Bashier et al. [7] considered two optimal control strategies, i.e., vaccination and treatment, delay differential equation with the SIR model. Recher et al. [8] assessed dengue vaccination impact on the model and also reduced the dengue mortality rate by the use of baseline at least 50% by 2020. Srivastav et al. [9] assessed the impact of treatment on dengue fever dynamics based on a case study in India. Rodrigues et al. [10] discussed a model with the use of vaccine control. Also, we formulated the optimal control policies to minimize the cost of interventions. Hamdan et al. [11] developed a deterministic mathematical model of the dengue transmission by considering the effect of temperature and hospitalization on the dengue dynamics. Khan et al. [12] discussed a dengue transmission model with hospitalization and estimated the model using the confirmed notified cases of the East Java Province, Indonesia, for 2018. The optimal control problem is solved numerically, and the results comprised of control system for different strategies.

In this article, a dengue SIR model is proposed, and it uses the two different optimal control approaches to determine the optimal strategy against dengue. Two controls, namely drug therapy and education control, are used cumulatively for the eradication of disease. The basic mathematical results are evaluated, and a global sensitivity analysis is performed to determine the most influential parameters of the model.

We give brief details about the mathematical modeling formulation and description of dengue in Sect. 2. Some fundamental properties of the model and their stability will be explored in Sect. 3. We then deal with the sensitivity analysis of the model in Sect. 4. The formulation of an optimal control problem and the associated results is shown in Sect. 5 also, we briefly discuss the numerical results with different costs. Concluding remarks will be provided in Sect. 6.

2 Mathematical Formulation and Its Description

In this section, we briefly describe the mathematical model which we discuss in this article. The model assumption consists of human and mosquito population in different compartments. This human and mosquitoes have been sub-divided into four and two sub-classes, respectively. Here, N_H be the total human population, at time $t > 0$, and there are $S_{H_1}(t)$ susceptible human at high risk and $S_{H_2}(t)$ susceptible human at low risk, $I_H(t)$ infected human, $R_H(t)$ immune (recovered) human, $S_m(t)$ susceptible mosquitoes, $I_m(t)$ infected mosquitoes. Immune (recovered) class does not exist in case of mosquito population. Let k be the total constant recruitment rate of mosquito population at time t. Total number of natural death rate of mosquitoes and human are δ and μ_H, respectively. It is admitted that the flow diagram in Fig. 1 from the susceptible to infected class, for each species, depends on the biting rate of mosquito, the transmission probabilities, as well the number of infected and susceptible of each species. Let β be denoted the biting rate of mosquito, which is the average number per mosquito per day, where m is number of hosts available as blood sources. α_H and α_m be the transmission potential from the human to mosquito and mosquito to human.

- Probability of human that mosquitoes choose humans is given by $\frac{N_H}{N_H+m}$
- Probability of human receives bites per unit time is given by $\frac{\beta N_m}{N_H}\frac{N_H}{N_H+m}$
- Probability of mosquitoes receives blood from human as meal per unit time is given by $\frac{\beta N_H}{N_H+m}$
- The infection rate per susceptible human is given by $\beta\alpha_H\frac{N_m}{N_H}\frac{N_H}{N_H+m}\frac{I_m}{N_m} = \frac{\beta\alpha_H}{N_H+m}I_m$
- The infection rate per susceptible mosquito is given by $\beta\alpha_m\frac{N_H}{N_H+m}\frac{I_H}{N_H} = \frac{\beta\alpha_H}{N_H+m}I_H$

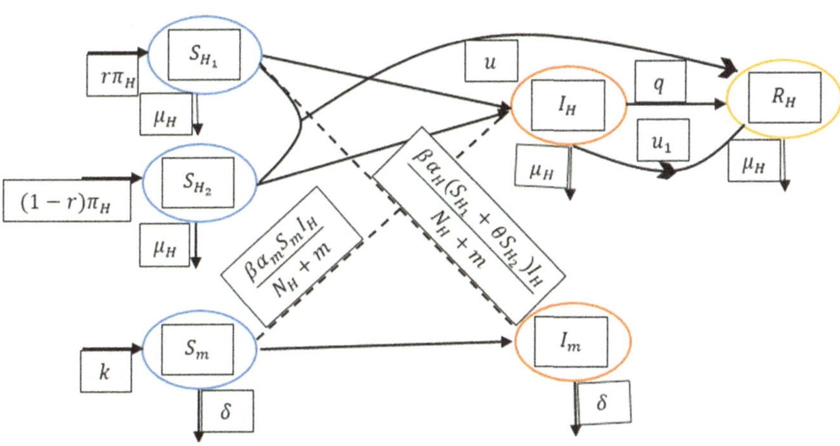

Fig. 1 Transmission diagram of dengue

At any time t, let $r\pi_H$ be constant rate in the region where r be the fraction of newly recruited population and join susceptible class. The persons in class S_{H_2} are less contact than S_{H_1} class person. It is assumed that θ is the relative chance of infection of low risk susceptible to high risk susceptible. Now, we applied control u $(0 \le u \le 1)$ as education campaigns to the susceptible individuals and u_1 $(0 \le u_1 \le 1)$ gives the drug therapy to the infected individuals; because of this, human goes to recovered class. Under these above assumptions, a dengue dynamical model has been developed as follows in Table 1:

$$N_H = S_{H_1}(t) + S_{H_2}(t) + I_H(t) + R_H(t)$$
$$N_m = S_m(t) + I_m(t)$$

Therefore, by the above assumptions, mathematical formulations for dengue transmission model are:

$$\frac{dS_m}{dt} = k - S_m\left(\frac{\beta\,\alpha_m I_H}{N_H + m}\right) - \delta S_m \tag{1}$$
$$\frac{dI_m}{dt} = S_m\left(\frac{\beta\,\alpha_m I_H}{N_H + m}\right) - \delta I_m \tag{2}$$
$$\frac{dS_{H_1}}{dt} = r\,\pi_H - S_{H_1}\left(\frac{\beta\alpha_H I_m}{N_H + m}\right) - (\mu_H + u)S_{H_1} \tag{3}$$
$$\frac{dS_{H_2}}{dt} = (1 - r)\pi_H - S_{H_2}\left(\frac{\beta\theta\alpha_H I_m}{N_H + m}\right) - (\mu_H + u)S_{H_2} \tag{4}$$
$$\frac{dI_H}{dt} = (S_{H_1} + \theta\, S_{H_2})\left(\frac{\beta\alpha_H I_m}{N_H + m}\right) - (q + \mu_H + u_1)I_H \tag{5}$$
$$\frac{dR_H}{dt} = u(S_{H_1} + S_{H_2}) + (q + u_1)I_H - \mu_H R_H \tag{6}$$

The system (1–6) can also be written as:

$$\frac{dS_m}{dt} = k - p_{11}S_m I_H - \delta S_m \tag{7}$$
$$\frac{dI_m}{dt} = p_{11}S_m I_H - \delta I_m \tag{8}$$
$$\frac{dS_{H_1}}{dt} = A - p_{12}S_{H_1}I_m - (\mu_H + u)S_{H_1} \tag{9}$$
$$\frac{HS_{H_2}}{Ht} = B - p_{12}\theta S_{H_2}I_m - (\mu_H + u)S_{H_2} \tag{10}$$
$$\frac{dI_H}{dt} = p_{12}(S_{H_1} + \theta\, S_{H_2})I_m - (q + \mu_H + u_1)I_H \tag{11}$$
$$\frac{dR_H}{dt} = u(S_{H_1} + S_{H_2}) + (q + u_1)I_H - \mu_H R_H \tag{12}$$

Table 1 Depiction of non-negative parameters

Parameters	Depiction
k	Total constant recruitment rate of mosquito population
δ	Natural death rate of mosquito
μ_H	Natural death rate of host
r	Newly recruited population and join susceptible class
π_H	Total recruitment rate of human population
q	Recovery rate of disease
β	Biting rate of mosquito which defines the average number of bites per day
u	Education control ratio per unit of time $0 \leq u \leq 1$
θ	Relative chance of infection of low risk susceptible to high risk susceptible
α_H & α_m	Transmission potential from the host to mosquito and mosquito to human
u_1	Drug therapy treatment to infected human $0 \leq u_1 \leq 1$

where

$$p_{11} = \left(\frac{\beta\alpha_m}{N_H + m}\right), \quad p_{12} = \left(\frac{\beta\alpha_H}{N_H + m}\right), \quad A = r\pi_H, \quad B = (1-r)\pi_H$$

with initial conditions $S_m(0) = S_m^0 > 0$, $I_m(0) = I_m^0 \geq 0$, $S_{H_1}(0) = S_{H_1}^0 > 0$, $S_{H_2}(0) = S_{H_2}^0 > 0$, $I_H(0) = I_H^0 \geq$ and $R_H(0) = R_H^0 \geq 0$ at time $t = 0$. All parameters in the system (7–12) are assumed to be non-negative and also the initial values taken to be non-negative values. The full descriptions of parameters are given in Table 1.

2.1 Positivity and Boundedness of the System

In this section, we discuss the positivity and boundedness of system (7–12) for mosquito and human population. For this, we have

$$\frac{dS_m}{dt}|_{S_m=0} = k > 0, \quad \frac{dI_m}{dt}|_{I_m=0} = p_{11}S_m I_H \geq 0, \quad \frac{dS_{H_1}}{dt}|_{S_{H_1}=0} = A > 0,$$

$$\frac{dS_{H_2}}{dt}|_{S_{H_2}=0} = B > 0, \quad \frac{dI_H}{dt}|_{I_H=0} = p_{12}(S_{H_1} + \theta S_{H_2}) > 0,$$

$$\frac{dR_H}{dt}|_{R_H=0} = u(S_{H_1} + S_{H_2}) + (q + u_1)I_H \geq 0.$$

Here, all the variables are non negative, so if the non-negative cone $\mathbb{R}_{6+}$ is in interior of domain, all the state variables are inward of domain in the vector field direction. Therefore, the system is non-negative in the bounded domain.

Theorem 1 *Let* $(S_m > 0, I_m > 0, S_{H_1} > 0, S_{H_2} > 0, I_H > 0, R_H > 0)$ *be the solution of the system (7)–(12) with the initial conditions* $(S_m^0, I_m^0, S_{H_1}^0, S_{H_2}^0, I_H^0, R_H^0)$ *are bounded in the domain D. where,* $D = \{(S_m, I_m, S_{H_1}, S_{H_2}, I_H, R_H) \in \mathbb{R}_+^6, (L_1(t), L_2(t)) \leq \left(\frac{k}{\delta}, \frac{\pi_H}{\mu_H}\right)\}$

Proof Let us construct a Lyapunov function of the form:

$$Ł(t) = (Ł_1(t), \quad Ł_2(t))$$

where

$$Ł(t) = (S_m(t) + I_m(t), \quad S_{H_1}(t) + S_{H_2}(t) + I_H(t) + R_H(t))$$

For the derivative with respect to t,

$$\begin{aligned}
Ł'(t) &= (Ł_1'(t), \ Ł_2'(t)) \\
&= (S_m'(t) + I_m'(t), \quad S_{H_1}'(t) + S_{H_2}'(t) + I_H'(t) + R_H'(t)) \\
&= (k - p_{11} S_m I_H - \delta S_m + p_{11} S_m I_H - \delta I_m, \quad A - p_{12} S_{H_1} I_H - (\mu_H + u) S_{H_1} \\
&\quad + B - p_{12}\theta S_{H_2} I_m - (\mu_H + u) S_{H_2} + p_{12}(S_{H_1} + \theta S_{H_2}) I_m - (q + \mu_H + u_1) I_H \\
&\quad + u(S_{H_1} + S_{H_2}) + (q + u_1) I_H - \mu_H R_H) \\
&= (k - \delta N_m, \quad A + B - \mu_H N_H)
\end{aligned}$$

Now, from the above discussion, it is easy to prove the above equation as follows:

$$\begin{aligned}
Ł_1'(t) &= k - \delta N_m \leq \quad \text{for } Ł_1 \geq \frac{k}{\delta} \\
Ł_2'(t) &= A + B - \mu_H N_m \leq 0 \quad \text{for } Ł_2 \geq \frac{A + B}{\mu_H}
\end{aligned}$$

Therefore, from the above equation, one has that $Ł'(t) \leq 0$.

On solving the above system, we have

$$0 \leq \left(Ł_1, \ L_2) \leq (\frac{k}{\delta} + Ł_1(0) + e^{-(\delta t)}, \ \frac{A + B}{\mu_H} + Ł_2(0) + e^{-(\mu_H t)}\right)$$

where $Ł_1(0)$ and $Ł_2(0)$ are, respectively, the intial conditions of $Ł_1(t)$ and $Ł_2(t)$. Thus, as $t \to \infty, 0 \leq (L_1(t), L_2(t)) \leq \left(\frac{k}{\delta}, \frac{A+B}{\mu_H}\right)$ or $0 \leq (L_1(t), L_2(t)) \leq \left(\frac{k}{\delta}, \frac{\pi_H}{\mu_H}\right)$. Therefore, D is attraction set [13].

3 Model Analysis

We have the following system of ODE for analysis of our model

$$\frac{dN_m}{dt} = k - \delta N_m \tag{13}$$

$$\frac{dI_m}{dt} = p_{11}(N_m - I_m)I_H - \delta I_m \tag{14}$$

$$\frac{dS_{H_1}}{dt} = A - p_{12}S_{H_1}I_m - (\mu_H + u)S_{H_1} \tag{15}$$

$$\frac{dS_{H_2}}{dt} = B - p_{12}\theta S_{H_2}I_m - (\mu_H + u)S_{H_2} \tag{16}$$

$$\frac{dI_H}{dt} = p_{12}(S_{H_1} + \theta\ S_{H_2})I_m - (q + \mu_H + u_1)I_H \tag{17}$$

$$\frac{dR_H}{dt} = u(S_{H_1} + S_{H_2}) + (q + u_1)I_H - \mu_H R_H \tag{18}$$

For the model system (13–18), we found the disease-free equilibrium point $D_E = (N_m^0, I_m^0, S_{H_1}^0, S_{H_2}^0, I_H^0, R_H^0) = \left(\frac{k}{\delta}, 0, \frac{A}{\mu_H+u}, \frac{B}{\mu_H+u}, 0, 0\right)$. Then, by next-generation matrix method from [9, 15], we have F and V as follows:

$$F = \begin{pmatrix} \frac{\partial F_1}{\partial I_m} & \frac{\partial F_1}{\partial I_H} \\ \frac{\partial F_2}{\partial I_m} & \frac{\partial F_2}{\partial I_H} \end{pmatrix} \quad \text{and} \quad V = \begin{pmatrix} \frac{\partial V_1}{\partial I_m} & \frac{\partial V_1}{\partial I_H} \\ \frac{\partial V_2}{\partial I_m} & \frac{\partial V_2}{\partial I_H} \end{pmatrix}$$

$$F = \begin{pmatrix} 0 & p_{11}(N_m - I_m) \\ p_{12}(S_{H_1} + \theta S_{H_2}) & 0 \end{pmatrix} \quad \text{and} \quad V = \begin{pmatrix} \delta & 0 \\ 0 & (q + \mu_H + u_1) \end{pmatrix}$$

then $(F \cdot V^{-1})$ are defined as:

$$F \cdot V^{-1} = \begin{pmatrix} 0 & \frac{p_{11}k}{\delta(q+\mu_H+u_1)} \\ \frac{p_{12}(A+\theta B)}{\delta(\mu_H+u)} & 0 \end{pmatrix}$$

Here, the largest eigenvalue of the matrix $(F \cdot V^{-1})$ is defined as basic reproduction number R_0 and possessed as follows:

$$R_0 = \frac{p_{11}\ p_{12}\ k(A + \theta B)}{\delta^2(\mu_H + u)(q + \mu_H + u_1)}$$

3.1 Endemic Equilibrium Point Existence

For the model system (13–18), we find the endemic equilibrium point which satisfies the following conditions:

$$\frac{dN_m}{dt} = 0, \frac{dI_m}{dt} = 0, \frac{dS_{H_1}}{dt} = 0, \frac{dS_{H_2}}{dt} = 0, \frac{dI_H}{dt} = 0, \frac{dR_H}{dt} = 0$$

Then, there exist a unique solution for the model system (13)–(18); we have following non-negative solution $E_E = (N_m^*, I_m^*, S_{H_1}^*, S_{H_2}^*, I_H^*, R_H^*)$ where

$$\begin{aligned}
N_m^* &= \frac{k}{\delta}, I_m^* = \frac{p_{11}(N_m^* - I_m^*)I_H^*}{\delta},\\
S_{H_1}^* &= \frac{A}{p_{12}I_m^* + (\mu_H + u)}, S_{H_2}^* = \frac{B}{p_{12}\theta I_m^* + (\mu_H + u)},\\
I_H^* &= \frac{p_{12}(S_{H_1}^* + \theta S_{H_2}^*)}{(q + \mu_H + u_1)}, R_H^* = \frac{u(S_{H_1}^* + S_{H_2}^*) + (q + u_1)I_H^*}{\mu_H}
\end{aligned}$$

Here, I_m^* roots satisfy the following quadratic equation:

$$f(I_m^*) = A_{11}I_m^{*2} + A_{12}I_m^* + A_{13} = 0$$

where

$$\begin{aligned}
A_{11} &= p_{12}^2\,\theta\,(p_{11}(A + B) + \delta(\mu_H + q + u_1))\\
A_{12} &= p_{12}(p_{11}(\theta B(\mu_H - kp_{12}\theta + u)) + \delta(1 + \theta)(\mu_H + u)(q + \mu_H + u_1))\\
A_{13} &= (\mu_H + u)^2\,\delta^2\,(\mu_H + q + u_1)(1 - R_0)
\end{aligned}$$

For I_m^* to be positive, we need I_H^* to be less than $\frac{p_{11}k}{\delta^2}$, i.e., $0 < I_m^* < \frac{p_{11}k}{\delta^2}$. So, here A_{12} is positive for $(\mu_H + u) > kp_{12}\theta$, and from the above equation $f(I_m^*) = 0$, we get

$$f(0) = (\mu_H + u)\,\delta^2\,(\mu_H + u)\,(\mu_H + q + u_1)(1 - R_0)$$

and $f(\frac{p_{11}k}{\delta^2}) < 0$ when $R_0 < 1$, $f(0) < 0$ and $f(\frac{p_{11}k}{\delta^2}) < 0$, so there is possibility of two roots or no roots in $0 < I_m^* < \frac{p_{11}k}{\delta^2}$. Also, we see that the term A_{12} is negative if $(\mu_H + u) > kp_{12}\theta$; then, we can say that the quadratic equation has unique positive root. So after combining the above equation, we have at most two positive roots of the quadratic equation Table 2.

Table 2 Existence of +ve roots

Cases	A_{11}	A_{12}	A_{13}	R_0	No. of sign change	No. of +ve roots
1.		++		$+R_0 < 1$	0	0
2.		+-	-	$R_0 < 1$	1	1
3.		+-		$+R_0 < 1$	2	0,2
4.		++	-	$R_0 < 1$	1	1

Theorem 2 *The disease-free equilibrium point (D_E) is asymptotically stable if $R_0 < 1$ and is unstable otherwise.*

Proof The Jacobian matrix of the system (1–6) at non-trivial disease-free equilibrium point $D_E = (N_m^0, I_m^0, S_{H_1}^0, S_{H_2}^0, I_H^0, R_H^0) = \left(\frac{k}{\delta}, 0, \frac{A}{\mu_H+u}, \frac{B}{\mu_H+u}, 0, 0\right)$ is obtained as follows:

$$J_{D_E} = \begin{pmatrix} -\delta & 0 & 0 & 0 & 0 & 0 \\ 0 & -\delta & 0 & 0 & 0 & 0 \\ 0 & -\frac{p_{12}A}{(\mu_H+u)} & -(\mu_H+u) & 0 & 0 & 0 \\ 0 & -\frac{p_{12}\theta B}{(\mu_H+u)} & 0 & -(\mu_H+u) & 0 & 0 \\ 0 & p_{12}(A+\theta B) & 0 & 0 & -(q+\mu_H+u_1) & 0 \\ 0 & 0 & u & u & (q+u_1) & -\mu_H \end{pmatrix}$$

From Jacobian matrix of the system, we found two eigenvalues are $-\delta$, $-\mu_H$ and other roots are solve by the given bi-quadratic equation.

$$\lambda^4 + H_1\lambda^3 + H_2\lambda^2 + H_3\lambda + H_4 = 0$$

where

$$\begin{aligned} H_1 &= (2\delta + 2\mu_H + q + u + u1), \\ H_2 &= (\delta^2 + (\mu_H + u)(\mu_H + q + u1) + 2\delta(2\mu_H + q + u + u_1)), \\ H_3 &= \delta(2(\mu_H + u)(\mu_H + q + u1) + \delta(2\mu_H + q + u + u1)), \\ H_4 &= \delta^2(\mu_H + u)(\mu_H + q + u1) = \frac{p_{11}p_{12}k(A + \theta B)}{R_0} > 0 \end{aligned}$$

For $0 < R_0 < 1$. According to Routh–Hurwitz criteria, the above equation has negative roots if the following condition holds :

$$H_1 > 0, \quad H_0 > 0, \quad H_1H_2 - H_3 > 0, (H_1H_2 - H_3) - H_1^2H_4 > 0, H_4 > 0.$$

So, after satisfying all these conditions, D_E is locally asymptotically stable provided $R_0 < 1$.

Remark 2.1 According to the above theorem, if R_0 is less than unity, a small influx of infected mosquitoes/humans into the population will not result in large outbreaks, and the disease will ultimately die out.

Theorem 3 *The endemic equilibrium* $E_E = (N_m^*, I_m^*, S_{H_1}^*, S_{H_2}^*, I_H^*, R_H^*)$ *is locally asymptotically stable when the parameters of the model satisfy Routh–Hurwitz criteria stated in the proof of this theorem; otherwise, it is unstable.*

Proof The Jacobian matrix of the system (7-12) at non- trivial equilibrium point $E_E = (N_m^*, I_m^*, S_{H_1}^*, S_{H_2}^*, I_H^*, R_H^*)$ is obtained as follows:

$$J_{E_E} = \begin{pmatrix} -\delta & 0 & 0 & 0 & 0 & 0 \\ p_{11}I_H & -\delta - p_{11}I_H & 0 & 0 & 0 & 0 \\ 0 & -p_{12}S_{H_1} & -\mu_H - p_{12}I_v u & 0 & 0 & 0 \\ 0 & -p_{12}S_{H_2}\theta & 0 & -\mu_H - p_{12}I_v\theta u & 0 & 0 \\ 0 & p_{12}(S_{H_1} + \theta S_{H_2}) & 0 & 0 & -(q + \mu_H + u_1) & 0 \\ 0 & 0 & u & u & (q + u_1) & -\mu_H \end{pmatrix}$$

From Jacobian matrix of the system, we found two eigenvalues $-\delta$, μ_H and other roots are solve by the given biquadratic equation:

$$\lambda^4 + G_1\lambda^3 + G_2\lambda^2 + G_3\lambda + G_4 = 0$$

where,

$$G_1 = \delta + 3\mu_H + p_{11}I_H + p_{12}I_v + q + p_{12}I_v\theta + 2u + u_1.$$

$$\begin{aligned} G_2 = {} & 3\mu_H^2 + p_{11}p_{12}I_H I_v + p_{12}I_v q + p_{11}p_{12}\theta I_H I_v + \theta p_{12}^2 I_v^2 + \\ & \theta q p_{12}I_v + 2up_{11}I_H + up_{12}I_v + 2qu + \theta u p_{12}I_v + u^2 + u_1 p_{11}I_H + u_1 p_{12}I_v + \\ & u_1\theta p_{12}I_v + 2uu_1 + \delta(3\mu_H + q + p_{12}(1+\theta)I_v + 2u + u_1) + \mu_H(3I_H p_{11} + 2(q + p_{12}I_v(1+\theta) + 2u + u_1)). \end{aligned}$$

$$\begin{aligned} G_3 = {} & \mu_H^3 + I_H I_v p_{11}p_{12}q + I_H I_v^2 p_{11}p_{12}^2\theta + I_H I_v p_{11}p_{12}q\theta + I_v^2 p_{12}^2 q\theta + I_H I_v p_{11}p_{12}u \\ & + 2I_H p_{11}qu + I_v p_{12}qu + I_H I_v p_{11}p_{12}\theta u + I_v p_{12}q\theta u + I_H p_{11}u^2 + qu^2 + I_H I_v \\ & p_{11}p_{12}u_1 + I_H I_v p_{12}\theta u_1 + I_v^2 p_{12}^2\theta u_1 + 2I_H p_{11}uu_1 + I_v p_{12}uu_1 + I_v p_{12}\theta uu_1 + \\ & u^2 u_1 + \mu_H^2(3I_H p_{11} + q + I_v p_{12}(1+\theta) + 2u + u_1) + \delta(3\mu_H^2 + I_v^2 p_{12}^2\theta + \\ & I_v p_{12}(1+\theta)(q + u + u_1) + 2\mu_H(q + I_v p_{12}(1+\theta) + 2u + u_1 + u(2q + u + 2u_1)) \\ & \mu_H(I_v^2 p_{12}^2\theta + I_v p_{12}(1+\theta)(q + u + u_1) + 2I_H p_{11}(q + I_v p_{12}(1+\theta) + 2u + u_1) + u(2q + u + 2u_1)). \end{aligned}$$

$$G_4 = (\delta + p_{11}I_H)(\mu_H + I_H p_{12}u)(\mu_H + I_v p_{12}\theta + u)(\mu_H + q + u_1).$$

Therefore, by Routh–Hurwitz criterion, the above biquadratic equation has negative roots if the following conditions hold:

$$G_1 > 0, \quad G_4 > 0, \quad G_1G_2 - G_3 > 0, \quad G_1G_2G_3 - G_3^2 - G_4G_1^2 \; > \; 0$$

So, after satisfying the above conditions, E_E is locally asymptotically stable [9, 16].

4 Sensitivity Analysis

In this section, we discuss the sensitivity analysis of the model on varying, $\beta, \alpha_H, \alpha_m$, k and μ_H parameters. These key parameters are regulate on the basic reproduction number R_0. Where,

$$R_0 = \frac{\beta^2 \alpha_H \alpha_m k(A + \theta B)}{\delta^2(\mu_H + u)(q + \mu_H + u_1)(N_H + m)^2}$$

Sensitivity parameters are not able to quantify the relative change in a variable when a parameter changes. The standard forward sensitivity index of variable ψ with respect to parameter ξ is the proportion of the relative change in ξ. Thus, the normalized forward sensitivity is defined as [9, 14]:

$$\Gamma_{\xi}^{\psi} = \frac{\partial \psi}{\partial \xi} \times \frac{\xi}{\psi}$$

On taking $\psi = R_0$ and $\xi = \{\beta, \alpha_H, \alpha_m, k, \mu_H\}$, the partial derivative is shown below:

$$\begin{aligned}
\frac{\partial \psi}{\partial \beta} &= \frac{2k\beta(A + \theta B)\alpha_H \alpha_m}{(N_H + m)^2 \delta^2 (\mu_H + u)(\mu_H + q + u_1)} \\
\frac{\partial \psi}{\partial \alpha_H} &= \frac{k\beta^2(A + \theta B)\alpha_m}{(N_H + m)^2 \delta^2 (\mu_H + u)(\mu_H + q + u_1)} \\
\frac{\partial \psi}{\partial \alpha_m} &= \frac{k\beta^2(A + \theta B)\alpha_H}{(N_H + m)^2 \delta^2 (\mu_H + u)(\mu_H + q + u_1)} \\
\frac{\partial \psi}{\partial k} &= \frac{\beta^2(A + \theta B)\alpha_H \alpha_m}{(N_H + m)^2 \delta^2 (\mu_H + u)(\mu_H + q + u_1)} \\
\frac{\partial \psi}{\partial \mu_H} &= -\left[\frac{k\alpha\beta^2(A + \theta B)\alpha_H}{(N_H + m)^2 \delta^2 (\mu_H + u)(q + u_1 + \mu_H)^2} \right. \\
&\quad \left. + \frac{k\alpha_m \beta^2(A + \theta B)\alpha_H}{(N_H + m)^2 \delta^2 (\mu_H + u)^2 (q + u_1 + \mu_H)}\right]
\end{aligned}$$

Here, we see that some of the partial derivatives are positive and some are negative. Then, we can say that if the results are positive, the basic reproduction number R_0 increases. If the partial derivative is negative, then R_0 decreases with the parameters' increase. To see the effects and proportional changes, we calculate elasticities and the parameters from Table 3.

$$\Im_{\beta} = \frac{\beta}{R_0}\frac{\partial\psi}{\partial\beta} = \frac{2k\beta^2(A+\theta B)\alpha_H\alpha_m}{R_0(N_H+m)^2\delta^2(\mu_H+u)(\mu_H+q+u_1)} = 2.00 > 0$$

$$\Im_{\alpha_H} = \frac{\alpha_H}{R_0}\frac{\partial\psi}{\partial\alpha_H} = \frac{\alpha_H k\beta^2(A+\theta B)\alpha_m}{R_0(N_H+m)^2\delta^2(\mu_H+u)(\mu_H+q+u_1)} = 0.9999 > 0$$

$$\Im_{\alpha_m} = \frac{\alpha_m}{R_0}\frac{\partial\psi}{\partial\alpha_m} = \frac{\alpha_m k\beta^2(A+\theta B)\alpha_H}{R_0(N_H+m)^2\delta^2(\mu_H+u)(\mu_H+q+u_1)} = 0.9999 > 0$$

$$\Im_{k} = \frac{k}{R_0}\frac{\partial\psi}{\partial k} = \frac{k\beta^2(A+\theta B)\alpha_H\alpha_m}{(N_H+m)^2\delta^2(\mu_H+u)(\mu_H+q+u_1)} = 0.9999 > 0$$

$$\Im_{\mu_H} = \frac{\mu_H}{R_0}\frac{\partial\psi}{\partial\mu_H} = -\frac{k\alpha_m\alpha_H\beta^2(A+\theta B)\mu_H(q+u+u_1+2\mu_H)}{R_0(N_H+m)^2\delta^2(\mu_H+u)^2(q+u_1+\mu_H)^2} = -0.05632 < 0$$

From the above expression, we see that $\Im_k = \Im_{\alpha_H} = \Im_{\alpha_m}$ which implies small changes in any of these parameters will have some impact on R_0. In Fig. 2, we have plotted contour plot of some key parameters on R_0. In plot (2a) and (2b), it is clear that the transmission probability α_H, α_m and biting rate β same effect on R_0, which means if we increase or decrease the value of α_m, α_H and β then the value of R_0 will increase or decrease respectively. From the Fig. (2c) and (2d) we can conclude that increase the value of k will increase the value of the R_0 but if we increase the value of μ_H, it will decrease the value of R_0, which is the death rate of human, and this is one of the best control policy to reduce the disease infection from the population. In Fig. 3, the parameter sensitivity shows the variation and plots the sensitivity indices of R_0 with respect to the parameters of interest.

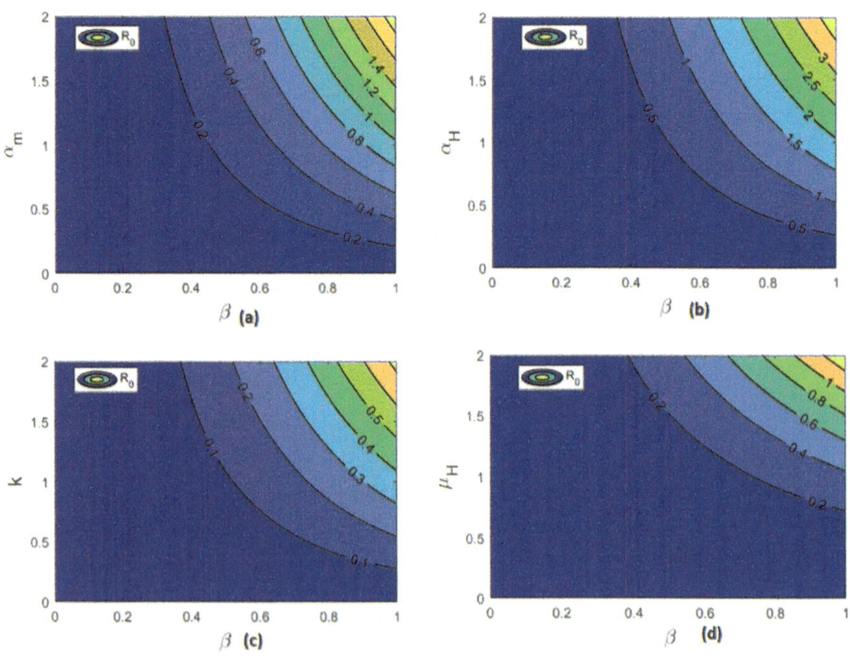

Fig. 2 Contour plots of R_0, with respect to the probability of disease transmission rate, death rate, recruitment rate and the biting rate of the mosquitoes and human

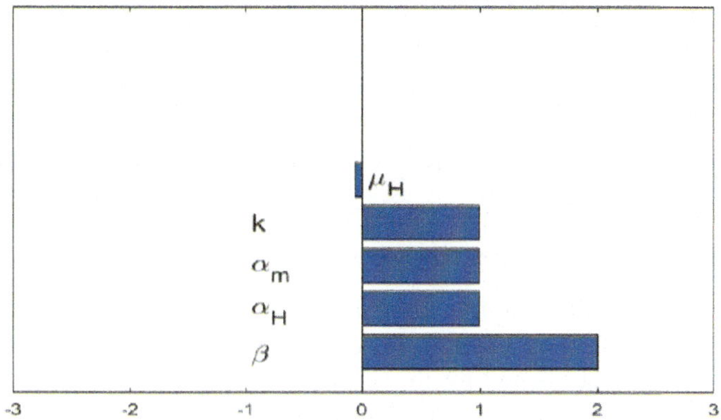

Fig. 3 Normalized forward sensitivity indices of R_0 with respect to model parameters. Parameter values:

Table 3 Parameter values

Symbol	Values	Dimension	Source
k	0–1000	–	Assumed
β	0.01	$per\ day^{-1}$	[9]
α_m	0.019	$per\ day^{-1}$	[9]
α_H	0.019	$per\ day^{-1}$	[9]
m	60	–	Assumed
δ	0.00002	$per\ day^{-1}$	[9]
u	$0 \leq u \leq 1$	–	Simulated
r	0.75	–	[18]
q	0.05977	$per\ day^{-1}$	[17]
u_1	$0 \leq u_1 \leq 1$	–	Simulated
θ	0.5	$per\ day^{-1}$	[18]
N_H	40	$per\ day^{-1}$	Assumed

5 Optimal Control Analysis

This section deals with dengue disease model to fix values of two methods of controls [4–6]. Let w_1 be weight constant and w_2, w_3 be positive constant aligned with the control square to balance the functional. Our aim is to minimize the infected human, cost of drug therapy, and education campaigns. We now use the fourth-order Runge–Kutta approach to solve the optimal system (1-6) in the finite time interval. With the help of the transversality and optimality conditions, we solve the state variables using the fourth-order Runge–Kutta forward method and the adjoint variables using the fourth-order Runge–Kutta backward method. (See, for details [4]). We define an objective functional to be minimized as follows:

$$J(u, u_1) = \min\left[\int_0^T (w_1 I_H + w_2 u^2 + w_3 u_1^2)\right] \mathrm{dt}$$

subject to the model (1–6). Our objective is to find controls (u^*, u_1^*) such that

$$J(u^*, u_1^*) = \min_{u, u_1 \in \Omega} J(u, u_1)$$

where $\Omega = u, u_1$: is Lebesque measurable and $0 \le u(t) \le 1$, $0 \le u_1(t) \le 1$ for $t \in [0, T]$ is the set for control space. Here, 0 and 1 are only bound for the controls. We can choose another bound for this interval. The square of the disease control parameter is taken to remove some unwanted side effects of the disease as well as to consider the over doses of the control [5].

Theorem 4 *: ∃ an optimal controls u^* and u_1^* that minimize $J(u, u_1)$ over the set Ω corresponding to the system (1)–(6). Then ∃ adjoint variables λ_i^s satisfying*

$$\frac{d\lambda}{dt} = -\frac{\partial H}{\partial V}.where V = (S_m, I_m, S_{H_1}, S_{H_2}, I_H, R_H)$$

with the transversality condition $\lambda_i(T) = 0$. The optimality condition are given as $\frac{\partial H^*}{\partial u} = 0$, $\frac{\partial H^*}{\partial u_1} = 0$

Furthermore, the controls u^* and u_1^* is given as

$$u^*(t) = \min\{1, \max[0, \hat{u}]\} \tag{19}$$

$$u_1^*(t) = \min\{1, \max[0, \hat{u_1}]\} \tag{20}$$

where $\hat{u} = \frac{(\lambda_3(t) - \lambda_6(t))S_{H_1} + (\lambda_4(t) - \lambda_6(t))S_{H_2}}{2w_2}$, $\hat{u_1} = \frac{(\lambda_5 - \lambda_6)I_H}{2w_3}$.

Proof The differential equations of the adjoint variables is obtained by the differentiation of the Hamiltonian function, $\frac{d\lambda}{dt} = -\frac{\partial H}{\partial V}$, where $V = (S_m, I_m, S_{H_1}, S_{H_2}, I_H, R_H)$. Thus, the adjoint system is given by

$$Ł = w_1 I_H + w_2 u^2 + w_3 u_1^2 + \lambda_1(t)\frac{dS_m}{dt} + \lambda_2(t)\frac{dI_m}{dt} + \lambda_3(t)\frac{dS_{H_1}}{dt} + \lambda_4(t)\frac{dS_{H_2}}{dt} + \lambda_5(t)\frac{dI_H}{dt} + \lambda_6(t)\frac{dR_H}{dt}$$

To minimize the Lagrangian L_1, let us consider the Hamiltonian of the problem as follows:

$$H^* = Ł + \lambda_1(t)\frac{dS_m}{dt} + \lambda_2(t)\frac{dI_m}{dt} + \lambda_3(t)\frac{dS_{HH_1}}{dt} + \lambda_4(t)\frac{dS_{H_2}}{dt} + \lambda_5(t)\frac{dI_H}{dt} + \lambda_6(t)\frac{dR_H}{dt}$$

where $\lambda_i(T)$, for $i = 1, 2, 3 \ldots 6$ are known as the adjoint variables, and it can be determined by solving the system of differential equation in the following form.

$$\lambda_1'(t) = -\frac{dH^*}{dS_m} = -\left\{-\left(\frac{\beta\alpha_m I_H}{N_H + m} + \delta\right)\lambda_1(t) + \left(\frac{\beta\alpha_m I_H}{N_H + m}\right)\lambda_2(t)\right\}$$

$$\lambda_2'(t) = -\frac{dH^*}{dI_m} = -\left\{-\left(\frac{\beta\alpha_H S_{H_1}}{N_H + m}\right)\lambda_3(t) - \left(\frac{\beta\theta\alpha_H S_{H_1}}{N_H + m}\right)\lambda_4(t) + \left(\frac{\beta\alpha_H (S_{H_1} + \theta S_{H_1})}{N_H + m}\right)\lambda_5(t)\right\}$$

$$\lambda_3'(t) = -\frac{dH^*}{dS_{H_1}} = -\left\{-\left(\frac{\beta\alpha_H I_m}{N_H + m} + (\mu_H + u)\right)\lambda_3(t) + \left(\frac{\beta\alpha_H I_m}{N_H m}\right)\lambda_5(t) + u\lambda_6(t)\right\}$$

$$\lambda_4'(t) = -\frac{dH^*}{dS_{H_2}} = -\left\{-\left(\frac{\beta\theta\alpha_H I_m}{N_H + m} + (\mu_H + u)\right)\lambda_4(t) + \left(\frac{\beta\theta\alpha_H I_m}{N_H m}\right)\lambda_5(t) + u\lambda_6(t)\right\}$$

$$\lambda_5'(t) = -\frac{dH^*}{dI_H} = -\left\{-\left(\frac{\beta\alpha_H S_m}{N_H + m}\right)\lambda_1(t) + \left(\frac{\beta\alpha_H S_m}{N_H m}\right)\lambda_2(t) - (q + \mu_H + u_1)\lambda_5(t) + (q + u_1)\lambda_6(t) + w_1\right\}$$

$$\lambda_6'(t) = -\frac{dH^*}{dR_H} = -\{-\mu_H \lambda_6(t)\}$$

Furthermore, differentiating the Hamiltonian function with respect to the control variable (u, u_1) to obtain:

$$\frac{\partial H^*}{\partial u} = -(\lambda_3(t) - \lambda_6(t))S_{H_1} - (\lambda_4(t) - \lambda_6(t))S_{H_2} + 2w_2 u(t) = 0$$

$$\frac{\partial H^*}{\partial u_1} = -(\lambda_5(t) - \lambda_6(t))I_H + 2w_3 u_1(t) = 0$$

Solving for u^* and u_1^*, we obtain:

$$u^* = \frac{(\lambda_3(t) - \lambda_6(t))S_{H_1} + (\lambda_4(t) - \lambda_6(t))S_{H_2}}{2w_2}, \quad u_1^* = \frac{(\lambda_5(t) - \lambda_6(t))I_H}{2w_3}.$$

Using the bounds of the control, we obtain the characterization given in Equation (19–20).

Figure 4 represents the proportion of dengue cases with and without control when we only give drug therapy to infected individuals. It shows that the proportion of dengue cases decreases with implementation of drug therapy. Furthermore, in Fig. 5, higher control (drug therapy and education campaigns) rate can be obtained if the cost of controls is cheaper. We use quadratic terminology in the control variables in the optimal control method to capture the nonlinear cost of control execution. This method is fairly standard and has been used in epidemiological modeling, including dengue modeling [2].

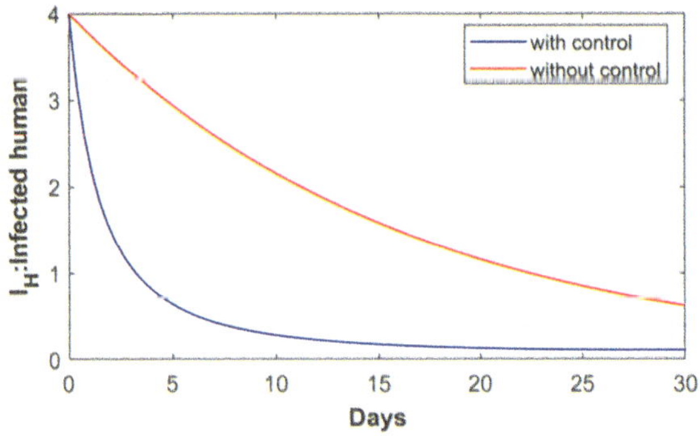

Fig. 4 Numerical simulations with and without control when the drug therapy given to only infected individuals

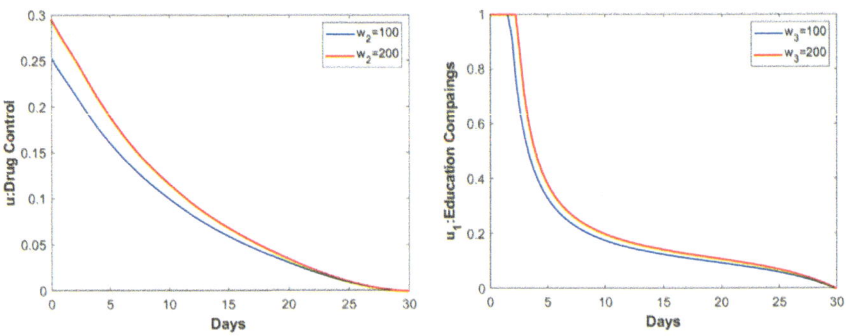

Fig. 5 Control profiles u and u_1 with different cost of drug and education controls $w_2, w_3 = 100, 200$

6 Discussion and Conclusion

This paper developed an epidemiological dengue mathematical model with two types of controls: drug therapy and education campaigns stability are discussed in detail. For this, we need to reduce R_0 less than one to get disease-free equilibrium and endemic equilibrium to be locally asymptotically stable. Sensitivity analysis of R_0 with various parameters has been discussed. It is sensitive to the parameter corresponding to transmission rate, death rate, recruitment rate and the biting rate of the mosquitoes and human. Next, we formulate an optimal control problem and solve it by using Pontrygin's principle. Finally, the optimal treatment strategy is caused. It is observed that drug control gives much better result than without control case also the cheap cost more effective. In [2, 4–6], the authors use many combination of controls. This could make the dynamics of dengue transmission more difficult to monitor. In addition, we use quadratic terms in the control variables in this article, which is popular in epidemiological model optimal control. Other terms in control variables should be considered for future research, as they may offer new insights into dengue transmission dynamics when controls are present. According to the medical point, our model is beneficial for all of us.

References

1. World Health Organization: Urgent need to develop vector tools. Available from: https://www.who.int/denguecontrol/Research/en
2. Ndii, M.Z., Mage, A.R., Messakh, J.J., Djahi, S.D.: Optimal vaccination strategy for dengue transmission in Kupang city. Heliyon **6**, e05345 (2020)
3. Iboi, E.A., Gumel, A.B.: Mathematical assessment of the role of Dengvaxia vaccine on the transmission dynamics of dengue serotypes. Math. Biosci. **304**, 25–47 (2018)
4. Kar, T.K., Jana, S.: A theoretical study on mathematical modelling of an infectious disease with application of optimal control. Biosystems **111**, 37–50 (2013)
5. Kar, T.K., Jana, S.: Application of three controls optimally in a vector-borne disease—a mathematical study. Commun. Nonli. Sci. Num. Simul. **18**, 2868–2884 (2009)
6. Kar, T.K., Batabyal, A.: Stability analysis and optical control of an SIR epidemic model with vaccination., Bio-system **104**, 127–135 (2011)
7. Bashier, E.B.M., Patidar, K.C.: Optimal control of an epidemiological model with multiple time delays. Appl. Math. Comput. **292**, 47–56 (2017)
8. Recher, M., Vaccine, K., Hombach, J., Jit., M.: Assessing dengue vaccination impact: model challenges and future directions. Vaccine **34**, 4461–4465 (2016)
9. Srivastav, A.K., Ghosh, M.: Assessing the impact of treatment on dynamics of dengue fever: a case study of India. Appli. Math. Comput. **362**, 124533 (2019)
10. Rodrigues, H.S., Teresa, M., Monteiro, T., Torres, D.F.M.: Vaccination model and optimal control strategies to Dengue. Math. BioSci. **247**, 1–12 (2014)
11. Hamdan, N.I.: The development of a deterministic dengue epidemic model with the influence of temperature: a case study in Malaysia. App. Math. Model. **90**, 547–567 (2021)
12. Khan, M.A., Fatmawati: Dengue Infection modelling and its optimal control analysis in East Java, Indonesia. Heliyon **7**, e06023 (2021)
13. Tewa, J.J., Dimi, L.J., Bowong, S.: Lyapunov function for a dengue disease transmission model. Chaos Soli. Fract. **39**, 936–941 (2007)
14. Sarkar, K., Khajanchi, S., Nieto, J.J.: Modelling and forecasting the Covid-19 pandemic in India. Chaos Sol. Frac. (2020). https://doi.org/10.1016/j.chaos.2020.110049
15. Baniya, V., Keval, R.: The influence of vaccination on the control of J.E. with standard incidence rate of mosquitoes, pigs and humans. J. App. Math.Comp. **65**, (2020).https://doi.org/10.1007/s12190-020-01367-2020
16. Baniya, V., Keval, R.: Mathematical modelling and stability analysis of Japanese Encephalitis, Add. Sci. Eng. Medi. **12** (2019).https://doi.org/10.1166/asem.2020.2528
17. Mishra, A., Ambrosio, B., Gakkhar, S., Aziz-Alaoui, M.: A network model for control of dengue epidemic using sterile insect technique. Math. Boisci. **15**, 441–460 (2018)
18. Ghosh, I., Tiwari P.K., Chattopadhyay, J.: Effect of active case finding on dengue control: implication from a mathematical model. J. Theor. Biol. **464**, 50–62 (2019)

Did the COVID-19 Lockdown in India Succeed? A Mathematical Study

Sandeep Sharma, Amit Sharma, and Fateh Singh

Abstract In this study, we estimate the basic reproduction number (R_0) for the ongoing COVID-19 pandemic for 10 seriously affected states and for the whole country for the lockdown period. For this, we formulate a SEIQHR mathematical model and fitted it to cumulative COVID-19 cases. The Government of India implemented the first phase of nationwide lockdown from March 25, 2020 to April 14, 2020 and extended the same from April 15, 2020 to May 3, 2020. We measure the effectiveness of the nationwide lockdown on the spread of COVID-19 in India. For this, we have estimated the basic reproduction number for three phases; namely March 14–31, 2020 (Phase I), April 1–15, 2020 (Phase II), and April 16–30, 2020 (Phase III). Our study finds that, in all the cases, the value of the R_0 is minimum at the end of phase III. This demonstrates the success of the implementation of lockdown in reducing the value of the basic reproduction number.

Keywords COVID-19 · The basic reproduction number · Mathematical model · Parameter estimation

1 Introduction

The ongoing pandemic of COVID-19 was originated in Wuhan (China) at the end of 2019. But, at the beginning of 2020, the disease started to spread to different parts of the globe. Due to its severity, the World Health Organization (WHO) first declared it a Public Health Emergency of International Concern on January 30, 2020, and subsequently a pandemic on March 11, 2020 [1]. Despite the serious preventive measures taken by health agencies of different countries, the cases are increasing at an exponential rate [2–5]. Health agencies and governments found it difficult to control the ongoing pandemic.

S. Sharma · F. Singh (✉)
Department of Mathematics, DIT University, Dehradun, Uttarakhand 248009, India

A. Sharma
Department of Mathematics, Shri P.N. Pandya Arts, M.P. Pandya Science & Smt. D.P. Pandya Commerce College, Lunawada, Gujarat 389230, India

P. Balasubramaniam et al. (eds.), *Mathematical Modelling and Computational Intelligence Techniques*, Springer Proceedings in Mathematics & Statistics 376,
https://doi.org/10.1007/978-981-16-6018-4_2

It is believed that coronavirus first transmitted to humans from bat species. Some studies [6–8] reported that the transmission of the virus from bat to human might be related to the seafood market, Wuhan, China [6, 9, 10]. Later, the disease spreads through human-to-human transmission mode. The primary symptoms of this deadly coronavirus include dry cough, abnormality in body temperature, breathing difficulty, and bilateral lung infiltration which are approximately 95 % similar to SARS-coronavirus [11, 12]. As per clinical research, direct contact with an infected individual and inhalation of droplets exhaled by an infected person are the main transmission pathways from infected individuals to healthy individuals [11]. The work carried out in [13] estimated that hospital-related transmission is also possible, and it is suspected to be the possible cause in 41% of patients. However, the root cause of the global spread of coronavirus is air travel of infected persons from one country to another.

Many studies performed on COVID-19 lockdown in India studied the impact of lockdown on different factors; e.g., improvement of air quality, reduction in environmental pollution, etc. [14–16]. Some studies estimated the values of the basic reproduction number (R_0) to investigate the impact of the lockdown [17, 18]. But we perform a comprehensive study and estimated the basic reproduction number for ten seriously affected states.

The chapter is organized in six sections. An overview of the lockdown scenario of India provided in Sect. 2. The formulation of the proposed model, expression of the basic reproduction number, and fitting of the model are discussed in Sect. 3. Estimation of the parameters and values of the basic reproduction number is given in Sect. 4. Finally, the discussion and conclusions of the present work are discussed in Sect. 5 and 6 respectively.

2 COVID-19 Lockdown: Indian Scenario

In India, the first case of COVID-19 was identified on January 30, 2020 in Kerala. Since then, it is spreading rapidly and invaded almost all the states of the country. The number of cases and mortality rate varies significantly from one state of the country to another. As of July 20, nearly 1.1 million COVID-19 cases have been reported in the country, and the number is increasing on a daily basis. All the major states are facing the stress caused by the ongoing COVID-19 pandemic. Some of the major cities (e.g., Delhi, Mumbai, Bengaluru, Kolkata, Indore, etc.) are among the worst affected cities.

Most of the reported cases have travel history from some COVID-19-affected countries or those who were in regular contact with COVID-19 patients (e.g., doctors, security personnel, or family members, etc.). Hence, there is no evidence of community spread (during the period considered in this work) of the disease [19].

Due to its huge population size and limited medical facilities, the country will not be able to control the disease once it starts to spread within the community. To avoid such a disastrous scenario, the Government of India implemented a nationwide lockdown of 21 days on March 25 and subsequently extended the same on April 15 for another 19 days. During the period of lockdown, people were not allowed to move from one place to another. Only the people indulge in necessary services (e.g., groceries, vegetables, and other daily needs items), media, the staff of the medical services and police were allowed to function. Researchers from Blavatnik School of Government at the University of Oxford created a tracker to calculate governments' response to COVID-19. The study, based on data from 73 countries, has identified India's response as one of the most stringent in the world [20]. Many experts termed the implementation of lockdown as the key to limiting the size of the pandemic in the country. However, despite this, the cases of coronavirus were increasing.

It is not a favorable situation for any country, and so for India, to be in lockdown as it hampers the growth and economy of the country. Lockdown in India had blended effects: negative as well as positive. A large number of skilled and working individuals lost their jobs during the COVID-19 lockdown while on the other side, lockdown in India was beneficial in reducing air pollution [15, 21, 22] and improved air quality significantly [16, 23, 24].

Another important issue related to the Indian system was its low capacity of testing in the beginning which, in turn, created a major hurdle in the country's fight against the pandemic [25]. At the time of implementation of its nationwide lockdown, India did not have enough testing kits. The highly heterogeneous pattern of COVID-19 across different states of the country making the phenomenon more complex [25, 26]. Comparing to the global scenario, conditions in India were better as the health agencies and authorities were successful in keeping the disease at a low scale [26, 27]. Despite all these good reasons, a clear understanding of the disease dynamics, which is the primary requirement to prevent infection, is still in its nascent state. The unavailability of previous literature on COVID-19 makes it a difficult task. In such a scenario, the use of mathematical models to gain insight about the key features of the disease transmission is a primary research area [18, 27–32]. The information provided by mathematical models can also help gauge the effect of preventive measures employed against COVID-19.

In the field of mathematical epidemiology, the basic reproduction number (R_0) maintains a crucial place. The value of this critical quantity hints at the success or failure of mitigation strategies applied for the prevention of a certain disease. In the present study, we formulate a mathematical model of the COVID-19 epidemic in India. Subsequently, we derive the expression of the basic reproduction number and by using the cumulative data of infected individuals, we, further, estimate its value for 10 states as well as for the whole country for the first three phases of lockdown (till April 30).

In short, the current work aimed to achieve the following two important issues

1. estimate the basic reproduction number for COVID 19 in India,
2. access the role of lockdown in controlling the spread of disease.

3 Methodology

3.1 Model Formulation

In the model formulation, we have divided the human population into six mutually exclusive compartments containing susceptible (S), exposed (E), infected (I), hospitalized (H), quarantined (Q), and recovered (R). It is assumed that susceptible individuals contracted the infection once they came into contact with an individual of exposed or infected class. Further, we also assume that the exposed individuals are not as capable as susceptible individuals in transmitting infection. The parameter δ represents the disease transmission rate from infected to susceptible individuals, while the parameter δ_0 is the reduction factor. The model also considers the constant recruitment rate Λ. This is realistic because despite the implementation of lockdown, the inter- and intramovement of the individuals was going on. This happened due to two factors: (i) the movement of daily wedges workers (Pravasi) (intramovement); (ii) the immigration of Indian nationals from different countries. A fraction, p, of exposed individual joins the infected class while the remaining $(1 - p)$ enters into the quarantine class. The parameter d represents the natural death rate for the human population. $\frac{1}{\mu}$ is the incubation period and α is the disease-induced death rate. r is the recovery rate for the individuals of the infected class, and ϕ is the rate at which infected individuals admitted to the hospital.

$$\begin{aligned}
\frac{\mathrm{d}S}{\mathrm{d}t} &= \Lambda - \frac{\delta S(I + \delta_0 E)}{N} - \mathrm{d}S \\
\frac{\mathrm{d}E}{\mathrm{d}t} &= \frac{\delta S(I + \delta_0 E)}{N} - \mathrm{d}E - \mu E \\
\frac{\mathrm{d}I}{\mathrm{d}t} &= \mu p E - (d + \alpha + r)I - \phi I \\
\frac{\mathrm{d}Q}{\mathrm{d}t} &= \mu(1 - p)E - (d + \alpha)Q - \sigma Q - \sigma_0 Q \\
\frac{\mathrm{d}H}{\mathrm{d}t} &= \phi I - (d + \alpha\alpha')H - \rho H + \sigma_0 Q \\
\frac{\mathrm{d}R}{\mathrm{d}t} &= \rho H + \sigma Q - \mathrm{d}R + rI
\end{aligned} \tag{1}$$

Further, we assume that the disease-induced death rate of hospitalized individuals is less than that of infected individuals and α' is the reduction factor corresponding the same. The quarantine individuals admitted to the hospital at a rate σ_0. The parameters

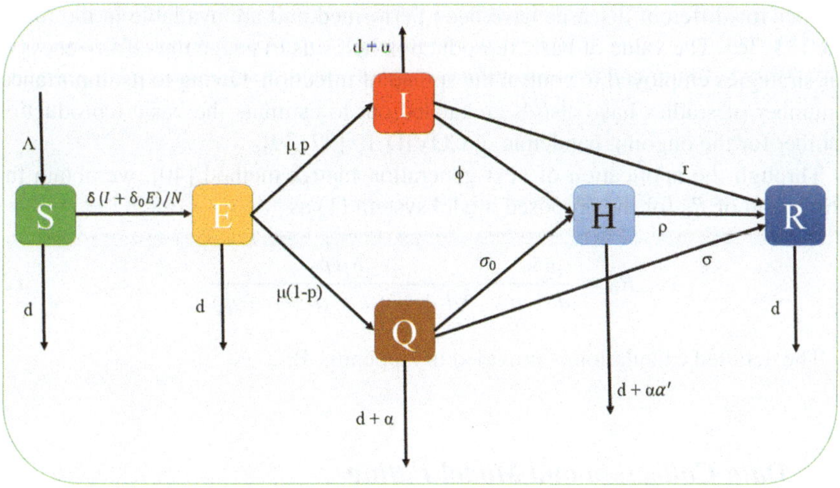

Fig. 1 Flow diagram of the model system (1).

ρ and σ are the recovery rate for the individuals of hospitalized and quarantine class, respectively.

All the parameters involved in the model are assumed positive. Further, the model is investigated under the following initial conditions

$$S(0) > 0, E(0) \geq 0, I(0) \geq 0, Q(0) \geq 0, H(0) \geq 0, R(0) \geq 0.$$

Since all the new recruits join the susceptible class, $S(0) > 0$. All other population depends upon the disease transmission rate, recovery rate, and death rates (natural as well as disease induced). If the disease transmission rate exceeds the sum of recovery rate and death rates, then $E(0)$, $I(0)$, $Q(0)$, $H(0)$, and $R(0)$ are positive else they are zero. Hence, all the populations except $S(0)$ are either positive or zero. Thus, the above initial conditions are biologically feasible.

The schematic diagram of the model system (1) is given in Fig. 1, and mathematical analysis is provided in Appendices.

3.2 *The Basic Reproduction Number*

The basic reproduction number (R_0) is a key concept in the study of infectious disease epidemiology. The value of the basic reproduction number measures the potential of an infection to spread in a region. In particular, the disease will continue to spread if $R_0 > 1$ and eliminated if $R_0 < 1$. Due to its importance, the estimation of the basic reproduction number is a very popular research area in infectious disease epidemiology. A number of studies dedicated to the estimation of the basic reproduction number for different diseases have been performed and are available in the literature [33–36]. The value of basic reproduction helps us to gauge the effectiveness of our strategies employed to control the spread of infection. Owing to its importance, a number of studies have also been carried out to estimate the basic reproduction number for the ongoing pandemic of COVID-19 [37–39].

Through the application of next generation matrix method [40], we obtain the expression of R_0 for the proposed model system (1) as

$$R_0 = \frac{\delta \delta_0}{d + \mu} + \frac{\delta \mu p}{(d + \mu)(d + \alpha + r + \phi)} \tag{2}$$

The detailed calculation is provided in Appendix B.

3.3 Data Collection and Model Fitting

In India, there was a total of four lockdowns had been implemented starting from March 25 to May 31. Out of these, the first two lockdowns were very strict and no relaxations were given to the people. In the last two lockdowns, implemented in the month of May, some relaxations were given in selected cities and regions [41]. To investigate the lockdown impacts on COVID-19, we divide the period of the first two lockdowns into three phases and estimate the basic reproduction number for 10 worst-affected states and for the whole country. We believe that such a comprehensive study will certainly give a clear picture of the outcomes of the COVID-19 lockdown in India.

To estimate the basic reproductive numbers for the COVID-19 pandemic in the 10 states and the whole country (India), we used the daily data on numbers of cases available on [26, 42] for the period from March 14, 2020 to April 30, 2020. We divide the data into three phases: phase I (from March 14–31), Phase II (from April 1–15), and phase III (from April 16–30). The main objective of dividing the data into three phases is to investigate the role of nationwide lockdown which has been implemented from March 25, 2020 to April 14, 2020 (first lockdown) and subsequently extended from April 15 to May 3, 2020 (second lockdown).

In the study of epidemiology, the disease transmission rate is a key parameter that is sensitive and varies from one place to another. Therefore, we use all the parameter values as given in Table 1 and estimate disease transmission rate and recruitment rate. The disease transmission rate δ estimated to match the reported infections in each state and for the whole country. We fit the model system (1) with daily active cases of COVID-19 for the whole country as well as for ten states by using in-built function *lsqcurvefit* in MATLAB (Mathworks, R2017a).

Table 1 Description of parameters used in the model

Parameters	Meaning	Value	Reference
Λ	Recruitment rate	See Tables 2, 3, and 4	Estimated
δ	Disease transmission rate from infected to susceptible individuals	See Tables 2, 3, and 4	Estimated
δ_0	Reduction in disease transmission for exposed individuals	0.2	[30]
d	Natural death rate	0.00003961	[43]
μ	Rate at which exposed individuals leaving exposed class to infected class	1/5.1	[27]
p	Fraction of exposed class joining infected class	0.95	assumed
α	Disease-related death rate	0.0175	[30, 42]
α'	Reduction in disease-induced mortality for hospitalized individuals	0.1	assumed
r	Recovery rate for infected class	1/14	[42]
ϕ	Rate at which infected individuals admitted to hospitals	0.2174	[7]
ρ	Recovery rate of hospitalized individuals	1/14	[30]
σ	Recovery rate of quarantine individuals	0.1162	[44]
σ_0	Rate by which quarantine individuals become infected	0.1429	[30]

Table 2 Estimated parameter values and initial populations for the period March 14–31, 2020

States	Λ	δ	$S(0)$	$E(0)$	$I(0)$	$H(0)$	$Q(0)$
Maharashtra	29.7384	0.5569	105529845.0	61.3108	20.9504	9.9560	12.6970
Gujarat	41.2333	0.5879	6971607.3	21.2616	0	0.6509	3.0959
Delhi	49.9969	0.6889	105529602.4	5.4318	7.8246	5.0649	48.6998
Madhya Pradesh	231.8286	0.8336	79445881.78	4.2344	0.0003	15.8794	3.2367
Rajasthan	6.3559	0.5766	105529859.6	24.4785	0.3092	43.0797	38.7481
Tamil Nadu	111.0181	0.9901	105529845.8	1.6910	0.8661	6.3692	24.2934
Uttar Pradesh	9.8381	0.5705	105529851.2	12.6078	13.4506	12.0693	21.0588
Andhra Pradesh	100.0000	0.8269	105529845.3	1.0309	1.0151	2.5159	30.5048
Telangana	485.2774	0.5520	103827062.4	30.3593	0.1665	12.2244	50.4436
West Bengal	154.0803	0.7505	105529846.3	3.4100	0	35.7535	18.1583

Table 3 Estimated parameter values and initial populations for the period April 1–15, 2020

States	Λ	δ	$S(0)$	$E(0)$	$I(0)$	$H(0)$	$Q(0)$
Maharashtra	35.1579	0.6242	164474.1	1023.6441	0.2358	84.2327	0.0833
Gujarat	92.7025	0.7994	105465645.8	100.3336	12.8877	50.0485	19.8812
Delhi	82.8778	0.5477	105493449.2	736.9751	52.5705	178.9264	329.7132
Madhya Pradesh	634.6673	0.6027	110786125.7	301.8412	15.2050	0.0906	56.6205
Rajasthan	284.4505	0.6002	14599523.01	405.0913	5.1981	0.3960	11.2410
Tamil Nadu	677.1277	0.4242	106592357.2	1014.7835	75.3869	15.6447	155.2254
Uttar Pradesh	10.4066	0.5164	131930985.4	358.6010	0.0133	38.7796	54.1280
Andhra Pradesh	611.8991	0.3892	105683447.9	479.0015	46.8980	58.7452	158.9297
Telangana	500.2755	0.4272	105529768.8	520.6687	12.5030	203.163	298.8051
West Bengal	10.0000	0.5879	105529990.8	69.9823	0.7996	19.0789	9.4820

Table 4 Estimated parameter values and initial populations for the period April 16–30, 2020.

States	Λ	δ	$S(0)$	$E(0)$	$I(0)$	$H(0)$	$Q(0)$
Maharashtra	789.1523	0.5233	106892013.6	4468.7449	83.3822	111.2058	242.6932
Gujarat	97.4263	0.4529	108682572.3	2947.1992	9.4044	1812.2461	3767.6181
Delhi	753.7834	0.5219	105785941.5	1177.7992	11.2122	500.2511	498.3525
Madhya Pradesh	2661.8965	0.4807	107922741.1	929.8147	254.8521	87.3521	2068.3516
Rajasthan	80.9682	0.4227	110741965.8	1412.8254	12.7402	212.1730	418.5005
Tamil Nadu	84.3871	0.5089	112780319.3	638.6235	2.2299	104.2202	103.9158
Uttar Pradesh	123.4160	0.4500	107546189.6	1269.2242	10.3975	103.4533	153.7916
Andhra Pradesh	1286.4031	0.5654	105911366.5	429.0830	6.6945	51.7046	29.7475
Telangana	788.4176	0.3095	107500310.7	631.0853	34.5328	2.7723	320.0567
West Bengal	738.8545	0.5387	97037035.3	291.1704	22.1629	85.5764	50.1882

Table 5 Estimated parameter values and initial populations of the model for whole India (March 14–April 30, 2020)

Phase	Λ	δ	$S(0)$	$E(0)$	$I(0)$	$H(0)$	$Q(0)$
Phase I	560.8891	0.5830	103783356.8	300.0346	99.8991	100.8935	476.3326
Phase II	3939.4788	0.4626	703191814.8	7456.8562	1936.0503	6961.4801	7888.9605
Phase III	6508.9350	0.3652	439415725.1	31286.5802	12986.4825	19727.8255	6099.6136

4 Results

The obtained phasewise estimates for δ and Λ are given in Table 2 (for the phase I, i.e., March 14–31, 2020), Table 3 (for the phase II, i.e., April 1–15, 2020), and in Table 4 (for the phase III, i.e., April 16–30, 2020), while the estimates for the same for the whole country are provided in Table 5. The estimated values of the basic reproduction number (R_0) for all the 10 states as well as for the whole country are

Table 6 Value of R_0 for different states

States	**Value of R_0**		
	For Phase I	For Phase II	**For Phase III**
Maharashtra	2.29	2.57	2.16
Gujarat	2.42	3.29	1.87
Telangana	2.27	1.76	1.28
Uttar Pradcsh	2.35	2.13	1.85
West Bengal	3.09	2.42	2.22
Madhya Pradesh	3.43	2.48	1.98
Andhra Pradesh	3.41	1.60	2.33
Delhi	2.84	2.26	2.15
Rajasthan	2.38	2.47	1.74
Tamil Nadu	4.08	1.75	2.10
India	2.40	1.91	1.50

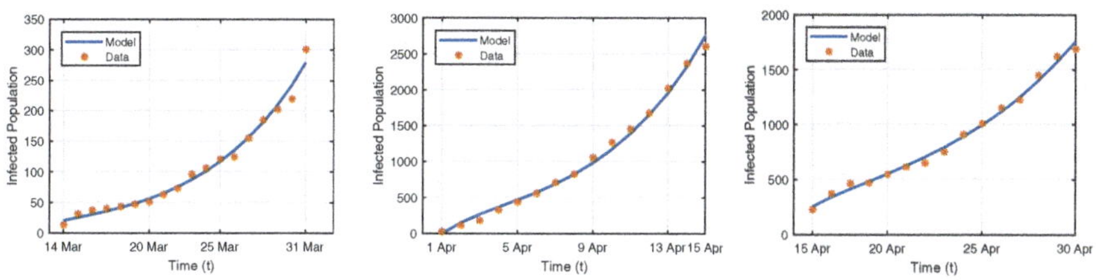

(a) Phase I (b) Phase II (c) Phase III

Fig. 2 Fitting of model system (1) with cumulative cases of COVID-19 (daily reported) for Maharashtra

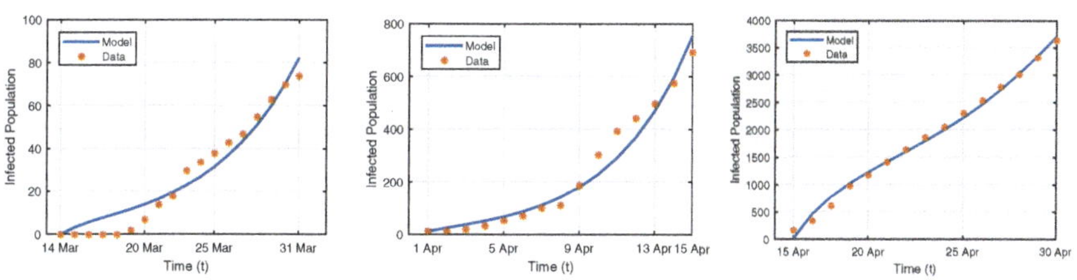

(a) Phase I (b) Phase II (c) Phase III

Fig. 3 Fitting of model system (1) with cumulative cases of COVID-19 (daily reported) for Gujarat

listed in Table 6. The plots for the data fitting of the cumulative COVID-19 cases (statewise and for the whole country) for all the three phases are shown in Figs. 2, 3, 4, 5, 6, 7, 8, 9, 10, 11, and 12. Further, the variation in basic reproduction number (R_0) with respect to various parameters (p, δ_0, r, ϕ, μ, α and d) is also obtained which are shown in Fig. 13.

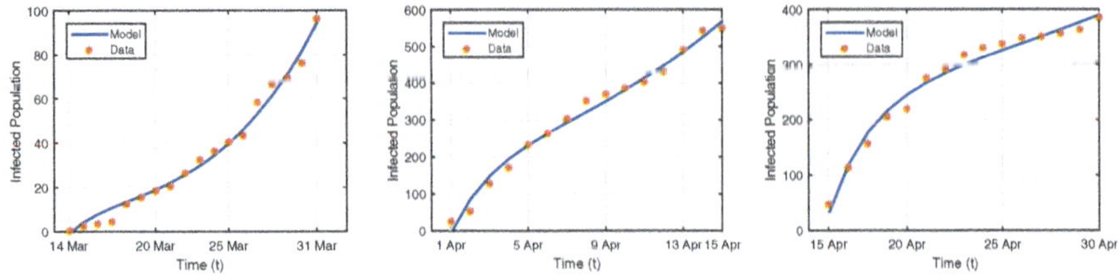

(a) Phase I (b) Phase II (c) Phase III

Fig. 4 Fitting of model system (1) with cumulative cases of COVID-19 (daily reported) for Telangana

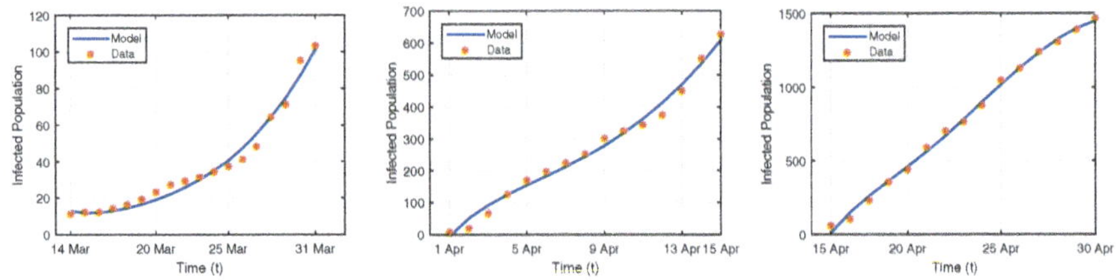

(a) Phase I (b) Phase II (c) Phase III

Fig. 5 Fitting of model system (1) with cumulative cases of COVID-19 (daily reported) for Uttar Pradesh

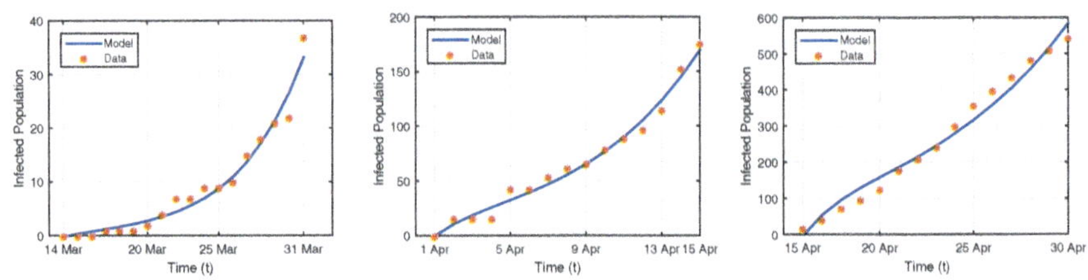

(a) Phase I (b) Phase II (c) Phase III

Fig. 6 Fitting of model system (1) with cumulative cases of COVID-19 (daily reported) for West Bengal

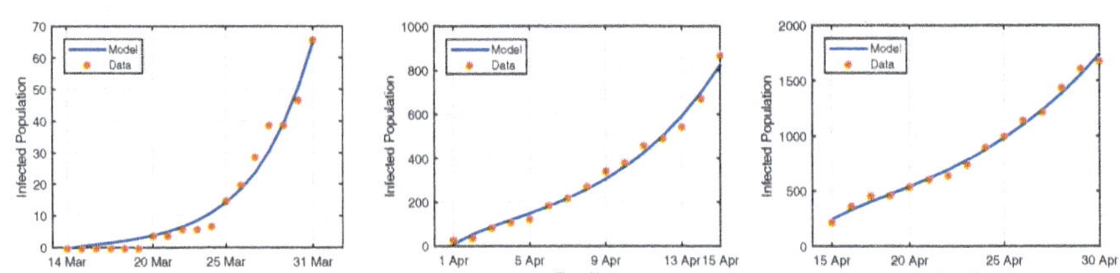

(a) Phase I (b) Phase II (c) Phase III

Fig. 7 Fitting of model system (1) with cumulative cases of COVID-19 (daily reported) for Madhya Pradesh

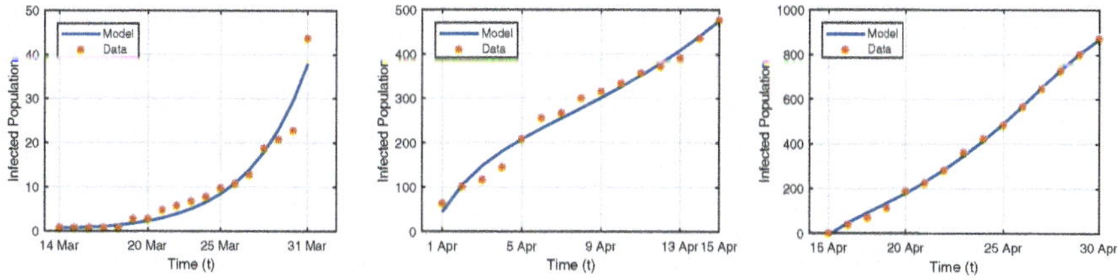

(a) Phase I (b) Phase II (c) Phase III

Fig. 8 Fitting of model system (1) with cumulative cases of COVID-19 (daily reported) for Andhra Pradesh

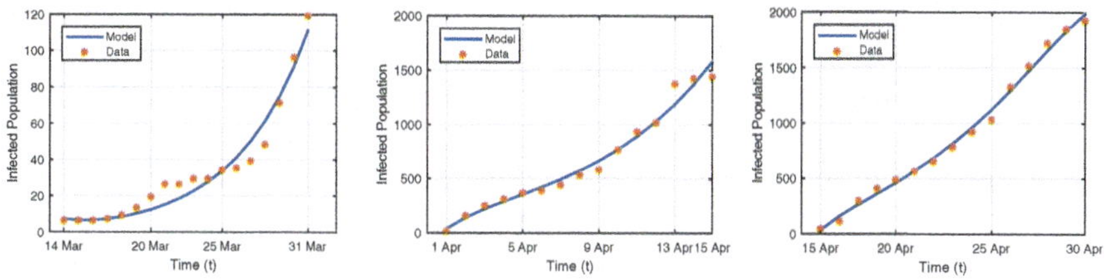

(a) Phase I (b) Phase II (c) Phase III

Fig. 9 Fitting of model system (1) with cumulative cases of COVID-19 (daily reported) for Delhi

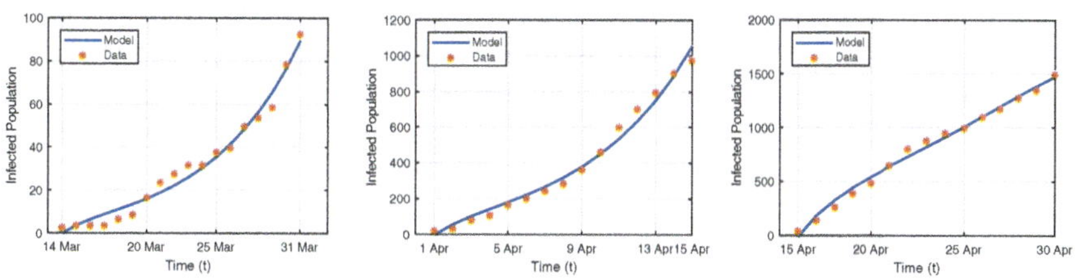

(a) Phase I (b) Phase II (c) Phase III

Fig. 10 Fitting of model system (1) with cumulative cases of COVID-19 (daily reported) for Rajasthan

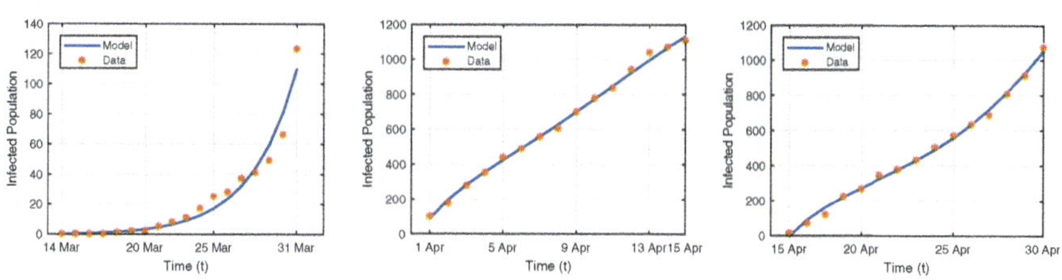

(a) Phase I (b) Phase II (c) Phase III

Fig. 11 Fitting of model system (1) with cumulative cases of COVID-19 (daily reported) for Tamil Nadu

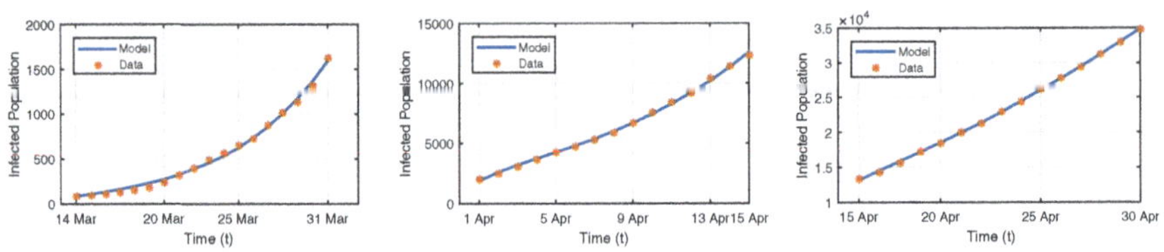

(a) Phase I (b) Phase II (c) Phase III

Fig. 12 Fitting of model system (1) with cumulative cases of COVID-19 (daily reported) for India

5 Discussion

The mathematical model proposed in the present work is fitted to real data of COVID-19, and parameters δ and Λ are estimated. However, other parameters involved in the model are selected from the literature (see Table 1). The expression of R_0, obtained in the Sect. 3.2, does not contain the recruitment rate (Λ) term. This suggests that the obtained R_0 is more sensitive to δ. For different states, values of R_0 are calculated which are shown in Table 6 for parameters value (in Table 1). From Table 6, it can be observed that Tamil Nadu registers the highest value (4.08) of R_0. This clearly highlights the fact that during the first phase, the COVID-19 was more severe in Tamil Nadu as compared to the other state involved in the current study. However, it is significantly reduced for Tamil Nadu in the second phase (1.75). During the second phase, the value of R_0 is highest for the state of Gujarat. Therefore, it implies that during the second phase the speed of the COVID-19 pandemic is highest in Gujarat. For the third phase, Andhra Pradesh and West Bengal are standing close in terms of the values of R_0 (2.33 for Andhra Pradesh and 2.22 for West Bengal). This also indicates that these two states may emerge as the next epicenter for the ongoing COVID-19 pandemic in India. Further, the estimated values of R_0 are greater than two for the five states (e.g., Maharashtra, Delhi, Tamil Nadu, Andhra Pradesh, and West Bengal) of India. Therefore, these five states need extra attention to the control of COVID-19 in India. Moreover, the estimated values of R_0 for Tamil Nadu and Andhra Pradesh depict a peculiar behavior as it first reduced from the first phase (4.08 for Tamil Nadu and 3.41 for Andhra Pradesh) to the second phase (1.75 for Tamil Nadu and 1.60 for Andhra Pradesh) and later increased during the third phase (2.10 for Tamil Nadu and 2.33 for Andhra Pradesh). This suggests that the implementation of lockdown either does not serve the purpose or is not followed properly in these two states.

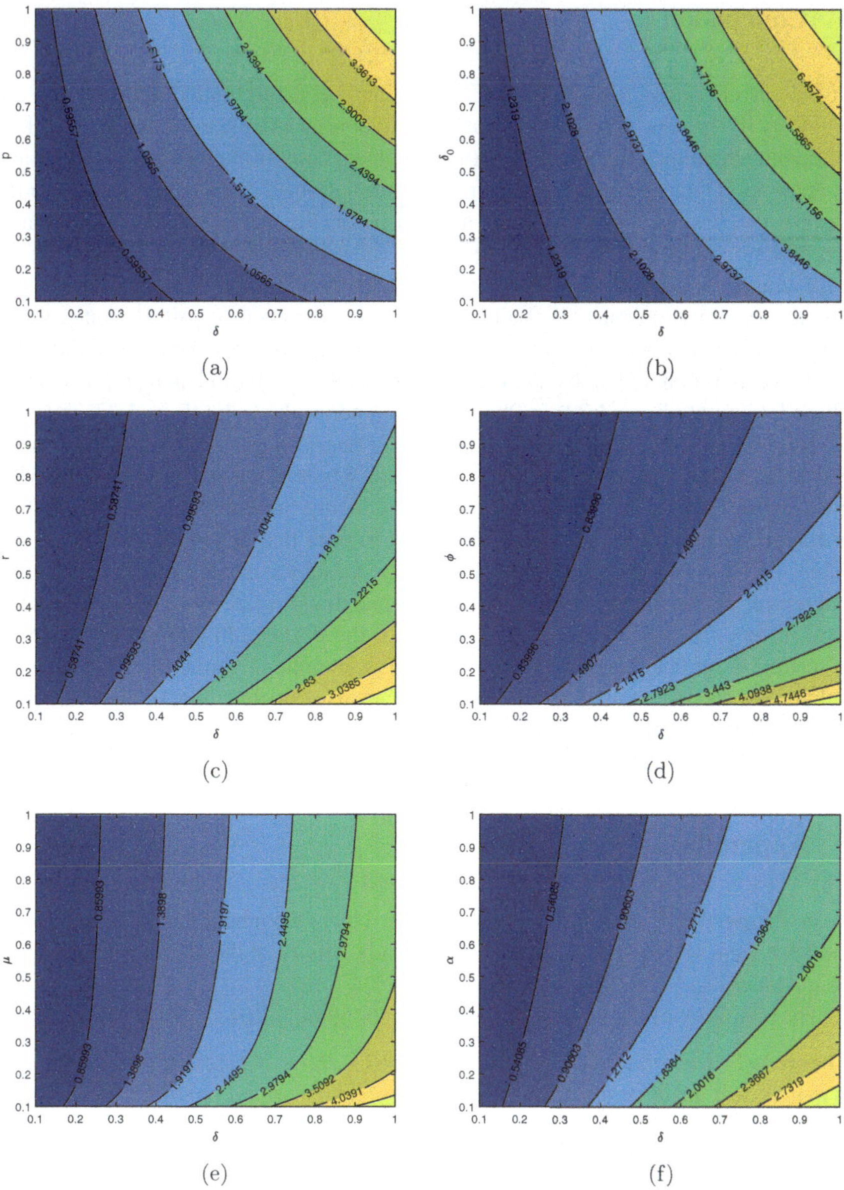

Fig. 13 Contour plots depicting the variation of R_0 with respect to various parameters

6 Conclusion

In the current work, a mathematical model has been formulated and fitted to the real COVID-19 data of India. The present model is an extension of the classic SEIR epidemic which includes two extra compartments; namely, the hospitalized (H) and quarantine (Q) compartments. After the formulation of the mathematical model, we use the next generation matrix method to obtain the expression of the basic reproduction number (R_0). Next, we fit the real data of COVID-19 cases using the least square method at the 95 % confidence interval to estimate the values of the disease transmission rate (δ). Values of R_0 are calculated for all 10 states under consideration and for the whole of India for all three phases. On the basis of the value of R_0, it can be concluded that the most affected state was Tamil Nadu ($R_0 = 4.08$) in Phase I, Gujarat ($R_0 = 3.29$) in Phase II, and Andhra Pradesh ($R_0 = 2.33$) in Phase III. It is observed that the value of R_0 is lower in Phase III as compared to the value of R_0 in Phase I and Phase II for all states and for the whole of India. This clearly signifies the success of the nationwide lockdown in reducing the burden of the COVID-19 pandemic in India. However, it is important to note that despite the decrease, the values of R_0 are significantly greater than one for all the cases under investigation. Therefore, in combination with nationwide lockdown, some other robust methods have to be employed to reduce the burden of the disease. In particular, the speed of testing should be increased to identify and quarantine potential infected cases. In light of the results obtained in this work, we can also conclude that in the absence of robust medical facilities, lockdown will be an effective measure to curtail the disease.

7 Appendix 1

Basic Properties Here, we will study the positivity and boundedness of the proposed model system (1). First, we state the result on positivity of the solution followed by boundedness.

The positivity of the solutions is established by the following theorem

Theorem 1 *The solution* $(S(t), E(t), I(t), Q(t), H(t), R(t))$ *of the proposed model system is non-negative for all* $t \geq 0$ *with non-negative initial condition.*

Proof From the first equation of the model system (1), we obtain

$$\frac{\mathrm{d}S}{\mathrm{d}t} \geq \frac{\Lambda}{2} - \left(\frac{\delta I}{N} + \frac{\delta\delta_0 E}{N} + d\right) S$$

Further, we can obtain

$$\frac{\mathrm{d}}{\mathrm{d}t}\left[S(t)\exp\left\{\int_0^t \left(\frac{\delta I}{N} + \frac{\delta\delta_0 E}{N}\right)\mathrm{d}\mathrm{d}\theta + \mathrm{d}t\right\}\right] \geq \frac{\Lambda}{2}\exp\left\{\int_0^t \left(\frac{\delta I}{N} + \frac{\delta\delta_0 E}{N}\right)\mathrm{d}\theta + \mathrm{d}t\right\}$$

on integration, we get

$$S(t)\exp\left\{\int_0^t\left(\frac{\delta I}{N}+\frac{\delta\delta_0 E}{N}\right)d\theta+dt\right\}-S(0)\geq\int_0^{t_1}\frac{\Lambda}{2}\exp\left\{\int_0^t\left(\frac{\delta I}{N}+\frac{\delta\delta_0 E}{N}\right)d\theta+dt\right\}d\vartheta$$

on rearranging the terms, we get

$$S(t_1)\geq S(0)\exp\left\{-\int_0^t\left(\frac{\delta I}{N}+\frac{\delta\delta_0 E}{N}\right)d\theta+dt\right\}$$
$$+\int_0^{t_1}\frac{\Lambda}{2}\exp\left\{\int_0^t\left(\frac{\delta I}{N}+\frac{\delta\delta_0 E}{N}\right)d\theta+dt\right\}d\vartheta\times exp\left\{-\int_0^t\left(\frac{\delta I}{N}+\frac{\delta\delta_0 E}{N}\right)d\theta+dt\right\}$$

From the above calculation, we obtain that $S(t) > 0$. The positivity of other components can be established similarly. This completes the proof.

Next, we will state the following result on the boundedness of the solutions

Theorem 2 *All solutions of the proposed model are bounded.*

Proof From the model system (1), we obtain

$$(S+E+I+Q+H+R)'=\Lambda-d(S+E+I+Q+H+R)-\alpha I-\alpha Q-\alpha\alpha' H$$

which, further, gives

$$(S+E+I+Q+H+R)'\leq\Lambda-d(S+E+I+Q+H+R)$$

which gives us

$$\lim_{t\to\infty} Sup(S+E+I+Q+H+R)\leq\frac{\Lambda}{d}$$

Further, from the first equation of the model system (1), we have

$$\frac{dS}{dt}\leq\Lambda-dS\Rightarrow S(t)\leq\frac{\Lambda}{d}$$

Similarly, we can obtain the bounds for other components of the solution. Thus, we obtain the feasible region (Ω) for the proposed model system as

$$\Omega=\left\{(S,E,I,Q,H,R)\in R_+^6|S+E+I+Q+H+R\leq\frac{\Lambda}{d}\right\}.$$

8 Appendix 2

Disease Free Equilibrium and Basic Reproduction Number It is trivial to note that the proposed model system (1) has a disease free equilibrium $E_0 = (\frac{A}{d}, 0, 0, 0, 0, 0)$.

Now, the progression from exposed class (E) to infected class (I) and Quarantine class (Q) are not considered to be new infections. Further, the infected compartments involve in the model are E and I. Combining all these information, the matrices F and V are given as

$$F = \begin{bmatrix} \delta\delta_0 & \delta \\ 0 & 0 \end{bmatrix}$$

$$V = \begin{bmatrix} (\mu + d) & 0 \\ -\mu p & (\alpha + d + r + \phi) \end{bmatrix}$$

Now, the expression of R_0 can be obtained by determining the largest eigenvalue of the matrix FV^{-1}, which is same as given in equation (2).

References

1. https://www.who.int/emergencies/diseases/novel-coronavirus-2019/events-as-they-happen (2020)
2. Ghosh, K., Sengupta, N., Manna, D., De, S.K.: Inter-state transmission potential and vulnerability of covid-19 in India. Progress Disaster Sci. **7**, 100114 (2020)
3. Livadiotis, G.: Statistical analysis of the impact of environmental temperature on the exponential growth rate of cases infected by covid-19. PLoS one **15**(5), e0233875 (2020)
4. Moftakhar, L., Seif, M.: The exponentially increasing rate of patients infected with covid-19 in Iran. Arch. Iran Med. **23**(4), 235–238 (2020)
5. Musa, S.S., Zhao, S., Wang, M.H., Habib, A.G., Mustapha, U.T., He, D.: Estimation of exponential growth rate and basic reproduction number of the coronavirus disease 2019 (covid-19) in Africa. Infect. Dis. Poverty **9**(1), 1–6 (2020)
6. Huang, C., Wang, Y., Li, X., Ren, L., Zhao, J., Hu, Y., Zhang, L., Fan, G., Xu, J., Gu, X., et al.: Clinical features of patients infected with 2019 novel coronavirus in Wuhan, China. Lancet **395**(10223), 497–506 (2020)
7. Li, Q., Guan, X., Wu, P., Wang, X., Zhou, L., Tong, Y., Ren, R., Leung, K.S., Lau, E.H., Wong, J.Y., Xing, X., Xiang, N., Wu, Y., Li, C., Chen, Q., Li, D., Liu, T., Zhao, J., Liu, M., Tu, W., Chen, C., Jin, L., Yang, R., Wang, Q., Zhou, S., Wang, R., Liu, H., Luo, Y., Liu, Y., Shao, G., Li, H., Tao, Z., Yang, Y., Deng, Z., Liu, B., Ma, Z., Zhang, Y., Shi, G., Lam, T.T., Wu, J.T., Gao, G.F., Cowling, B.J., Yang, B., Leung, G.M., Feng, Z.: Early transmission dynamics in Wuhan, China, of novel coronavirus-infected pneumonia. New Engl. J. Med. **382**(13), 1199–1207 (2020). PMID: 31995857
8. Zhou, P., Yang, X.L., Wang, X.G., Hu, B., Zhang, L., Zhang, W., Si, H.R., Zhu, Y., Li, B., Huang, C.L., et al.: A pneumonia outbreak associated with a new coronavirus of probable bat origin. Nature **579**(7798), 270–273 (2020)
9. Chan, J.F.W., Yuan, S., Kok, K.H., To, K.K.W., Chu, H., Yang, J., Xing, F., Liu, J., Yip, C.C.Y., Poon, R.W.S., et al.: A familial cluster of pneumonia associated with the 2019 novel coronavirus indicating person-to-person transmission: a study of a family cluster. Lancet **395**(10223), 514–523 (2020)

10. Zhu, N., Zhang, D., Wang, W., Li, X., Yang, B., Song, J., Zhao, X., Huang, B., Shi, W., Lu, R., Niu, P., Zhan, F., Ma, X., Wang, D., Xu, W., Wu, G., Gao, G.F., Tan, W.: A novel coronavirus from patients with pneumonia in China, 2019. New Engl. J. Med. **382**(8), 727–733 (2020). PMID: 31978945
11. World Health Organization. https://www.who.int/health-topics/coronavirus
12. Gralinski, L.E., Menachery, V.D.: Return of the coronavirus: 2019-ncov. Viruses **12**(2), 135 (2020)
13. Wang, D., Hu, B., Hu, C., Zhu, F., Liu, X., Zhang, J., Wang, B., Xiang, H., Cheng, Z., Xiong, Y., et al.: Clinical characteristics of 138 hospitalized patients with 2019 novel coronavirus-infected pneumonia in Wuhan, China. Jama **323**(11), 1061–1069 (2020)
14. Mahato, S., Pal, S., Ghosh, K.G.: Effect of lockdown amid covid-19 pandemic on air quality of the megacity delhi, india. Sci. Total Environ. **730**, 139086 (2020)
15. Naqvi, H.R., Datta, M., Mutreja, G., Siddiqui, M.A., Naqvi, D.F., Naqvi, A.R.: Improved air quality and associated mortalities in India under covid-19 lockdown. Environ. Pollut. **268**, 115691 (2021)
16. Singh, R.P., C.A.: Impact of lockdown on air quality in India during covid-19 pandemic. Air Quality Atmos. Health **13**, 921–928 (2020)
17. Pai, C., Bhaskar, A., Rawoot, V.: Investigating the dynamics of covid-19 pandemic in India under lockdown. Chaos Solitons Fractals **138**, 109988 (2020)
18. Sardar, T., Nadim, S.S., Chattopadhyay, J.: Assessment of 21 days lockdown effect in some states and overall India: a predictive mathematical study on covid-19 outbreak. arXiv preprint arXiv:2004.03487 (2020)
19. Indian Council of Medical Research. https://icmr.nic.in/
20. Hale, T., Angrist, N., Goldszmidt, R., Kira, B., Petherick, A., Phillips, T., Webster, S., Cameron-Blake, E., Hallas, L., Majumdar, S., et al.: A global panel database of pandemic policies (oxford covid-19 government response tracker). Nat Hum Behav 1–10 (2021)
21. Mor, S., Kumar, S., Singh, T., Dogra, S., Pandey, V., Ravindra, K.: Impact of covid-19 lockdown on air quality in Chandigarh, India: understanding the emission sources during controlled anthropogenic activities. Chemosphere **263**, 127978 (2021)
22. Sathe, Y., Gupta, P., Bawase, M., Lamsal, L., Patadia, F., Thipse, S.: Surface and satellite observations of air pollution in India during covid-19 lockdown: implication to air quality. Sustain. Cities Soc. **66**, 102688 (2021)
23. Bhadra, A., Mukherjee, A., Sarkar, K.: Impact of population density on covid-19 infected and mortality rate in india. Model. Earth Syst. Environ. **7**(1), 623–629 (2021)
24. Panda, S., M.C.N.J.e.a.: A study on variation of atmospheric pollutants over Bhubaneswar during imposition of nationwide lockdown in India for the covid-19 pandemic. Air Quality Atmos. Health **14**, 97–108 (2021)
25. Sahasranaman, A., Kumar, N.: Network structure of covid-19 spread and the lacuna in India's testing strategy. Available at SSRN 3558548 (2020)
26. https://www.mygov.in/covid-19 (2020)
27. Chatterjee, K., Chatterjee, K., Kumar, A., Shankar, S.: Healthcare impact of covid-19 epidemic in India: a stochastic mathematical model. Med. J. Armed Forces India **76**(2), 147–155 (2020)
28. Batista, M.: Estimation of the final size of the second phase of coronavirus epidemic by the logistic model. medrxiv (2020)
29. Giordano, G., Blanchini, F., Bruno, R., Colaneri, P., Di Filippo, A., Di Matteo, A., Colaneri, M.: Modelling the covid-19 epidemic and implementation of population-wide interventions in Italy. Nat. Med. **26**(6), 855–860 (2020)
30. Senapati, A., Rana, S., Das, T., Chattopadhyay, J.: Impact of intervention on the spread of covid-19 in India: a model based study. arXiv preprint arXiv:2004.04950 (2020)
31. Tanvi, Sajid, M., Aggarwal, R., Rajput, A.: Assessing the impact of transmissibility on a cluster-based covid-19 model in India. Int. J. Model. Simul. Sci. Comput. (2020)
32. Tanvi, Rajiv Aggarwal, A.R.M.S.: Modelling the optimal interventions to curtail the cluster based covid-19 pandemic in India: Efficacy of prevention measures. Appl. Comput. Math. **20**(1), 70–94 (2021)

33. Marques, C., Forattini, O., Massad, E.: The basic reproduction number for dengue fever in Sao Paulo State, Brazil: 1990–1991 epidemic. Trans. R. Soc. Trop. Med. Hygiene **88**(1), 58–59 (1994)
34. Mukandavire, Z., Liao, S., Wang, J., Gaff, H., Smith, D.L., Morris, J.G.: Estimating the reproductive numbers for the 2008–2009 cholera outbreaks in Zimbabwe. Proc. Nat. Acad. Sci. **108**(21), 8767–8772 (2011)
35. Mukandavire, Z., Smith, D.L., Morris, J.G., Jr.: Cholera in Haiti: reproductive numbers and vaccination coverage estimates. Sci. Rep. **3**(1), 1–8 (2013)
36. Nishiura, H., Mizumoto, K., Villamil-Gómez, W.E., Rodríguez-Morales, A.J.: Preliminary estimation of the basic reproduction number of Zika virus infection during Colombia epidemic, 2015–2016. Travel Med. Infect. Disease **14**(3), 274–276 (2016)
37. Cao, Z., Zhang, Q., Lu, X., Pfeiffer, D., Jia, Z., Song, H., Zeng, D.D.: Estimating the effective reproduction number of the 2019-ncov in china. medRxiv (2020)
38. Liu, Y., Gayle, A.A., Wilder-Smith, A., Rocklv, J.: The reproductive number of COVID-19 is higher compared to SARS coronavirus. J. Travel Med. **27**(2) (2020)
39. Zhou, T., Liu, Q., Yang, Z., Liao, J., Yang, K., Bai, W., Lu, X., Zhang, W.: Preliminary prediction of the basic reproduction number of the Wuhan novel coronavirus 2019-ncov. J. Evidence-Based Med. **13**(1), 3–7 (2020)
40. Van den Driessche, P., Watmough, J.: Reproduction numbers and sub-threshold endemic equilibria for compartmental models of disease transmission. Math. Biosci. **180**(1–2), 29–48 (2002)
41. https://www.mha.gov.in/ (2020)
42. https://www.covid19india.org/ (2020)
43. https://data.worldbank.org/indicator/sp.dyn.le00.in?locations=in (2020)
44. Tang, B., Wang, X., Li, Q., Bragazzi, N.L., Tang, S., Xiao, Y., Wu, J.: Estimation of the transmission risk of the 2019-ncov and its implication for public health interventions. J. Clin. Med. **9**(2), 462 (2020)

Tumour Growth and Its Treatment Response Delineate with Mathematical Models

Bhavyata Patel, Rhydham Karnik, and Dhanesh Patel

Abstract Cancer is characterized by abnormal cell growth and forms the extra mass of tissue known as a tumour. It leads to many serious health problems even to death. There are many techniques available to diagnose cancer, and several cancers have not been cured yet as treatment needs to be improved. In this paper, we have made an attempt to develop mathematical models of tumour growth and its treatment processes for more agreeable solution by considering some possible factor in Hahnfeldt et.al. model. The models were studied both analytically and numerically, predicting tumour growth and response to treatment processes.

Keywords Modelling of tumour growth · Differential equations · Angiogenesis inhibitor · Qualitative aspect · Stability

1 Introduction

Cancer is a group of diseases which involves sporadic cell growth with the potential to spread to other parts of the body which is defined as metastasis. Cancer involves a large number of complex disease in nature and its side effects. It has become second in terms of rates of mortality in the world today and approximately 9.6 million people died due to cancer in the year 2018 [1]. A lot of research is going on to understand the abnormality in behaviour of cancer cells. Normal, healthy cells differentiate and stop cell division. They are replaced by newly formed daughter cells, while, on the other hand, cancer cells keep on dividing uncontrollably unlike normal cells. Most cancer cells form a mass of tissue that is called a tumour [1].

B. Patel · D. Patel (✉)
Department of Applied Mathematics, Faculty of Technology and Engineering, The Maharaja Sayajirao University of Baroda, Vadodara, Gujarat, India
e-mail: patel.bhavyata-appmath@msubaroda.ac.in

R. Karnik
Department of Zoology, Faculty of science, The Maharaja Sayajirao University of Baroda, Vadodara, Gujarat, India

P. Balasubramaniam et al. (eds.), *Mathematical Modelling and Computational Intelligence Techniques*, Springer Proceedings in Mathematics & Statistics 376,
https://doi.org/10.1007/978-981-16-6018-4_3

There are several treatments available to curb sporadic tumour growth. Among them a lot of interdisciplinary research is going on in the field of chemotherapy and antiangiogenesis therapy. Many treatments also involves a combination of such therapies.

Tumour angiogenesis (process of creation of new blood vessels from an existing vasculature) plays a major role in the growth of tumour. Due to high rates of cell division, the tissue requires more oxygen and nutrients. This increase in demand leads to formation of new vasculature, which is necessary for tumour growth [2, 3]. One of the major pathways in the angiogenesis involve vasculature endothelial growth factor (VEGF) and its receptor [4].

Several mathematical models have been initiated to describe this process among which differential equation model gives a way into a quantitative approach for cancer biology [5, 6]. Among this Malthusian growth model is one of the basic models that describes the tumour growth in terms of growth rate and death rate. The solution of this model predicted an exponential growth. The disadvantage of the Malthusian model is that it does not take into account a number of factors such as increased competition for limited resources and environmental constraints. It would be adequate to slow down the growth rate. In the initial stage, this factor will not affect the tumour growth, and therefore, this model is used only for early stage of cancer [6].

The logistic growth model is a generalization of this model by considering carrying capacity to introduce the concept that the pair interaction decreases the tumour growth due to a limited availability of resources and space.

$$\frac{dp}{dt} = \alpha p(1 - \frac{p}{k}) \quad p(0) = p_0 \tag{1}$$

where α is proportionality constant, k is carrying capacity, p is tumour volume and p_0 is tumour volume at time $t = 0$. This equation shows that when tumour volume is near to its carrying capacity, i.e. $\frac{dp}{dt}$ tends to 0 then tumour growth tends to be constant. If tumour volume is very small to the carrying capacity ($p \ll k$) then the tumour grows exponentially.

The Gompertz growth model emerged from an actuarial count number of living tumour cell as a function of their age. Let us consider the tumour volume is proportional to exponentially decaying birth rate $\beta(t)$.

$$\frac{dp}{dt} = -\beta(t)p(t) \quad p(0) = p_0 \tag{2}$$

$$\frac{d\beta}{dt} = -\gamma\beta(t) \quad \beta(0) = \beta_0 \tag{3}$$

where $\beta(0) = \beta_0$ be the initial tumour volume net fecundity, $p(0) = p_0$ is initial tumour volume and γ is a proportionality constant. Tumour vasculature controls the tumour volume through its supply of nutrients [7]. We know that derivative of Gompertzian function is used to observe tumour growth slowdown with size [6].

Hahnfeldt defined carrying capacity as the maximum tumour volume that can be sustained in a tissue at a given time. During the initial stage of development, tumour uses the nutrients and oxygen provided by the vasculature until tumour volume is equal to carrying capacity. Once the tumour volume is higher than carrying capacity, it releases stimulatory and inhibitory signals for tumour angiogenesis which leads to formation of new vasculature, and this results into change in carrying capacity with respect to time. Hahnfeldt model takes into consideration the concept that tumour growth process depends on development of vascular and can be controlled by antiangiogenesis treatment. This model considers the Gompertz model as a tumour growth model and further includes time-dependent carrying capacity in terms of stimulation and inhibition and also confirmed validity of model in vivo lab experiments[6].

$$\frac{\mathrm{d}p}{\mathrm{d}t} = -\gamma p ln(\frac{p}{k}) \tag{4}$$

$$\frac{\mathrm{d}k}{\mathrm{d}t} = \eta p - \theta k p^{\frac{2}{3}} - \delta k - f k D_r(t) \tag{5}$$

Equation (4) shows that the change in the number of tumour cells is dependent on the ratio of $\frac{p}{k}$, i.e. number of cells present to the carrying capacity. If the ratio of tumour Volume to carrying capacity is greater than 1, then there will be inhibition of tumour, and if it is less than 1, then there will be stimulation.

Equation (5) shows the change in carrying capacity with time due to various factors taken into consideration. From the results, it can be concluded that Observed values of this treatment were approximately similar with expected values from the derived model [6].

In 2004, Alberto d'Onofrio and Alberto Gandolf considers that the potential doubling time of the vasculature to be constant overall, further also divided the endothelial cells into three subdivisions [8]. In 2009 Alberto d'Onofrioa, Paola Cerrai further extended the Hahnfeldt model by considering two different types of drugs (vessel-disrupting antiangiogenic drugs and drugs inhibiting exponential cell division of endothelial cells)[8]. Jan Poleszczuk, Marek Bodnar and Urszula Fory proposed that Halfeldt model is valid only for antiangiogenic treatments destroying the endothelial cells. They further modified the model which is only valid for the antiangiogenic treatment focusing on blocking the VEGF signalling [9]. Benzekry et al. numerically investigate the behaviour and convergence of Hahnfeldt model [2].

In 2014 Heiko Enderling et al. considered various approaches towards cancer therapy, which included antitumour treatments (chemotherapy or immunotherapy) and antiangiogenic cancer therapy. Further analyses of different mathematical models and Hahnfeldt model by considering the fact that antitumour treatment would negatively affect change in cell population with time [5]. Heiko Enderling model

$$\frac{\mathrm{d}p}{\mathrm{d}t} = -\gamma p \ln\left(\frac{p}{k}\right) - \omega p \tag{6}$$

$$\frac{dk}{dt} = \eta p - \theta k p^{\frac{2}{3}} - f k D_r(t) \tag{7}$$

The term ω defined as killing strength of the antitumour drug that is being used. The range of ω is from $0 \leq \omega \leq 1$.

The main purpose of this paper is to modify the mathematical model of Hahnfeldt et.al. [6] and analyse stability of the modified Hahnfeldt model. Comparing both the models with experimental data, in the mathematical model from Hahnfeldt et al. the spontaneous loss of vasculature (δ) is considered negligible and hence taken as zero, whereas Heiko Enderling removed the term in his mathematical model [5]. Here we propose that this term cannot be removed and cannot be taken as zero even if the value of δ is almost negligible. We further define the range of δ from 0.005 to 0.05.

2 Tumour Growth Treatment Model

Tumour growth reduction depends upon various treatments.

1. Antitumour treatment.
2. Antiangiogenesis treatment.
3. Immunotherapy
4. Radiation
5. Surgery.

In this model, we have considered only antitumour treatment and antiangiogenesis treatment. We have defined spontaneous loss of carrying capacity, which happens due to natural death of endothelial cells.

$$\frac{dp}{dt} = -\gamma p \ln\left(\frac{p}{k}\right) - \omega p \tag{8}$$

This equation is according to Enderling et al. where $-\omega p$ is added for chemotherapy treatment. Here, ω is defined as killing strength in the range of [0, 1]

$$\frac{dk}{dt} = \eta p - \theta k p^{\frac{2}{3}} - \delta k - f k D_r(t) \tag{9}$$

The change in carrying capacity with time will depend on this four assumptions:

1. ηp ; increase in carrying capacity due to stimulatory signals from tumour cells (angiogenesis) where η is the rate at which stimulator is released per day and p is the tumour volume at a given time.
2. $\theta k p^{\frac{2}{3}}$; decrease in the carrying capacity with time due to inhibitory signals from the tumour cells. The term $\frac{2}{3}$ was derived by Hahnfeldt using diffusion-consumption equation.
3. δk; decrease in the carrying capacity due to natural death of endothelial cells.

4. $fkD_r(t)$; decrease in the carrying capacity due to antiangiogenic drugs. f is rate at which drug is administered. D_r can be further defined as

$$D_r(t) = \int_0^t A(T)\mathrm{e}^{(-c_r(t-T))}\,\mathrm{d}T$$

where T is the time at which drug is administered, c_r is clearance rate of any drug and $A(T)$ is dosage given during treatment.

Hahnfeldt et al. experimented different treatment modules of TNP-470, angiostatin or endostatin.

We introduce spontaneous loss as a δk in this equation. Here, f is the rate at which the drug is administered and $D_r(t)$ is the drug concentration at a given time inside the body of the patient.

3 Stability Analysis

3.1 *Qualitative Analysis of Hahnfeldt Et. Al. Model and Modified Hahnfeldt Model Without Treatment*

We have a system of nonlinear ordinary differential equations in the modified Hahnfeldt model, and therefore, first we must linearize this system and find eigenvalues of the Jacobian matrix at the equilibrium point [10].

We have analysed the stability for untreated tumours. i.e. $D_r(t) = 0$ and $\omega = 0$.

The equilibrium is found when system is independent of time
i.e. $\frac{\mathrm{d}p}{\mathrm{d}t} = 0$ and $\frac{\mathrm{d}k}{\mathrm{d}t} = 0$.

The equilibrium point of giving nonlinear system of ordinary differential equation is:

$$E(p,k) = \left(\left(\frac{\eta-\delta}{\theta}\right)^{\frac{3}{2}}, \left(\frac{\eta-\delta}{\theta}\right)^{\frac{3}{2}}\right) \tag{10}$$

There is no trivial equilibrium which means:

$$E(p,k) \neq (0,0) \tag{11}$$

From equation (9) and (10), at an equilibrium point i.e. $\frac{\mathrm{d}p}{\mathrm{d}t} = 0$ and $\frac{\mathrm{d}k}{\mathrm{d}t} = 0$. which yields

$$q1 = -\gamma p \ln\left(\frac{p}{k}\right) = 0 \tag{12}$$

$$q2 = \eta p - \theta k p^{\frac{2}{3}} - \delta k = 0 \tag{13}$$

The Jacobian matrix is as follows

$$J = \frac{\partial(q1, q2)}{\partial(p, k)} = \begin{bmatrix} \frac{\partial q1}{\partial p} & \frac{\partial q1}{\partial k} \\ \frac{\partial q2}{\partial p} & \frac{\partial q2}{\partial k} \end{bmatrix}$$

$$J = \begin{bmatrix} -\gamma & \gamma \\ \frac{1}{3}\eta + \frac{2}{3}\delta & -\eta \end{bmatrix}$$

Eigenvalues of the Jacobian matrix at its equilibrium point is:

$$E_1 = \frac{1}{2}(-(\gamma + \eta) + \sqrt{\left(\gamma^2 + \eta^2) - \frac{2}{3}\eta\gamma + \frac{8}{3}\gamma\delta\right)} \tag{14}$$

$$E_1 = \frac{1}{2}(-(\gamma + \eta) - \sqrt{\left(\gamma^2 + \eta^2) - \frac{2}{3}\eta\gamma + \frac{8}{3}\gamma\delta\right)} \tag{15}$$

Using the eigenvalues, we have predicted the stability of given system.

Qualitative analysis of both the models is the same. However, the eigenvalues will be different because δ is considered 0 in Hahnfeldt et al. model.

3.2 *Result and Discussion*

The experimental data for simulation of the Hahnfeldt model as well as modified Hahnfeldt model was extracted from various in vivo experiments carried in [6, 11–13] using Webplotdigitizer. The extracted data was plotted scattered diagrams. Both the models were simulated using the parameters mentioned in Table 1. For Hahnfeldt model, δ is taken as zero [6].

As all the eigenvalues are real and negative, it is concluded that

the system is asymptotically stable. This means that the values will remain stable with time.

Table 1 Parameter values

Data set or Cell line	γ	η	θ	δ	
				MHM	HM
Lewis lung carcinoma (Hahnfeldt et al.) [6]	0.192	5.85	0.00873	0.0295	0.0
Breast adeno-carcinoma xenografts-1[14]	4.7192	0.1519	0.0015	0.0215	0.0
Breast adeno-carcinoma xenografts-2 [14]	0.1340	1.1503	0.0088	0.0053	0.0
H226 (lung cancer cell line) [14]	0.1028	0.7264	0.0075	0.0053	0.0
HT29 (Human colon cancer cell line) [14]	0.2138	0.5061	0.0017	0.0191	0.0

HM Hahnfeldt et al. *Model MHM* Modified Hahnfeldt Model

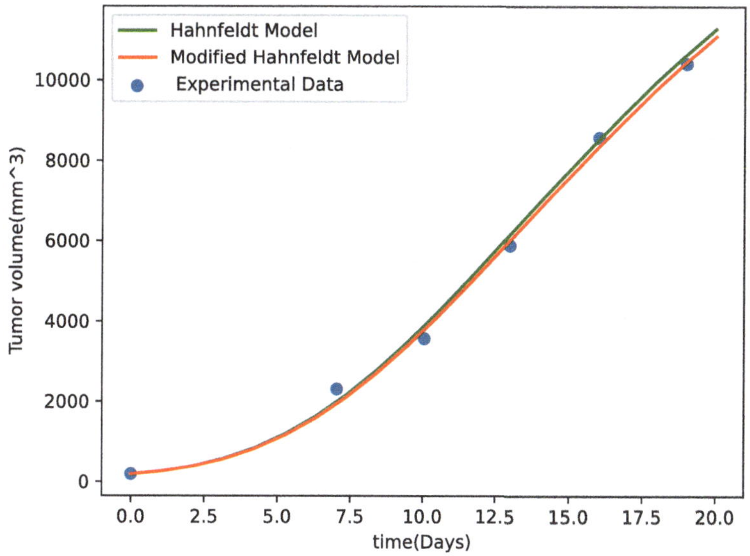

Fig. 1 Tumour volume versus time plotted from experimental data of Lewis lung carcinoma extracted from Hahnfeldt et al. This figure shows the comparison between the Hahnfeldt model and the modified Hahnfeldt model [6]

Table 2 Eigenvalue with corresponding parameter (Modified Hahnfeldt Model) taken from Table 1

Data set or Cell line	Eigenvalue
Lewis lung carcinoma (Hahnfeldt et al.)	[−0.1259323 −5.9160677]
Breast adenocarcinoma xenografts-1	[−4.78536872 −0.08573128]
Breast adenocarcinoma xenografts-2	[−0.08149756 −1.20280244]
H226 (lung cancer cell line)	[−0.06463747 −0.76456253]
HT29 (Human colon cancer cell line)	[-0.11469448 -0.60520552]

Table 3 Eigenvalue with corresponding parameter (Hahnfeldt et al. Model) taken from Table 1

Data set or Cell line	Eigenvalue
Lewis lung carcinoma (Hahnfeldt et al.)	[−0.12658451 −5.91541549]
Breast adenocarcinoma xenografts-1	[−4.77093137 −0.10016863]
Breast adenocarcinoma xenografts-2	[−0.08573605 −1.19856395]
H226 (lung cancer cell line)	[−0.0651568 −0.7640432]
HT29 (human colon cancer cell line)	[−0.12030885 −0.59959115]

In Fig. 1, we have plotted tumour volume vs time from experimental data, Hahnfeldt et al. model as well as modified Hahnfeldt et al. model, where the parameters are taken as per Table 2 and value of δ is 0.05. The experimental data of Lewis lung carcinoma is taken from [6]. The simulations of Hahnfeldt et al. model and modified Hahnfeldt et al. model were compared with experimental data, and we see that the modified Hahnfeldt et al. model gives more precise results and overlaps with the experimental data compared to Hahnfeldt et al. model (Table 3).

In Fig. 2 the simulations of Hahnfeldt model and the modified Hahnfeldt model are compared with other in vivo experiments. In Fig. 2a large difference can be seen between both the simulations where the Hahnfeldt model shows more growth and the modified Hahnfeldt model shows similarity to the experimental data. Figure 2b and 2c both the models showed similar results, but the root mean squared error for the modified Hahnfeldt model was lower. Figure 2d shows similar results as Fig. 2a where the Hahnfeldt model estimates more growth as compared to the modified Hahnfeldt model.

In Fig. 3, the bar graph indicates the root mean square error of both the models in comparison with the experimental data.

4 Numerical Simulation

The experimental data for simulation of the Hahnfeldt model as well as modified Hahnfeldt model was extracted from various in vivo experiments carried in [6, 11–13] using Webplotdigitizer. The extracted data was plotted scattered diagrams.

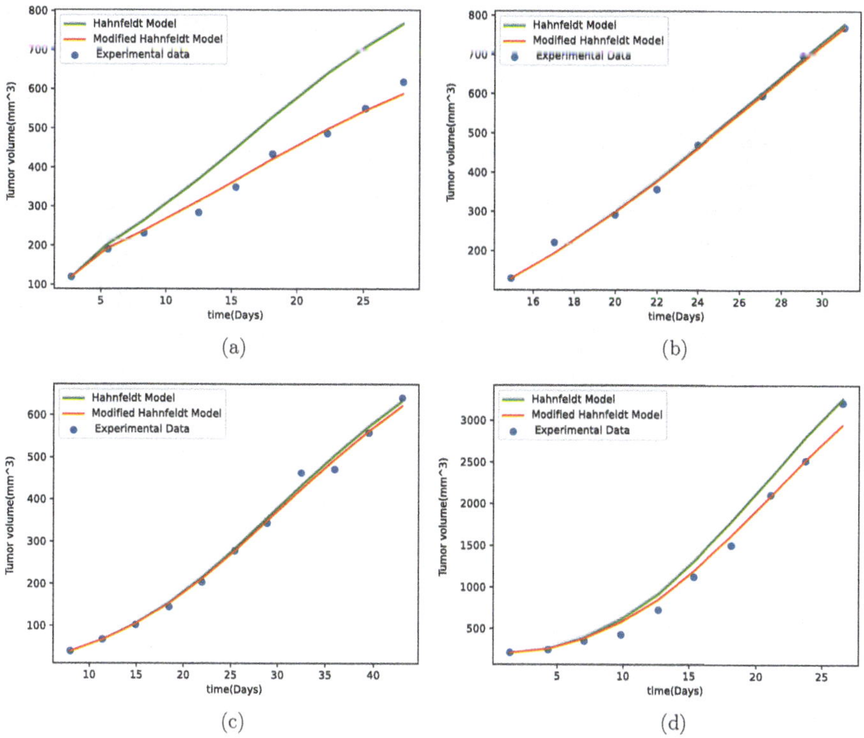

Fig. 2 **a** Experimental data from Higgins et al. is used to compare both models. Tumour volume vs time is plotted which shows the comparison between them[11]. **b** The experimental data from Higgins et al. is used to compare both models. Tumour volume vs time graph is plotted which shows the comparison between them [11]. **c** The experimental data from Hoang et al. is used to compare both models. Tumour volume vs time graph is plotted which shows the comparison between them [12]. **d** The experimental data from Selvakumaran et al. is used to compare both models. Tumour volume vs time graph is plotted which shows the comparison between them [13]

Numerical simulations were carried out for the data without treatment and solved using PYTHON built in function *odeint*.

4.1 Root Mean Square Error

Here we have calculated root mean square error to analyse the accuracy of both the models compared to experimental data

$$RMSE = \sqrt{\frac{\Sigma(x_1 - x_2)^2}{N}} \tag{16}$$

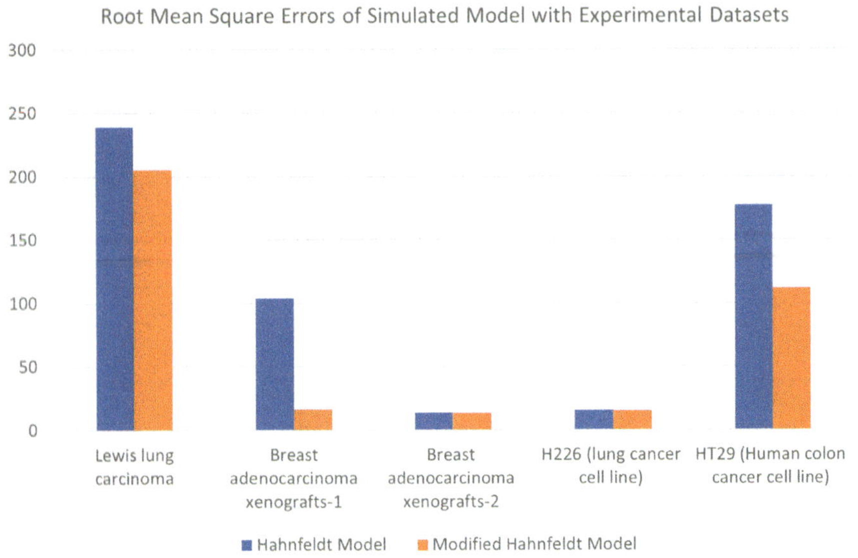

Fig. 3 Graph shows the root mean square error comparison between the Hahnfeldt model and the modified Hahnfeldt model for different data sets

where RMSE is root mean square error, N is total number of data, x_1 is simulated value and x_2 is estimated value of tumour volume.

5 Conclusion

We have developed a model of tumour growth, tumour treatment and explained how model can be used to simulate complex biological systems. Also, we have performed stability analysis of Hahnfeldt model as well as modified Hahnfeldt model with different values of parameters to recognize the dynamics of tumour growth without treatment. It can also be seen that the eigenvalues are all real and negative, hence the system is asymptotically stable and value of parameters remain stable with time. That means this holds true in a biological system.

The spontaneous loss of vasculature due to natural death of endothelial cells inhibits the tumour growth. Tumour angiogenesis results in proliferation of vasculature and can be stopped by various antiangiogenic drugs. Apart from drug induced death or inhibition of proliferation of endothelial cells, natural death due to apoptosis is also one aspect during tumour angiogenesis. This spontaneous loss of vasculature due to natural death of endothelial cells cannot be considered negligible and according to our assumption is should be between 0.005 and 0.05 which provides minimum error and simulation is also similar to that of experimental data. In this paper, we have only taken into consideration in vivo experiments without any treatment to first understand the functioning of both the models in the absence of treatment. Further work needs to be done experiments with antiangiogenesis as well as chemotherapeutic treatments.

References

1. World Health Organization Cancer: http://www.who.int/health-topics/cancer#tab=tab_1. Last accessed 18 May 2021
2. Benzekry, S., Gandolfi, A., Hahnfeldt, P.: A mathematical model of systemic inhibition of angiogenesis in metastatic development. Congress SMAI **2013**, 75–87 (2013)
3. Dołbniak, M., Świerniak, A.: Comparison of simple models of periodic protocols for combined anticancer therapy. Comput. Math. Methods Med. **2013**, 11 (2013)
4. Folkman, J.: Clinical applications of research on angiogenesis. New Engl. J. Med. **333**, 1757–1763 (1995)
5. Enderling, H., Chaplain, M.: Mathematical modeling of tumor growth and treatment. Curr. Pharm. Design **20**(30), 4934–40 (2014)
6. Hahnfeldt, P., Panigrahy, D., Folkman, J., Hlatky, L.: Tumor development under angiogenic signaling: a dynamical theory of tumor growth, treatment response, and postvascular dormancy. Cancer Res. **59**, 4770–4775 (1999)
7. Stepien, T., Kostelich, E., Kuang, Y. : Mathematics + cancer: An undergraduate "Bridge" course in applied mathematics. SIAM Rev. **62**(1), 244–263 (2020)
8. d'Onofrio, A., Gandolfi, A.: Tumour eradication by antiangiogenic therapy: analysis and extensions of the model by Hahnfeldt et al. Math. Biosci. **191**(2), 159–84 (2004)
9. Poleszczuk, J., Bodnar, M., Foryś, U.: New approach to modeling of antiangiogenic treatment on the basis of Hahnfeldt et. al. model. Math. Biosci. Eng. **8**(2), 591–603 (2011)
10. Sari, E., Lestari, D., Yulianti, E., Subekti, S.: Stability analysis of a mathematical model of tumor with chemotherapy. J. Phys. Conf. Ser. **1321**(2), 022072 (2019)
11. Higgins, B., Kolinsky, K., Linn, M., Adames, V., Zhang, Y., Moisa, C., Dugan, U., Heimbrook, D., Packman, K.: Antitumor activity of capecitabine and bevacizumab combination in a human estrogen receptor-negative breast adenocarcinoma xenograft model. Anticancer Res. **27**(4), 2279–87 (2007)
12. Hoang, T., Huang, S., Armstrong, E., Eickhoff, J., Harari, P.M.: Enhancement of radiation response with bevacizumab. J. Experim. Clin. Cancer Res. **31**(1), 37 (2012)
13. Selvakumaran, M., Yao, K., Feldman, M., O'Dwyer, P.: Antitumor effect of the angiogenesis inhibitor bevacizumab is dependent on susceptibility of tumors to hypoxia-induced apoptosis. Biochem. Pharmacol. **75**(3), 627–38 (2008)
14. Argyri, K., Dionysiou, D., Misichroni, F., Stamatakos, G.: Numerical simulation of vascular tumour growth under antiangiogenic treatment: addressing the paradigm of single-agent Bevacizumab therapy with the use of experimental data. Biol. Direct **11**(1), 12 (2016)

A Computational Approach to the Three-Body Coulomb Problem: Positron Scattering from Atomic Systems

Kuru Ratnavelu and Jia Hou Chin

Abstract The fundamental three-body Coulomb system has been intractable. Nonetheless, there were quantum leaps in this research field over the last 3 decades, especially in the simplest system of electron and positron-hydrogen atomic systems. Generally, the dynamics of positron-atom scattering systems can be considered similar to the electron-atom scattering systems. However, these early approximations which extended the electron-atom scattering codes, ranging from the born-based approximation methods to the close-coupling (CC) and the powerful R-Matrix methods, failed to consider the positronium formation (Ps) channel and seemed valid only for high-energy regimes. In the last 25 years, many theoretical approaches have been developed to provide a realistic treatment for positron scattering from atoms or molecules with the incorporation of the Ps channel. Here, we will discuss a theoretical method that incorporates the optical potential formalism to treat the three-body collisions involving these scattering systems. Computational details are described, and various test studies are presented.

Keywords Three-body Coulomb system · Positron-atom scattering · Optical potential

---

K. Ratnavelu (✉)
Institute of Computer Science and Digital Innovation, UCSI University, 56000 Kuala Lumpur, Malaysia
e-mail: kurunathan@ucsiuniversity.edu.my

Institute of Mathematical Sciences, Faculty of Science, University of Malaya, 50603 Kuala Lumpur, Malaysia

J. H. Chin
Department of Actuarial Science and Applied Statistics, UCSI University, 56000 Kuala Lumpur, Malaysia
e-mail: chinjh@ucsiuniversity.edu.my

P. Balasubramaniam et al. (eds.), *Mathematical Modelling and Computational Intelligence Techniques*, Springer Proceedings in Mathematics & Statistics 376,
https://doi.org/10.1007/978-981-16-6018-4_4

1 Introduction

The positron (e^+) is an antiparticle of the electron (e^-). Essentially, it has the same mass and magnitude of charge as the electron but it differs with its positive charge. The study in anti-matter has pique the interest of physicists since the first postulation of positron by Dirac.

Positron and electron are particles with opposite sign in the charges and the collision of these particles with atoms bring about some interesting physical differences. The static interaction is distinctly different with being repulsive between atom and the positron and attractive for the electron case. Whereas the polarization effects are the same, the distinct difference in the sign of the charge leads to the positron–electron correlations in the form of the virtual and real positronium (Ps) formation. The earlier theoretical works that used the electron-atom codes to extend to the study of positron-atom collisions with only the change in the magnitude of the sign completely ignored these physical interactions. However, this leads to an incomplete theoretical calculations in the measured physical parameters such as Ps formation cross sections.

In the last three decades, technological advancements in the development of positron sources have provided further impetus to theoreticians to calculate accurate physical observables such as total cross sections and differential cross sections [1]. Here, we will not review the state of theoretical and experimental works, and the reader is referred to various reviews such as Ratnavelu et al. [2] and Chiari and Zecca [3].

The present work was motivated by the theoretical work of Mitroy [4] in e^+-H which explicitly used the Ps states in the close-coupling expansion of the wavefunction together with the atomic states. The e^+-H is one of the fundamental 3-body Coulombic systems. It involves the positron, an electron and the proton. The e^+-hydrogenic atom is similar to this fundamental system with its inert core as a charged ion, the atomic electron and the probing positron.

But there is a drawback in the use of the close-coupling method that expands the wave function in terms of the set of hydrogenic and positronioum wave functions [4, 5]. There is a possibility that expanding the wave functions in both centers of mass would lead to overcompleteness. This is overcomed by the prescription of Mitroy [5]. Mitroy [6] and Kernoghan et al. [7] performed large-scale close-coupling calculations on the positron-hydrogen system in the 1990s. Thus, we pursued the optical potential formalism of McCarthy and Stelbovics [8] within the close-coupling formalism of Mitroy [4] to the positron-hydrogen system [9]. Other theoretical works using the present optical potential method can be found in [10–13].

In the Sect. 2, the Feshbach formalism of the optical potential method of as applied to the positron-hydrogenic system is described and in Sect. 3, the computational details are described. In Sect. 3.4, we outline the computational code and present some calculations in Sect. 4. The conclusions are given in Sect. 5.

2 Optical Potential Formalism

For positron scattering from H atom (or the corresponding hydrogenic target), the system is essentially (or reduced) a simple three-body Coulombic system. Thus, the Hamiltonian for the positron-hydrogen (or hydrogenic) atom can be approximated to

$$\mathrm{H} = K_1 + K_2 + v_1 + v_2 + v_3 = K + v \tag{1}$$

where K_1 and K_2 stand for the kinetic energy operators of the incoming positron and the valence electron, respectively. v_1 and v_2 are the positron-proton (or core) and the electron-proton (or core) potential operators, respectively, while v_3 is the positron–electron potential operator. For simplicity, let $K = K_1 + K_2$ and $v = v_1 + v_2 + v_3$. The spin–orbit coupling is ignored, as the electron spin plays a role only at the application of the Pauli's exclusion principle.

Thus, the Schrödinger equation is written for the scattering system

$$(E^{(+)} - H)\psi(r_1, r_2) = 0 \tag{2}$$

where $\Psi(r_1,r_2)$ is the trial wave function of the collision system and (+) denotes the outgoing spherical-wave boundary conditions. The coordinates r_1 and r_2 represent the valence electron and positron with respect to the proton, respectively.

Using an eigen function expansion of the positron scattering states $F_\alpha\ (r_2)$ and the Ps states $\phi_\beta(\rho)$ which are coupled to the atomic hydrogenic states $\psi_\alpha(r_1)$, we obtain

$$\Psi(r_1, r_2) = \sum_\alpha \psi_\alpha(r_1)F_\alpha(r_2) + \sum_\beta \varphi_\beta(\rho)G_\beta(R) \tag{3}$$

where ρ is the relative coordinate and R is the center of mass of outgoing Ps.

The hydrogenic state $\psi_\alpha(r_1)$, and the positronium state $\varphi_\beta(\rho)$, satisfy these conditions:

$$\left\langle \psi_\alpha(r_1) \middle| \left(-\frac{1}{2}\nabla_1^2 - \frac{1}{r_1} - \varepsilon_\alpha \right) \middle| \psi_\alpha(r_1) \right\rangle = 0 \tag{4}$$

$$\left\langle \varphi_\beta(\rho) \middle| \left(-\nabla_\rho^2 - \frac{1}{\rho} - \varepsilon_\beta \right) \middle| \varphi_\beta(\rho) \right\rangle = 0 \tag{5}$$

Thus, Eq. (2) can be rewritten as

$$(E - H)\left\{ \sum_\alpha \psi_\alpha(r_1)F_\alpha(r_2) + \sum_\beta \varphi_\beta(\rho)G_\beta(R) \right\} = 0 \tag{6}$$

The Hamiltonian, H is defined as

$$H = -\frac{1}{2}\nabla_1^2 - \frac{1}{2}\nabla_2^2 - \frac{1}{r_1} + \frac{1}{r_2} - \frac{1}{r_{12}} \tag{7}$$

Following standard integrating procedures, Eq. (6) can be rewritten as;

$$\begin{aligned}\left(E + \frac{1}{2}\nabla_2^2 - \varepsilon_{\alpha'}\right)F_{\alpha'}(r_2) &= \sum_{\alpha}\left\langle\psi_{\alpha'}\left|\left(\frac{1}{r_2} - \frac{1}{r_{12}}\right)\right|\psi_{\alpha}\right\rangle F_{\alpha}(r_2)\\ &\quad + \sum_{\beta}\langle\psi_{\alpha'}|(H - E)|\varphi_{\beta}G_{\beta}\rangle\end{aligned} \tag{8}$$

$$\begin{aligned}\left(E + \frac{1}{4}\nabla_R^2 - \varepsilon_{\beta'}\right)G_{\beta'}(R) &= \sum_{\beta}\left\langle\varphi_{\beta'}\left|\left(\frac{1}{r_2} - \frac{1}{r_1}\right)\right|\varphi_{\beta}\right\rangle G_{\beta}(R)\\ &\quad + \sum_{\alpha}\langle\varphi_{\beta'}|(H - E)|\psi_{\alpha}F_{\alpha}\rangle\end{aligned} \tag{9}$$

By standard techniques [4], these equations can be written as the momentum-space Lippmann–Schwinger (LS) equations for a positron with the momentum k incident on a hydrogenic atom in ψ_α state. The T-matrix of the positron-atom scattering is

$$\begin{aligned}\langle k'\psi_{\alpha'}|T|k\psi_{\alpha}\rangle &= \langle k'\psi_{\alpha'}|V|k\psi_{\alpha}\rangle + \sum_{\alpha'}\int d^3k''\frac{\langle k'\psi_{\alpha'}|V|k''\psi_{\alpha'}\rangle\langle k''\psi_{\alpha'}|T|k\psi_{\alpha}\rangle}{\left(E^{(+)} - \varepsilon_{\alpha''} - \frac{1}{2}k''^2\right)}\\ &\quad + \sum_{\beta''}\int d^3k''\frac{\langle k'\psi_{\alpha'}|V|k''\varphi_{\beta'}\rangle\langle k''\varphi_{\beta''}|T|k\psi_{\alpha}\rangle}{\left(E^{(+)} - \varepsilon_{\beta''} - \frac{1}{4}k''^2\right)}\end{aligned} \tag{10}$$

$$\begin{aligned}\langle k'\varphi_{\beta'}|T|k\psi_{\alpha}\rangle &= \langle k'\varphi_{\beta'}|V|k\psi_{\alpha}\rangle + \sum_{\alpha''}\int d^3k''\frac{\langle k'\varphi_{\beta'}|V|k''\varphi_{\alpha'}\rangle\langle k''\psi_{\alpha''}|T|k\psi_{\alpha}\rangle}{\left(E^{(+)} - \varepsilon_{\alpha''} - \frac{1}{2}k''^2\right)}\\ &\quad + \sum_{\beta''}\int d^3k''\frac{\langle k'\varphi_{\beta'}|V|k''\varphi_{\beta''}\rangle\langle k''\varphi_{\beta'}|T|k\psi_{\alpha}\rangle}{\left(E^{(+)} - \varepsilon_{\beta''} - \frac{1}{4}k''^2\right)}\end{aligned} \tag{11}$$

Ratnavelu and collaborators [9, 11, 13] had extended the formalism of Feshbach [14] to the positron-hydrogenic atoms with some approximations. Here, the space of collision reaction can be divided into two orthogonal or complementary subspaces by means of the projection operators P and Q. The P-space describes the finite set of the discrete channels considered including the ground state, while Q-space includes the continuum and the remaining discrete states that are not explicitly coupled in the coupled-channel calculation. In principle, the P-space should include the H (or hydrogenic) atomic states as well as the Ps states but these lead to over-completeness, and an approximation is made within the atomic (H) center-of-mass and ignore the Ps center-of-mass. These will remove the over-completeness that afflicts calculations that uses expansions about the two centers of mass. Details can be found in Ref. [9].

The Schrodinger Eq. (6) can be written in the operator form as

$$E - K - v = 0 \tag{12}$$

where $K = K_1 + K_2$ and $v = v_1 + v_2 + v_3$. Applying the properties of P and Q in Eq. (12)

$$\sum_{\alpha'} \langle \psi_\alpha | (E - K - v)(P + Q) | \psi_{\alpha'} \rangle F_{\alpha'} = 0 \tag{13}$$

Hence, the optical potential [9] is defined as

$$V^{(Q)} = v_1 + v_3 + (v_1 + v_3) Q \frac{1}{Q(E - K - v)Q} Q (v_1 + v_3) \tag{14}$$

Equation (14) is an approximation to the original Schrodinger equation appearing in Eq. (13). The second term of the optical potential $V^{(Q)}$ formula is called the complex non-local polarization potential $W^{(Q)}$ and $V^{(Q)}$ can be simplified as

$$V^{(Q)} = V + W^{(Q)} \tag{15}$$

Using Faddeev's spectral representation [15], the Green's function in Eq. (14) becomes

$$\frac{1}{Q(E - K - v)Q} = Q \sum_n \left| \Psi_n^{(-)} \right\rangle \frac{1}{E - E_n} \left\langle \Psi_n^{(-)}(q) \right| Q \tag{16}$$

The spectral index, n is a discrete notation for remaining discrete channels including the target continuum states. It defines the asymptotic partition of the three-body system into bound or continuum states and specifies the quantum numbers and momenta within each partition. The optical potential is written explicitly by substituting $Q = \sum_{c \in Q} |\psi_c\rangle\langle\psi_c|$ into Eq. (16):

$$\begin{aligned} V^{(Q)} &= (v_1 + v_{12}) + (v_1 + v_{12}) \\ &\times \sum_n \sum_{m \in Q} |\psi_m\rangle \langle \psi_m | \Psi_n^{(-)} \rangle \frac{1}{E - E_n} \langle \Psi_n^{(-)} | \psi_m \rangle \langle \psi_m | (v_1 + v_{12}) \end{aligned} \tag{17}$$

where $\Psi_n^{(-)}$ is the three-body wave function for ingoing spherical-wave boundary conditions. Thus, optical potentials can be decomposed into the first- and second-order terms [16] where q and q' are the wavenumbers associated with the incoming channel i and outgoing channel j.

$$V_{ij}(q', q) = V_{ij}^{(1)}(q', q) + V_{ij}^{(2)}(q', q)$$

$$V_{ij}^{(1)}(q', q) = \langle q'\psi_i|v_1 + v_3|\psi_j q\rangle$$
$$V_{ij}^{(2)}(q', q) = \sum_k \langle q'\psi_i|v_1 + v_3\left|\psi_k \chi_k^{(-)}\right\rangle \frac{1}{E - E_k}\left\langle \chi_k^{(-)}\psi_k\right|v_1 + v_3|\psi_j q\rangle \quad k \in Q \tag{18}$$

where $\chi_k^{(-)}$ is a distorted-wave. The second-order term $V_{ij}^{(2)}$ is separated into Hermitian and anti-Hermitian parts

$$V_{ij}^{(2)} = -U_{ij} - iW_{ij} \tag{19}$$

where U_{ij} is the real polarization term involving virtual (off-energy-shell) excitations into Q space. W_{ij} is the absorptive term involving on-shell excitations. The extreme screening approximation [17] is used to calculate the polarization potential for the continuum. In this approximation, the full Coulomb wave is used for the slower of two particles (either the bound electron or the incoming positron), and a plane wave is used for the faster of the two particles. This partially satisfies the requirement that P and Q-space are orthogonal to the ground state of hydrogen. This is represented as

$$\begin{aligned} |\ell\rangle &= |\psi^{(-)}(q')\rangle \quad \left|\chi_\ell^{(-)}(q)\right\rangle = |q\rangle \quad q \ge q' \\ |\ell\rangle &= |q'\rangle \quad \left|\chi_\ell^{(-)}(q)\right\rangle = |\psi^{(-)}(q)\rangle \quad q < q' \end{aligned} \tag{20}$$

Then, the continuum potential is written as

$$V_{ij}^{(2)}(p) = \frac{1}{2}\int_{-1}^{1} du \int d^3q'' \int d^3p'' \langle (K + p)\psi_i|v_3|\Psi^{(-)}(q_<)q_>\rangle \times \frac{1}{E^{(+)} - \frac{1}{2}(q''^2 + p''^2)}\langle q_>\psi^{(-)}(q_<)|v_3|\psi_j K\rangle \tag{21}$$

where $p = q - q'$ and $|K| = \sqrt{2E_0}$. This is justified on the basis of the optical model to calculate ionization cross sections which are in reasonably good agreement with available experimental data [17].

The total ionization cross section is related to the polarization potential for the continuum [17] by

$$\sigma_I = \left(\frac{2}{k}\right)(2\pi)^3 W(0) \tag{22}$$

$V_c = U + iW$, where k is the momentum of the incident positron and $W(0)$ is the imaginary part of the polarization potential V_c when $|k_i| = |k_f|$. McCarthy and Stelbovics [16] have shown that ionization cross sections for the electron hydrogen

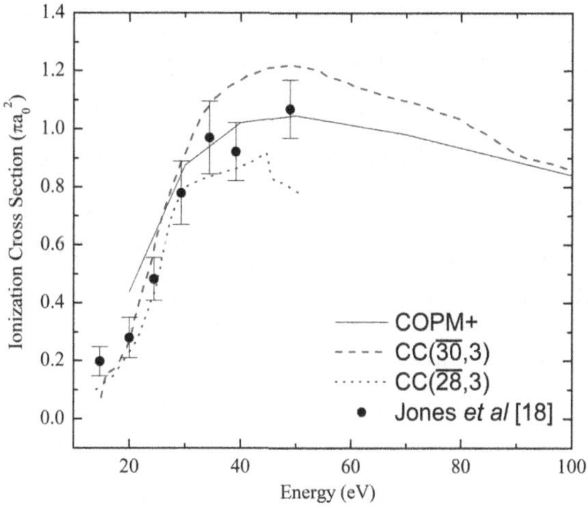

Fig. 1 Total ionization cross section for e^+-H scattering

atom calculated using the extreme screening approximation agrees well with experimental data above 100 eV. Their theoretical ionization cross sections (denoted as continuum optical potential (COPM-)) also show fair agreement for energies as low as 50 eV.

In Fig. 1, the total ionization cross sections for the positron-hydrogen atom (COPM+), calculated using Eq. (19) is compared with other theoretical calculations (CC $(\overline{28}, 3)$[6]), (CC $(\overline{30}, 3)$ [7]) and experimental measurements [18]. The COPM+ is excellent agreement with other theoretical and experimental measurements.

3 Computational Details

3.1 Computational Details of the Optical Potentials

The non-local continuum polarization potential, V_c, can be written as

$$V_c = \int \mathrm{d}q_1 \int \mathrm{d}q_2 f^*(k, q_1, q_2)(E - K)^{-1} f(k', q_1, q_2) \tag{23}$$

where $f(k, q_1, q_2)$ is the direct breakup matrix element. Equation (23) can be integrated by decomposing the Green's function into its principal value and imaginary parts. By using the hyperspherical transformation, the equation can be written as the real and imaginary parts of V_c where

$$\mathrm{Re}\, V_c = \int \mathrm{d}\hat{q}_1 \int \mathrm{d}\hat{q}_2 \int_0^\infty \mathrm{d}\sigma \int_0^{\pi/4} \mathrm{d}\alpha J_\alpha \left(\frac{\sigma^4 F - W^2 F_{\mathrm{on}}}{W - \sigma^2} \right) \tag{24}$$

$$\text{Im}\, V_c = -\pi \int d\hat{q}_1 \int d\hat{q}_2 \int_0^{\infty} d\sigma \int_0^{\pi/4} d\alpha\, J_\alpha W^2 F_{\text{on}} \tag{25}$$

$$\sigma_W = (2E)^{\frac{1}{2}},\ W = 2E,\ J = \sigma^5 J_\alpha,\ J_\alpha = (\sin\alpha\cos\alpha)^2,\ F_{\text{on}} = F\left(k, k', \sigma_W, \sigma_W\right) \tag{26}$$

As the functions in the integrands in Eq. (24) and (25) are functions of one-dimensional variables, the integrals can be integrated using the Diophantine method. By introducing the Diophantine [19] variables $x_1, x_2, x_3, x_4, x_5, x_6$ and x_7, Eq. (25) can be written as

$$\text{Im}\, W_{\text{LC}} = -8\pi^4 \int_0^1 dx_1 \int_0^1 dx_2 \int_0^1 dx_3 \int_0^1 dx_4 \int_0^1 dx_5 \times \int_0^1 dx_7 F_{\text{on}} E_p^2 \left[\cos\left(\frac{\pi}{4}x_5\right) \sin\left(\frac{\pi}{4}x_5\right)\right]^2 \tag{27}$$

which can be computed using the Diophantine multidimensional integration method.

Table 1 shows the convergence of the continuum optical potential for 1 s–1 s coupling of H atom for various momentum transfers at selected energies. In general, the imaginary part of V_c converged rapidly as expected. The integration of the real part is problematic due to the existence of the Green's function. However, large

Table 1 Convergence of the continuum polarization potentials for the ground state of hydrogen atom with different quadrature points for momentum transfer $P = 0.0$ a.u and $P = 0.5$ a.u

$P = 0$	$E = 20$ eV		$E = 50$ eV		$E = 150$ eV	
Q.P	Real	Imag	Real	Imag	Real	Imag
10,000	−0.1047	-0.1193^{-1}	-0.6712^{-1}	-0.4461^{-1}	-0.3529^{-1}	-0.5618^{-1}
60,000	−0.1310	-0.1193^{-1}	−0.1069	-0.4486^{-1}	-0.4510^{-1}	-0.4789^{-1}
120,000	−0.6520	-0.1194^{-1}	-0.7685^{-1}	-0.4480^{-1}	-0.2833^{-1}	-0.4701^{-1}
200,000	−0.2173	-0.1194^{-1}	-0.8661^{-1}	-0.4524^{-1}	-0.3159^{-1}	-0.5009^{-1}
400,000	−0.2173	-0.1194^{-1}	-0.7819^{-1}	-0.4500^{-1}	-0.4742^{-1}	-0.4879^{-1}
P = 0.50	E = 20 eV		E = 50 eV		E = 150 eV	
Q.P	Real	Imag	Real	Imag	Real	Imag
10,000	-0.6110^{-1}	-0.9450^{-2}	-0.5834^{-1}	-0.2670^{-1}	-0.1612^{-1}	-0.2510^{-1}
60,000	-0.7170^{-1}	-0.9410^{-2}	-0.4262^{-1}	-0.2961^{-1}	-0.8790^{-2}	-0.2857^{-1}
120,000	-0.8820^{-1}	-0.9310^{-2}	-0.4646^{-1}	-0.2901^{-1}	-0.2119^{-1}	-0.2945^{-1}
200,000	-0.8110^{-1}	-0.9380^{-2}	-0.4628^{-1}	-0.2903^{-1}	-0.1615^{-1}	-0.2817^{-1}
400,000	-0.7860^{-1}	-0.9990^{-2}	-0.4260^{-1}	-0.2920^{-1}	-0.1704^{-1}	-0.2844^{-1}

number of quadrature points is needed for a good convergence (at about 400,000 quadrature points).

3.2 Numerical Solutions of Lippmann–Schwinger Equations

The abbreviated partial-wave form of the coupled LS equations [4] is

$$T_{ij}(q,k_j) = V_{ij}(q,k_j) + \sum_l \int_0^\infty dq' q'^2 V_{il}(q,q') G_l(q'^2) \times T_{lj}(q',k_j) \tag{28}$$

where q are the momenta for off shell and can take arbitrary values while k_l, the on-shell momenta are the momenta of the external electron for the channels l.

The equations can be solved numerically by using Gaussian quadrature method by transform the LS equations into an algebraic form in term of the coordinate x_{ln}

$$x_{ln} = \begin{cases} k_l & n = 1 \\ q_{n-1} & n = 2, 3, \ldots, N+1 \end{cases} \tag{29}$$

Hence, the coupled LS equations can be written as

$$T_{ij}(x_{in},k_j) = V_{ij}(x_{in},k_j) + \sum_l \sum_{n'=1}^{N+1} K_{il}(x_{in},x_{ln'}) \times T_{lj}(x_{ln'},k_j) \tag{30}$$

in which the kernels are

$$K_{il}(x_{in},x_{ln'}) = W_{ln} V_{il}(x_{in},x_{ln'}) \tag{31}$$

where W_{ln} are the superweights'

$$W_{ln} = \begin{cases} x_{ln}^2 w_{n-1}\left[\frac{1}{2}\left(k_l^2 - x_{ln}^2\right)\right]^{-1} & , \ n = 2, 3, \ldots, N+1 \\ -k_l^2 \sum\limits_{n'=2}^{N+1} w_{n'-1}\left[\frac{1}{2}\left(k_l^2 - x_{ln'}^2\right)\right]^{-1} - i\pi k_l & , \ n = 1 \end{cases} \tag{32}$$

Solving the LS equations using Gaussian quadrature method has been well detailed in [8]. In general, the quadrature points are distributed in such a way that they cover the regions of large and small k, aside from including closely spaced points near the on-shell values of k. Since there are atomic on-shell and Ps on-shell coordinates in the positron cases, a five-panel composite mesh is implemented for the distribution of the quadrature points [20]. As shown in Fig. 2, there are 5 regions and the on-shell momenta of the atomic and Ps are located on the second and fourth regions. The mid-point and the widths of Region 2 and Region 4 are adjusted in such

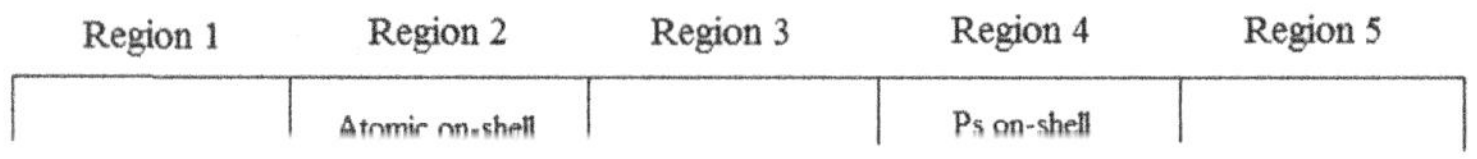

Fig. 2 Five-panel composite mesh. The quadrature points are distributed among the 5 regions

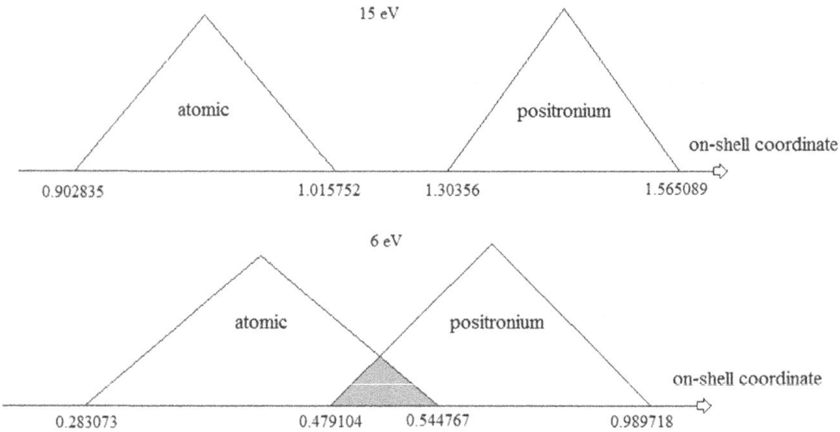

Fig. 3 The illustration of the on-shell coordinates: (Top panel) CC (5,3) for positron-rubidium scattering at 15 eV. (Bottom panel) CC (5,3) for positron-rubidium scattering at 6 eV. The gray area is the overlapping region of atomic and Ps on-shell coordinates

a way that the quadrature points in Region 2 are close to the atomic channels on-shell coordinates, and the quadrature points in Region 4 will be close to the Ps channels on-shell coordinates. It is crucial to have sufficiently dense meshes near the on-shell momenta in order to discretize the kernel.

At lower energy in the positron-hydrogenic atom scattering systems, the on-shell values of the atomic and Ps states can overlap. Region 3 vanishes when the atomic on-shell coordinates (Region 2) are overlapping with the Ps on-shell coordinates (Region 4). The scenario is depicted in Fig. 3. In such cases, the five-panel composite mesh will be less effective in distributing the quadrature points. The best solution for this problem is to resort to single Gaussian mesh [8] for those lower energies in which the on-shell coordinates overlap.

3.3 *Convergence of the Cross Section*

In the calculations, the convergence of the cross section is ensured by using sufficiently large total angular momentum, JMAX. Generally, higher JMAX is required for better convergence at higher incident energies. In Table 2, a large difference in the partial-wave sum and extrapolated partial-wave sum in the 5p channel when JMAX

Table 2 Total cross section for 5s and 5p channel in the unit of πa_0^2 obtained from the CC8 calculation for positron-rubidium scattering at 100 eV. PW refers to the partial-wave

Channel	5s		5p	
JMAX	60	180	60	180
PW sum	7.242	10.328	16.942	26.355
Integral	7.243	7.285	24.226	24.697
Extrapolated PW sum	7.257	10.328	25.959	27.458

= 60 indicates that the cross sections of the 5p scattering has not converged. This issue is rectified by using JMAX = 180.

However, it can be observed from Table 2 that there is a big difference between the partial-wave sum and extrapolated partial-wave sum of 5s and 5p channels when JMAX = 60 and JMAX = 180. Further investigation shows that the CC8 calculation is inconsistent for $J > 60$, as depicted in Fig. 4 as there are fluctuations in the CC8 calculation when $J > 60$. This issue is rectified by merging the CC calculation with the unitarized born approximation (UBA). By using a merging program, the T-matrices of the CC and UBA calculations are merged into a new T-matrix where the new JMAX is the combination of J of the CC before the fluctuations and JMAX of the UBA.

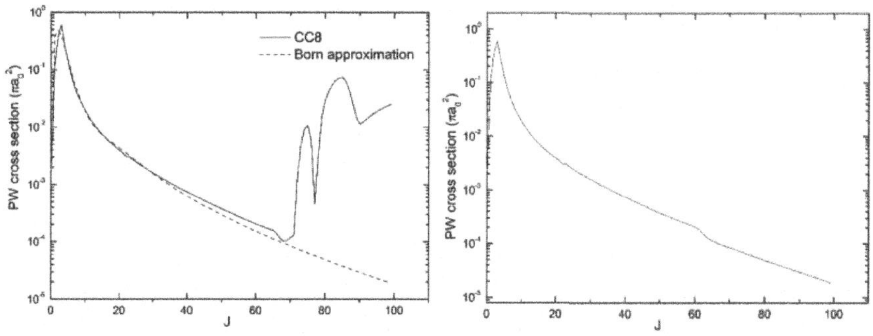

Fig. 4 (Left panel) Partial-wave cross section for 5 s channel in the unit of πa_0^2 obtained from the CC8 calculation of positron-rubidium scattering at 100 eV. The CC8 calculation shows fluctuations around $J = 65$. (Right panel) Partial-wave cross section by using the merged T-matrix and new JMAX

3.4 An Overview of the Coupled-Channel Optical Method (CCOM) Computational Codes

In Sects. 3.1 and 3.2, we have detailed the calculation details of the optical potential and the two-centre Lippman-Schwinger Equations for the scattering of positron from hydrogenic atoms.

Here, we outline the computational code for CCOM method. The CCOM code developed from [4] has been extended to include the continuum optical potentials into the code. The main code CCOM contains various modules to prepare the calculation.

The main input data is contained in a file called CC. ATS and has the necessary information to prepare for the calculation of the positron scattering from a hydrogenic atom.

The following orbital data, channel data and Voptical data are used in the following codes:

1. Orbital Code—Use the input orbital data for hydrogenic atom and Ps-atom to construct the orbital functions for the calculation.
2. Channel Code—Uses the input channel data information to construct the channel information for the calculation.
3. Core Code—Use the core orbital data to construct the core-orbitals for the calculation.
4. Voptical Code—Uses the optical potential input data to construct the optical potential for the calculation.

The other major modules are involved in the construction of various potentials that are needed in the formalism discussed in Sect. 2. These modules are:

1. VDA Code—This module is used for the construction and usage of the 1st Born Momentum-space matrix element.
2. VDR Code—This module is used to construct the radial potential matrix elements for partial waves.
3. VTAIL Code—This module consists of various subroutines:
   a. To compute the value of the integral of two continuum functions and an inverse power of R.
   b. To compute dipole and quadrupole coulomb radial tail integrals.
   c. To computer the value of the integral of two spherical Bessel functions and an inverse power of R.
4. VOPT—Consists of various subroutines and functions such as:
   a. To set up the 3-j and 6-j Wigner for the polar integration of the optical potential.
   b. To evaluation of real or imaginary part of the optical potential for a given channel pair at arbitrary K.
   c. To integrate optical potential matrix elements.

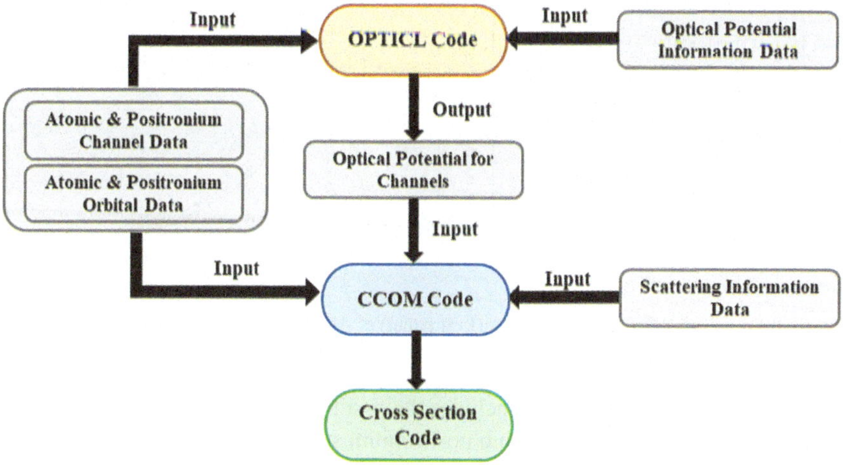

Fig. 5 Flowchart of the CCOM computational codes

5. There are various modules in the CCOM (m, n) code that is needed to calculate Coulomb functions, Legendre polynomials, Wigner elements, density matrices for the orbital functions, grid and quadrature modules for the integration of the LS- Equations and Offdiag optical potentials. The set of numerical packages such as EISPACK to solve the matrix equations among other numerical routines.
6. The OPTICL CODE is a separate program to generate the optical potentials for the positron-hydrogenic atoms and used similar orbital, channel and core data. The code calculates the required optical potential as described in the previous section, and it serves as the input for the CCOM code.

The flowchart of the CCOM method is depicted in Fig. 5.

4 Results

In this section, we report the total cross sections (TCS) and total positronium formation cross sections (TPCS) of the positron-hydrogen (e^+-H) and positron-rubidium (e^+-Rb) atom systems. The following calculations were performed in the energy region of 4 eV to 100 eV:

- Positron-hydrogen atom system
  - CCO (9,6): The $n = 1, 2, 3$ and 4 (1s, 2s, 2p, 3s, 3p, 3d, 4s, 4p and 4d) hydrogen states are included in the expansion together with 6 positronium states (1s, 2s, 2p, 3s, 3p, 3d). The continuum optical potentials for the 1s-1s, 1s-2s, 1s-2p, 2s-2s and 2s-2p and 2p-2p are used.
- Positron-rubidium atom system [11]

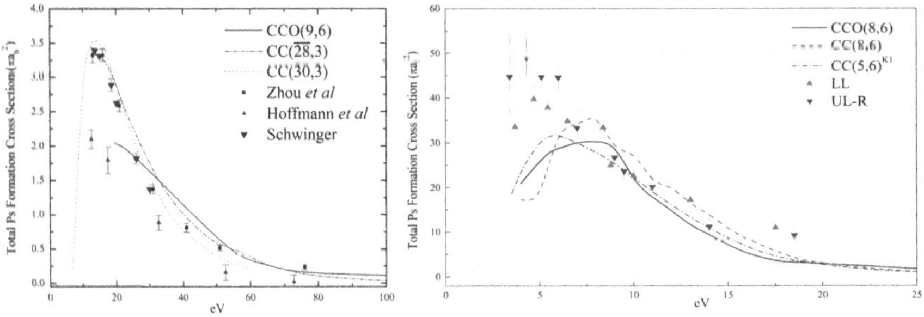

Fig. 6 Total Ps formation cross sections: (Left panel) e^+-H. (Right panel) e^+-Rb

- CC (8,6): CC method that includes the first 8 Rb atomic states (5s, 5p, 4d, 6s, 6p, 5d, 7s, 7p) together with 6 positronium states (1s, 2s, 2p, 3s, 3p, 3d).
- CCO(8,6): The 14 states in CC(8,6) are used together with the continuum optical potentials in the 5s-5s, 5s-5p and 5p-5p coupling.

The TPCS of the e^+-H and e^+-Rb systems are shown in Fig. 6. For e^+-H, the present CCO (9,6) shows overall qualitative agreement with the CC ($\overline{28}$,3) [6], CC($\overline{30}$,3) [7] and Schwinger model [21], as well as the experiments [22, 23] above 20 eV. Below 20 eV, the CCO (9,6) is closer to the Hoffman data but underestimating the other measurement data. It is plausible that the use of $n = 3$ Ps states in the present calculation has led to these differences as we expect $n = 3$ Ps states to play a significant role in the lower and low-intermediate energy region.

For e^+-Rb, we compared the TPCS of CC (8,6) and CCO (8,6) with the CC $(5,6)^{K1}$ R-matrix calculation of Kernoghan et al. [24] (which used 5 atomic states with 6 Ps states in the calculation) as well as the experimental measurement of Surdutovich et al. [25] (LL and UL-R). In general, all theoretical models show good qualitative agreement with the experimental data. However, at energies below 10 eV, our calculations show some differences in structures with the R-matrix calculations. All methods are not in agreement with the measurements below 8 eV. As seen in the positron-H case, it is very plausible that the inclusion of $n = 3$ Ps states contribute to these differences.

The TCS of the e^+-H and e^+-Rb systems are depicted in Fig. 7. For e^+-H, the present CCO (9,6) shows only qualitative agreement with the other calculations and the experiment data of Zhou et al. [22]. The quantitative differences are within 10–15% with other methods.

For e^+-Rb, the CCO (8,6) and CC (8,6) calculations were compared with the R-matrix method (CC $(5,6)^K$ [24]), 5-state CC (CC5^E [26]), frozen core approach (MG3 [27]) and 3-state relativistic CC (RCC [28]). The experimental results of the Detroit group [29] is also included. At first glance, CC (8,6) has better agreement with the theoretical and experimental results than CCO(8,6). However, it must be noted that CCO (8,6) calculation takes into account the continuum effect. The higher TCS of CCO (8,6) especially when the incident energy is larger than ~20 eV indicates that continuum effect is significant at intermediate and high energies.

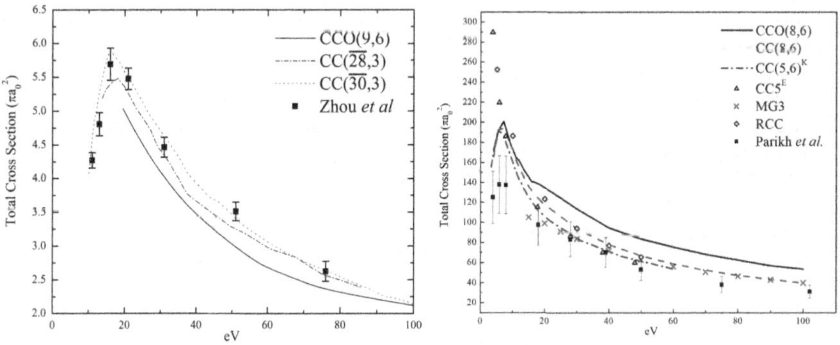

Fig. 7 Total cross sections: (Left panel) e^+-H. (Right panel) e^+-Rb

5 Conclusion

Here, the present CCO (*m*, *n*) method has been fairly successful in demonstrating its application to most hydrogenic atoms such as H and Rb. Other earlier works for Na and Li have been reported. The differences with other large and expensive calculations for some of the physical quantities cannot be gauged with the lack of agreement among other measurements. In general, we are confident that this method has potential to be extended to positron scattering from large atomic systems and molecules.

Acknowledgements KR is grateful to the late Prof Ian E. McCarthy and the late Prof Jim Mitroy for their generosity in sharing their knowledge over their three decades of collaboration from the early days at Flinders University in the late 1980s. The later collaboration with Professors Stephen J Buckman and Prof M. J. Brunger is also greatly valued. The authors also thank Dr. Mohd Zahurin bin Mohamed Kamali of University of Malaya for providing data for this paper.

References

1. Charlton, M., Humberston, J.W.: Positron Physics, vol. 11. Cambridge University Press (2001)
2. Ratnavelu, K., Brunger, M.J., Buckman, S.J.: Recommended positron scattering cross sections for atomic systems. J. Phys. Chem. Ref. Data **48**(2), 023102 (2019)
3. Chiari, L., Zecca, A.: Recent positron-atom cross section measurements and calculations. Eur. Phys. J. D. **68**(10), 1–25 (2014)
4. Mitroy, J.: Close coupling theory of positron? Hydrogen scattering. Aust. J. Phys. **46**(6), 751–772 (1993)
5. Mitroy, J.: An calculation of positron-hydrogen scattering at intermediate energies. J. Phys. B: At. Mol. Opt. Phys. **29**(7), L263 (1996)
6. Mitroy, J.: Positron–hydrogen scattering at intermediate energies. Aust. J. Phys. **49**(5), 919–936 (1996)
7. Kernoghan, A.A., Robinson, D.J.R., McAlinden, M.T., Walters, H.R.J.: Positron scattering by atomic hydrogen. J. Phys. B: At. Mol. Opt. Phys. **29**(10), 2089 (1996)

8. McCarthy, I.E., Stelbovics, A.T.: Momentum-space coupled-channels optical method for electron-atom scattering. Phys. Rev. A **28**(5), 2693 (1983)
9. Ratnavelu, K., Rajagopal, K.K.: Continuum effects on positron scattering of atomic hydrogen at intermediate energies. J. Phys. B: At. Mol. Opt. Phys. **32**(14), L381 (1999)
10. Kamali, M.Z.M., Ratnavelu, K.: Positron-hydrogen scattering at low intermediate energies. Phys. Rev. A **65**(1), 014702 (2001)
11. Chin, J.H., Ratnavelu, K., Zhou, Y.: Positron-rubidium atom scattering at intermediate and high energies. Eur. Phys. J. D **66**(3), 1–8 (2012)
12. Ratnavelu, K., Ng, S.Y.: Positron–lithium atom and electron–lithium atom scattering systems at intermediate and high energies. Chin. Phys. Lett. **23**(7), 1753 (2006)
13. Ratnavelu, K., Ong, W.E.: Electron and positron scattering from atomic potassium. Eur. Phys. J. D **64**(2), 269–285 (2011)
14. Feshbach, H.: Ann. Phys. (NY) **5**, 357 (1958)
15. Faddeev, L.D.: Mathematical Aspects of the Three-Body Problem in the Quantum Scattering Theory, vol. 69. Israel Program for Scientific Translations (1965)
16. McCarthy, I.E., Stelbovics, A.T.: Continuum in the atomic optical model. Phys. Rev. A **22**(2), 502 (1980)
17. Ratnavelu, K.: Positron impact ionisation of H and He atoms: the continuum model. Aust. J. Phys. **44**(3), 265–270 (1991)
18. Jones, G.O., Charlton, M., Slevin, J., Laricchia, G., Kover, A., Poulsen, M.R., Chormaic, S.N.: Positron impact ionization of atomic hydrogen. J. Phys. B: At. Mol. Opt. Phys. **26**(15), L483 (1993)
19. Haber, S.: Numerical evaluation of multiple integrals. SIAM Rev. **12**(4), 481–526 (1970)
20. Ratnavelu, K., Mitroy, J., Stelbovics, A.T.: Positron-hydrogen and positronium-proton scattering at intermediate and high energies. J. Phys. B: At. Mol. Opt. Phys. **29**(13), 2775 (1996)
21. Kar, S., Mandal, P.: Positronium formation in positron-hydrogen scattering using Schwinger's principle. Phys. Rev. A **62**(5), 052514 (2000)
22. Zhou, S., Li, H., Kauppila, W.E., Kwan, C.K., Stein, T.S.: Measurements of total and positronium formation cross sections for positrons and electrons scattered by hydrogen atoms and molecules. Phys. Rev. A **55**(1), 361 (1997)
23. Hofmann, A., Falke, T., Raith, W., Weber, M., Becker, D.P., Lynn, K.G.: Ionization of atomic hydrogen by positrons. J. Phys. B: At. Mol. Opt. Phys. **30**(14), 3297 (1997)
24. Kernoghan, A.A., McAlinden, M.T., Walters, H.R.J.: Positron scattering by rubidium and caesium. J. Phys. B: At. Mol. Opt. Phys. **29**(17), 3971 (1996)
25. Surdutovich, A., Jiang, J., Kauppila, W.E., Kwan, C.K., Stein, T.S., Zhou, S.: Measurements of positronium-formation cross sections for positrons scattered by Rb atoms. Phys. Rev. A **53**, 2861 (1996)
26. McEachran, R.P., Horbatsch, M., Stauffer, A.D.: Positron scattering from rubidium. J. Phys. B: At. Mol. Opt. Phys. **24**, 1107 (1991)
27. Gien, T.T.: Total cross sections for electron and positron collisions with rubidium in a model-potential approach. J. Phys. B: At. Mol. Opt. Phys. **26**(20), 3653 (1993)
28. Feng, X., McEachran, R.P., Stauffer, A.D.: Relativistic close-coupling calculations of positron scattering from the heavy alkalis. Nucl. Instrum. Methods Phys. Res., Sect. B **143**(1–2), 27–31 (1998)
29. Parikh, S.P., Kauppila, W.E., Kwan, C.K., Lukaszew, R.A., Przybyla, D., Stein, T.S., Zhou, S.: Toward measurements of total cross sections for positrons and electrons scattered by potassium and rubidium atoms. Phys. Rev. A **47**(2), 1535 (1993)

Common Best Proximity Points for Some Contractive Type Mappings

M. Sankara Narayanan and M. Marudai

Abstract The aim of this article is to establish the existence of Common Best Proximity Point (CBPP) for two non-self-mappings on a closed subsets of a metric space having weak $\mathcal{P}$-property. Further, we established the same for Kannan and Chatterjea type non-self-contractive mappings with $k = 1$ satisfying demicompact property on a convex bounded closed subset of a Banach space having weak $\mathcal{P}$-property. We also present an example to strengthen our main result.

Keywords Weak $\mathcal{P}$-property · CBPP · Demicompact · Contractive mappings

1 Introduction

Let $\mathcal{A} \neq \emptyset, \mathcal{B} \neq \emptyset$ be subsets of a complete metric space (X, ρ) which are closed. A contraction map $T : \mathcal{A} \to \mathcal{B}$ will not give assurance that $Tx = x$. It is possible that $\rho(x, Tx) > 0$. In this situation, it is necessary to find an element x so that $\rho(x, Tx)$ is minimum. For this one would compute a solution $x \in \mathcal{A}$ approximately with $\rho(x, Tx) \geq \rho(\mathcal{A}, \mathcal{B})$ is minimum and it was the beginning of Best Proximity Point (BPP) theory. BPP results give existence of x with $\rho(x, Tx) = \rho(\mathcal{A}, \mathcal{B})$ which is the global minimum value of $\rho(x, Tx)$. This x is BPP of T. BPP becomes a fixed point, when the considered map becomes a self-map. Many researchers namely Eldred et al. [1, 2] Karapinar [3], Prolla [4], Reich [5], Sehgal, and Singh [6, 7] have investigated BPP results on various mappings. Common fixed point theorem induces the authors [8–14] to investigate CBPP results for non-self-mappings. The authors [8] proved CBPP theorems in a complete metric space using weak $\mathcal{P}$-property for $k \in [0, 1)$.

M. S. Narayanan (✉)
Department of Mathematics, The Gandhigram Rural Institute (Deemed to be University), Gandhigram, Tamil Nadu 624302, India

M. Marudai
Department of Mathematics, Bharathidasan University, Tiruchirappalli, Tamil Nadu 620024, India

P. Balasubramaniam et al. (eds.), *Mathematical Modelling and Computational Intelligence Techniques*, Springer Proceedings in Mathematics & Statistics 376,
https://doi.org/10.1007/978-981-16-6018-4_5

The paper is arranged in the following way. In Sect. 2, we recall some definitions and results needed in the rest of the paper. The last section is devoted to establish **CBPP** results.

2 Preliminaries

In this section, we list out some definitions and results which are useful to prove our main results.

Definition 1 A mapping $T : X \to X$ is said to be demicompact if for every bounded sequence $\{x_n\}$ in X such that $x_n - Tx_n \to x \in X$, there is a convergent subsequence of $\{x_n\}$.

The idea of demicompactness was introduced in order to discuss fixed points. Further examples of demicompact mappings were introduced by Petryshyn [15].

Definition 2 Let $\mathcal{A} \neq \emptyset, \mathcal{B} \neq \emptyset$ be subsets of a metric space. We define $\mathcal{A}_0, \mathcal{B}_0$ and $\rho(\mathcal{A}, \mathcal{B})$ as follows.

(i) $\mathcal{A}_0 := \{x \in \mathcal{A} :$ for some $y \in \mathcal{B}, \rho(x, y) = \rho(\mathcal{A}, \mathcal{B})\}$;
(ii) $\mathcal{B}_0 := \{y \in \mathcal{B} :$ for some $x \in \mathcal{A}, \rho(x, y) = \rho(\mathcal{A}, \mathcal{B})\}$;
(iii) $\rho(\mathcal{A}, \mathcal{B}) := inf\{\rho(x, y) : x \in \mathcal{A}$ and $y \in \mathcal{B}\}$.

Definition 3 ([12]). Let $x \in \mathcal{A}$. x is said to be **CBPP** of the non-self-mappings $T_2, T_1 : \mathcal{A} \to \mathcal{B}$ if $\rho(x, T_2x) = \rho(x, T_1x) = \rho(\mathcal{A}, \mathcal{B})$.

Definition 4 ([16]). Let $\mathcal{A} \neq \emptyset, \mathcal{B} \neq \emptyset$ be subsets of (X, ρ), where (X, ρ) is a metric space such that $\mathcal{A}_0 \neq \emptyset$. The pair $(\mathcal{A}, \mathcal{B})$ is said to satisfy the $\mathcal{P}$-**property** if whenever $x_1, x_2 \in \mathcal{A}_0$, $y_1, y_2 \in \mathcal{B}_0$, then $\rho(x_1, y_1) = \rho(x_2, y_2) = \rho(\mathcal{A}, \mathcal{B}) \Rightarrow \rho(x_1, x_2) = \rho(y_1, y_2)$.

Definition 5 ([14]). Let $\mathcal{A} \neq \emptyset, \mathcal{B} \neq \emptyset$ be two subsets of (X, ρ), where (X, ρ) is a metric space such that $\mathcal{A}_0 \neq \emptyset$. Then the pair $(\mathcal{A}, \mathcal{B})$ is said to satisfy the **weak** $\mathcal{P}$-**property** if whenever $x_1, x_2 \in \mathcal{A}_0$, $y_1, y_2 \in \mathcal{B}_0$, then $\rho(x_1, y_1) = \rho(x_2, y_2) = \rho(\mathcal{A}, \mathcal{B}) \Rightarrow \rho(x_1, x_2) \leq \rho(y_1, y_2)$.
Note 1. $\mathcal{P}$-property implies weak $\mathcal{P}$-property.

Lemma 1 *([17]). Let $\mathcal{A} \neq \emptyset, \mathcal{B} \neq \emptyset$ be closed subsets of (X, ρ), where (X, ρ) is a complete metric space. Let $T : \mathcal{A} \to \mathcal{B}$ be a function with $\mathcal{A}_0 \neq \phi$. Then $T(\mathcal{A}_0) \subseteq \mathcal{B}_0$.*

Definition 6 ([18]). Let (X, ρ) be a metric space and let $T : X \to X$ be a function. Then T is said to be **Kannan mapping** if there exists $\alpha \in [0, \frac{1}{2})$ such that

$$\rho(Tx, Ty) \leq \alpha[\rho(x, Tx) + \rho(y, Ty)], \forall x, y \in X.$$

Theorem 1 *([18]). Let $T : X \to X$ be a function, such that $\rho(Tx, Ty) \leq \frac{\alpha}{2}[\rho(x, Tx) + \rho(y, Ty)], \forall x, y \in X$ where $\alpha \in [0, 1)$ and (X, ρ) is a complete metric space. Then, T has a unique $x \in X$ such that $Tx = x$*

Lemma 2 *([19]). Let $\mathcal{A} \neq \emptyset, \mathcal{B} \neq \emptyset$ be two subsets of a complete metric space (X, ρ) which are closed. Assume that the following axioms are true.*

(i) $\mathcal{A}_0 \neq \phi$;
(ii) Weak $\mathcal{P}$-property is satisfied by the pair.

Then, $\mathcal{B}_0$ is closed.

Lemma 3 *([19]). Let $\mathcal{A} \neq \emptyset, \mathcal{B} \neq \emptyset$ be closed subsets of (X, ρ), where (X, ρ) is a complete metric space. Suppose that the following axioms are true.*

(i) $\mathcal{A}_0 \neq \phi$;
(ii) Weak $\mathcal{P}$-property is satisfied by the pair;
(iii) $T : \mathcal{A} \to \mathcal{B}$ is a continuous function with $T(\mathcal{A}_0) \subseteq \mathcal{B}_0$.

Then, $T(\overline{\mathcal{A}_0}) \subseteq \mathcal{B}_0$.

In 1972, S. K. Chatterjea introduced the concept of C-contraction.

Definition 7 ([20]). Let $T : X \to X$ be a function, where (X, ρ) is a metric space. Then, T is **C-contraction** if there exists $\alpha \in (0, \frac{1}{2})$ with

$$\rho(Tx, Ty) \leq \alpha[\rho(x, Ty) + \rho(y, Tx)], \quad \forall x, y \in X.$$

Theorem 2 *([20]). Let (X, ρ) be a complete metric space and let $T : X \to X$ be a* ***C-contraction****. Then, T has a unique fixed point.*

3 Main Results

In this section, We prove some results on **CBPP** for two non-self-mappings by using demicompact property.

Theorem 3 *Let $\mathcal{A} \neq \emptyset, \mathcal{B} \neq \emptyset$ be subsets of a metric space (X, ρ), which are closed. Let $T_1, T_2 : \mathcal{A} \to \mathcal{B}$ be mappings with $\mathcal{A}_0 \neq \phi$ and compact. Suppose that the following axioms are true.*

(i) Weak $\mathcal{P}$-property is satisfied by the pair $(\mathcal{A}, \mathcal{B})$;
(ii) $\rho(T_1x, T_2y) \leq \rho(x, y) \forall x, y \in \mathcal{A}$.

Then, there exists $x \in \mathcal{A}$ such that $\rho(x, T_1x) = \rho(x, T_2x) = \rho(\mathcal{A}, \mathcal{B})$.

Proof Let $x_0 \in \mathcal{A}_0$. As $\mathcal{A}_0 \neq \phi$, Lemma 1 assures that $T_1x_0 \in T_1(\mathcal{A}_0) \subseteq \mathcal{B}_0$. Then by definition of $\mathcal{A}_0$ we can find $x_1 \in \mathcal{A}_0$ with $\rho(x_1, T_1x_0) = \rho(\mathcal{A}, \mathcal{B})$. Again $T_2x_1 \in T_2(\mathcal{A}_0) \subseteq \mathcal{B}_0$, we find $x_2 \in \mathcal{A}_0$ with $\rho(x_2, T_2x_1) = \rho(\mathcal{A}, \mathcal{B})$. Since $x_2 \in \mathcal{A}_0$, and

$T_1x_2 \in T_1(\mathcal{A}_0) \subseteq \mathcal{B}_0$, we find $x_3 \in \mathcal{A}_0$ with $\rho(x_3, T_1x_2) = \rho(\mathcal{A}, \mathcal{B})$. Continuing in this way, we get a sequence $\{x_n\}$ in $\mathcal{A}_0$ with

$$\begin{aligned}\rho(x_{2n-1}, T_1x_{2n-2}) &= \rho(\mathcal{A}, \mathcal{B})\\ \rho(x_{2n}, T_2x_{2n-1}) &= \rho(\mathcal{A}, \mathcal{B})\\ \rho(x_{2n+1}, T_1x_{2n}) &= \rho(\mathcal{A}, \mathcal{B})\\ \rho(x_{2n+2}, T_2x_{2n+1}) &= \rho(\mathcal{A}, \mathcal{B}).\end{aligned}$$

Since $(\mathcal{A}, \mathcal{B})$ has weak $\mathcal{P}$-property, it follows that

$$\begin{aligned}\rho(x_{2n}, x_{2n-1}) &\leq \rho(T_1x_{2n-2}, T_2x_{2n-1})\\ \rho(x_{2n+1}, x_{2n}) &\leq \rho(T_1x_{2n}, T_2x_{2n-1})\end{aligned}$$

$$\rho(x_{2n+1}, x_{2n}) \leq \rho(x_{2n}, x_{2n-1}), \text{ by (ii).} \tag{1}$$

$\rho(x_{2n+2}, x_{2n+1}) \leq \rho(T_1x_{2n}, T_2x_{2n+1})$

$$\rho(x_{2n+2}, x_{2n+1}) \leq \rho(x_{2n+1}, x_{2n}), \text{ by (ii).} \tag{2}$$

Thus, we get $\rho(x_{n+1}, x_n) \leq \rho(x_n, x_{n-1})$, $\forall n \in N$, from (1) and (2). Therefore $\rho(x_{n+1}, x_n)$ is convergent. Since $\mathcal{A}_0$ is compact, $\{x_n\}$ has a subsequence $\{x_{n_k}\}$ with $x_{n_k} \to x$ (say) in $\mathcal{A}_0$ as $k \to \infty$. Now

$$\begin{aligned}\rho(T_1x_{n_k}, T_1x) &\leq \rho(T_1x_{n_k}, T_2x_{n_k}) + \rho(T_2x_{n_k}, T_1x)\\ &\leq \rho(x_{n_k}, x_{n_k}) + \rho(x_{n_k}, x) \to 0 \text{ as } k \to \infty.\end{aligned}$$

Therefore, $T_1x_{n_k} \to T_1x$. Similarly, $T_2x_{n_k} \to T_2x$
$\rho(\mathcal{A}, \mathcal{B}) = \rho(x_{2_{n_k-1}}, T_1x_{2_{n_k-2}}) \to \rho(x, T_1x)$
$\rho(\mathcal{A}, \mathcal{B}) = \rho(x_{2_{n_k}}, T_2x_{2_{n_k-1}}) \to \rho(x, T_2x)$. Thus, $\rho(x, T_1x) = \rho(x, T_2x) = \rho(\mathcal{A}, \mathcal{B})$.
This completes the proof.

Now we prove the following two results which will be used to prove the subsequent results.

Lemma 4 *Let C be a convex bounded closed subset of a Banach space. Let $T : C \to C$ be a function satisfying $\rho(Tx, Ty) \leq \frac{1}{2}[\rho(x, Tx) + \rho(y, Ty)]$. If T is demicompact, then T has a point $x \in C$ such that $Tx = x$.*

Proof Fix $x_0 \in C$, let $T_n : C \to C$ be defined as $T_n(x) = (1 - \alpha_n)x_0 + \alpha_n Tx$ where $\alpha_n \in [0, 1)$ such that $\alpha_n \to 1$ as $n \to \infty$.

$$\begin{aligned}\|T_n(x) - T_n(y)\| &= \alpha_n \|Tx - Ty\|\\ &\leq \frac{\alpha_n}{2}[\|x - Tx\| + \|y - Ty\|]\end{aligned}$$

Therefore, T_n is a Kannan mapping. Since C is complete, there exists $x_n \in C$ so that $x_n = T_n x_n$. Therefore, $x_n = (1-\alpha_n)x_0 + \alpha_n T x_n$. Since C is bounded

$$\begin{aligned}\|x_n - Tx_n\| &= \|(1-\alpha_n)x_0 + \alpha_n T x_n - T x_n\| \\ &= (1-\alpha_n)\|x_0 - T x_n\| \to 0\end{aligned}$$

as $n \to \infty$ so that $(I-T)(x_n) \to 0$. Due to the demicompactness of T, $\{x_n\}$ has a subsequence $\{x_{n_k}\}$ with $x_{n_k} \to x$ as $k \to \infty$.

$$\begin{aligned}\|x - Tx\| &= \|x - x_{n_k} + x_{n_k} - Tx\| \\ &\le \|x - x_{n_k}\| + \|T_{n_k} x_{n_k} - Tx\| \\ &= \|x - x_{n_k}\| + \|(1-\alpha_{n_k})x_0 + \alpha_{n_k} T x_{n_k} - Tx\| \\ &\le \|x - x_{n_k}\| + (1-\alpha_{n_k})\|x_0 - Tx\| + \alpha_{n_k}\|T x_{n_k} - Tx\| \\ &= \|x - x_{n_k}\| + (1-\alpha_{n_k})\|x_0 - Tx\| + \frac{\alpha_{n_k}}{2}[\|x_{n_k} - T x_{n_k}\| + \|x - Tx\|] \\ &\to \frac{1}{2}\|x - Tx\| \text{ as } k \to \infty\end{aligned}$$

Thus $\|x - Tx\| \le \frac{1}{2}\|x - Tx\|$.

Hence, $x = Tx$. This completes the proof.

Lemma 5 *Let C be a convex bounded closed subset of a Banach space. Let $T : C \to C$ be a function satisfying $\rho(Tx, Ty) \le \frac{1}{2}[\rho(x, Ty) + \rho(y, Tx)]$. If T is demicompact, then T has a point $x \in C$ such that $Tx = x$.*

Proof Fix $x_0 \in C$, let $T_n : C \to C$ be defined as $T_n(x) = (1-\alpha_n)x_0 + \alpha_n T x$ where $\alpha_n \in [0, 1)$ such that $\alpha_n \to 1$ as $n \to \infty$.

$$\begin{aligned}\|T_n(x) - T_n(y)\| &= \alpha_n \|Tx - Ty\| \\ &\le \frac{\alpha_n}{2}[\|x - Ty\| + \|y - Tx\|]\end{aligned}$$

Therefore, T_n is a $C-$contraction mapping on C. Since C is complete, there exists $x_n \in C$ so that $x_n = T_n x_n$. Since C is bounded

$$\begin{aligned}\|x_n - Tx_n\| &= \|(1-\alpha_n)x_0 + \alpha_n T x_n - T x_n\| \\ &= (1-\alpha_n)\|x_0 - T x_n\| \to 0\end{aligned}$$

as $n \to \infty$ so that $(I-T)(x_n) \to 0$. Due to the demicompactness of T, $\{x_n\}$ has a subsequence $\{x_{n_k}\}$ such that $x_{n_k} \to x$ as $k \to \infty$. It follows that $\lim_{k\to\infty}[(1-\alpha_{n_k})(x_0 - x) + \alpha_{n_k}(T x_{n_k} - x)] = 0$, that is, $\lim_{k\to\infty}(T x_{n_k} - x) = 0$

$$\begin{aligned}\|x - Tx\| &= \|x + x_{n_k} - x_{n_k} + Tx_{n_k} - Tx_{n_k} - Tx\| \\ &\leq \|x - x_{n_k}\| + \|x_{n_k} - Tx_{n_k}\| + \frac{1}{2}[\|x_{n_k} - Tx\| + \|x - Tx_{n_k}] \\ &\rightarrow \frac{1}{2}\|x - Tx\| \text{ as } k \rightarrow \infty\end{aligned}$$

Thus $\|x - Tx\| \leq \frac{1}{2}\|x - Tx\|$

Hence, $x = Tx$. This completes the proof.

The proof of the following result is analogous to the proof of Theorem 3.11 of [8].

Theorem 4 *Let $\mathcal{A} \neq \emptyset, \mathcal{B} \neq \emptyset$ be two subsets of a Banach space X, which are convex, closed and bounded. Let $T_1, T_2 : \mathcal{A} \rightarrow \mathcal{B}$ be functions with $\mathcal{A}_0 \neq \phi$. Suppose that the following axioms are true.*

(i) Weak $\mathcal{P}$-property is satisfied by the pair $(\mathcal{A}, \mathcal{B})$;
(ii) T_1 and T_2 are demicompact and are continuous ;
(iii) T_1 and T_2 satisfies
$\rho(T_1x, T_1y) \leq \frac{1}{2}[\rho(x, T_1x) + \rho(y, T_1y) - 2\rho(\mathcal{A}, \mathcal{B})]$;
$\rho(T_2x, T_2y) \leq \frac{1}{2}[\rho(x, T_2x) + \rho(y, T_2y) - 2\rho(\mathcal{A}, \mathcal{B})]$;
$\rho(T_1x, T_2y) \leq \frac{1}{2}[\rho(x, T_1x) + \rho(y, T_2y) - 2\rho(\mathcal{A}, \mathcal{B})]$.

Then, there exists $x \in \mathcal{A}$ such that $\rho(x, T_1x) = \rho(x, T_2x) = \rho(\mathcal{A}, \mathcal{B})$.

Proof Since $\mathcal{A}_0 \neq \phi$ and since the pair $(\mathcal{A}, \mathcal{B})$ having weak $\mathcal{P}-$ property. As a consequence of Lemma 2, $\mathcal{B}_0$ is closed. Since $S(\mathcal{A}_0) \subset \mathcal{B}_0$, $T_1(\overline{\mathcal{A}_0}) \subset \overline{\mathcal{B}_0} = \mathcal{B}_0$. Now define the operator $P_{\mathcal{A}_0} : T_1(\overline{\mathcal{A}_0}) \rightarrow \mathcal{A}_0$ by $P_{\mathcal{A}_0}(y) = \{x \in \mathcal{A}_0 : \rho(x, y) = \rho(\mathcal{A}, \mathcal{B})\}$, $P_{\mathcal{A}_0}$ is well defined, since if there exists $x_0 \in \mathcal{A}_0$ such that $\rho(x_0, y) = \rho(\mathcal{A}, \mathcal{B})$ then by weak $\mathcal{P}-$property, $\rho(x, x_0) \leq \rho(y, y) = 0 \Rightarrow x = x_0$.
Claim: $P_{\mathcal{A}_0} \circ T_1$ is demicompact
Suppose x_n bounded such that $x_n - P_{\mathcal{A}_0} \circ T_1(x_n) \rightarrow l$.

$$\begin{aligned}\|x_n - T_1x_n - (\rho(\mathcal{A}, \mathcal{B}) + l)\| &= \|x_n - T_1x_n + P_{\mathcal{A}_0} \circ T_1(x_n) - P_{\mathcal{A}_0} \circ T_1(x_n) - (\rho(\mathcal{A}, \mathcal{B}) + l)\| \\ &\leq \|x_n - P_{\mathcal{A}_0} \circ T_1(x_n) - l\| + \|P_{\mathcal{A}_0} \circ T_1(x_n) - T_1x_n - \rho(\mathcal{A}, \mathcal{B})\| \\ &\rightarrow 0\end{aligned}$$

as $n \rightarrow \infty \Rightarrow (I - T_1)(x_n) \rightarrow \rho(\mathcal{A}, \mathcal{B}) + l$ as $n \rightarrow \infty$. Since T_1 is demicompact there exists $\{x_{n_k}\} \subset \{x_n\}$ such that $x_{n_k} \rightarrow x$ as $n \rightarrow \infty$. Thus $(I - P_{\mathcal{A}_0} \circ T_1)(x_n) \rightarrow l \Rightarrow \{x_n\}$ has a convergent subsequence. Therefore $P_{\mathcal{A}_0} \circ T_1$ is demicompact.
Again by applying the condition of weak $\mathcal{P}$-property of $(\mathcal{A}, \mathcal{B})$, it can be obtained that $\rho(P_{\mathcal{A}_0}(T_1x), T_1x) = \rho(\mathcal{A}, \mathcal{B})$ and $\rho(P_{\mathcal{A}_0}(T_1y), T_1y) = \rho(\mathcal{A}, \mathcal{B})$ which in turn imply that

$$\begin{aligned}\rho(P_{\mathcal{A}_0}(T_1x), P_{\mathcal{A}_0}(T_1y)) &\leq \rho(T_1x, T_1y)\\ &\leq \frac{1}{2}[\rho(x, T_1x) + \rho(y, T_1y) - 2\rho(\mathcal{A}, \mathcal{B})]\\ &\leq \frac{1}{2}[\rho(x, P_{\mathcal{A}_0}(T_1x)) + \rho(P_{\mathcal{A}_0}(T_1x), T_1x)\\ &\quad + \rho(y, P_{\mathcal{A}_0}(T_1y)) + \rho(P_{\mathcal{A}_0}(T_1y), T_1y) - 2\rho(\mathcal{A}, \mathcal{B})]\\ &\leq \frac{1}{2}[\rho(x, P_{\mathcal{A}_0}(T_1x)) + \rho(y, P_{\mathcal{A}_0}(T_1y))]\end{aligned}$$

for any $x, y \in \overline{\mathcal{A}_0}$. Thus Lemma 4 confirms $P_{\mathcal{A}_0} \circ T_1$ has a fixed point x_1. This implies $\rho(x_1, T_1x_1) = \rho(\mathcal{A}, \mathcal{B})$. Define $P_{\mathcal{A}_0} : T_2(\overline{\mathcal{A}_0}) \to \mathcal{A}_0$. It can be easily seen that x_2 is fixed point of $P_{\mathcal{A}_0} \circ T_2$ so that $\rho(x_2, T_2x_2) = \rho(\mathcal{A}, \mathcal{B})$.
Claim: $x_1 = x_2$. By weak $\mathcal{P}$-property, $\rho(x_1, T_1x_1) = \rho(\mathcal{A}, \mathcal{B})$ and $\rho(x_2, T_2x_2) = \rho(\mathcal{A}, \mathcal{B})$, imply that

$$\begin{aligned}\rho(x_1, x_2) &\leq \rho(T_1x_1, T_2x_2)\\ &\leq \frac{1}{2}[\rho(x_1, T_1x_1) + \rho(x_2, T_2x_2) - 2\rho(\mathcal{A}, \mathcal{B})]\\ &= 0.\end{aligned}$$

Therefore, $x_1 = x_2$ and hence $\rho(x, T_1x) = \rho(x, T_2x) = \rho(\mathcal{A}, \mathcal{B})$. This completes the proof.

Example 1 Let $X := R^2$ with respect to a metric
$\rho((a_1, b_1), (a_2, b_2)) = |a_1 - a_2| + |b_1 - b_2|$ for all $(a_1, b_1), (a_2, b_2) \in X$.
Let $\mathcal{A} = \{(0, 0), (0, 3), (0, 6), (0, 9)\}$ and $\mathcal{B} = \{(1, -1), (1, 2), (1, 5), (1, 8)\}$. Then $(\mathcal{A}, \mathcal{B})$ satisfies weak P-property and $\rho(\mathcal{A}, \mathcal{B}) = 2$.
Let $T_1, T_2 : \mathcal{A} \to \mathcal{B}$ be defined as follows:
$T_1a = \begin{cases}(1, 5) \text{ if } a = (0, 0)\\ (1, 8) \quad \text{otherwise}\end{cases}$ and
$T_2a = (1, 8)$ for all $a \in \mathcal{A}$.
Then T_1 and T_2 satisfy demicompact property, and it is easy to see that T_1 and T_2 satisfy all other conditions of Theorem 4. It is easy to check that $(0, 9)$ is the **CBPP** of T_1 and T_2.

The proof of the following result is analogous to the proof of Theorem 3.14 of [8].

Theorem 5 *Let $\mathcal{A} \neq \emptyset, \mathcal{B} \neq \emptyset$ be two convex subsets of a Banach space X, which are closed and bounded. Let $T_1, T_2 : \mathcal{A} \to \mathcal{B}$ be functions with $\mathcal{A}_0 \neq \phi$. Suppose that the following axioms are true.*

(i) Weak $\mathcal{P}$-property is satisfied by the pair $(\mathcal{A}, \mathcal{B})$;
(ii) T_1 and T_2 are demicompact and are continuous;
(iii) T_1 and T_2 satisfies
$\rho(T_1x, T_1y) \leq \frac{1}{2}[\rho(x, T_1y) + \rho(y, T_1x) - 2\rho(\mathcal{A}, \mathcal{B})]$;

$\rho(T_2x, T_2y) \leq \frac{1}{2}[\rho(x, T_2y) + \rho(y, T_2x) - 2\rho(\mathcal{A}, \mathcal{B})]$;
$\rho(T_1x, T_2y) < \frac{1}{2}[\rho(x, T_2y) + \rho(y, T_1x) - 2\rho(\mathcal{A}, \mathcal{B})]$ *and*
(iv) $\rho(T_1x, T_2y) < \rho(x, y)$ *if* $x \neq y$.

Then there exists $x \in \mathcal{A}$ *such that* $\rho(x, T_1x) = \rho(x, T_2x) = \rho(\mathcal{A}, \mathcal{B})$.

Proof The closedness of $\mathcal{B}_0$ and $T_1(\overline{\mathcal{A}_0}) \subseteq \mathcal{B}_0$, $T_2(\overline{\mathcal{A}_0}) \subseteq \mathcal{B}_0$ are already proved in Lemmas 2 and 3. Let $P_{\mathcal{A}_0} : T_1(\overline{\mathcal{A}_0}) \to \mathcal{A}_0$ be defined by

$$P_{\mathcal{A}_0}y = \{x \in \mathcal{A}_0 : \rho(x, y) = \rho(\mathcal{A}, \mathcal{B})\}.$$

Claim: $P_{\mathcal{A}_0} \circ T_1$ is demicompact
Suppose x_n bounded such that $x_n - P_{\mathcal{A}_0} \circ T_1(x_n) \to l$.

$$\begin{aligned}\|x_n - T_1x_n - (\rho(\mathcal{A}, \mathcal{B}) + l)\| &= \|x_n - T_1x_n + P_{\mathcal{A}_0} \circ T_1(x_n) - P_{\mathcal{A}_0} \circ T_1(x_n) - (\rho(\mathcal{A}, \mathcal{B}) + l)\| \\ &\leq \|x_n - P_{\mathcal{A}_0} \circ T_1(x_n) - l\| + \|P_{\mathcal{A}_0} \circ T_1(x_n) - T_1x_n - \rho(\mathcal{A}, \mathcal{B})\| \\ &\to 0\end{aligned}$$

as $n \to \infty \Rightarrow (I - T_1)(x_n) \to \rho(\mathcal{A}, \mathcal{B}) + l$ as $n \to \infty$. Since T_1 is demicompact, there exists $\{x_{n_k}\} \subset \{x_n\}$ such that $x_{n_k} \to x$ as $n \to \infty$. Thus $(I - P_{\mathcal{A}_0} \circ T_1)(x_n) \to l \Rightarrow \{x_n\}$ has a convergent subsequence. Therefore, $P_{\mathcal{A}_0} \circ T_1$ is demicompact.
Since weak $\mathcal{P}$-property is satisfied by the pair $(\mathcal{A}, \mathcal{B})$, so that $\rho(P_{\mathcal{A}_0}(T_1x), T_1x) = \rho(\mathcal{A}, \mathcal{B})$ and
$\rho(P_{\mathcal{A}_0}(T_1y), T_1y) = \rho(\mathcal{A}, \mathcal{B})$.
Now,

$$\begin{aligned}\rho(P_{\mathcal{A}_0}(T_1x), P_{\mathcal{A}_0}(T_1y)) &\leq \rho(T_1x, T_1y) \\ &\leq \frac{1}{2}[\rho(x, T_1y) + \rho(y, T_1x) - 2\rho(\mathcal{A}, \mathcal{B})] \\ &\leq \frac{1}{2}[\rho(x, P_{\mathcal{A}_0}(T_1y)) + \rho(P_{\mathcal{A}_0}(T_1y), T_1y) \\ &\quad + \rho(y, P_{\mathcal{A}_0}(T_1x)) + \rho(P_{\mathcal{A}_0}(T_1x), T_1x) - 2\rho(\mathcal{A}, \mathcal{B})] \\ &\leq \frac{1}{2}[\rho(x, P_{\mathcal{A}_0}(T_1y)) + \rho(y, P_{\mathcal{A}_0}(T_1x))]\end{aligned}$$

for any $x, y \in \overline{\mathcal{A}_0}$. Lemma 5 confirms $P_{\mathcal{A}_0} \circ T_1$ has a fixed point x_1 this implies $\rho(x_1, T_1x_1) = \rho(\mathcal{A}, \mathcal{B})$. Define $P_{\mathcal{A}_0} : T_2(\overline{\mathcal{A}_0}) \to \mathcal{A}_0$. It can be easily see that x_2 is fixed point of $P_{\mathcal{A}_0} \circ T_2$ so that $\rho(x_2, T_2x_2) = \rho(\mathcal{A}, \mathcal{B})$.
Claim: $x_1 = x_2$. By weak $\mathcal{P}$-property, $\rho(x_1, T_1x_1) = \rho(x_2, T_2x_2) = \rho(\mathcal{A}, \mathcal{B})$ and hence it follows that

$$\begin{aligned}\rho(x_1, x_2) &\leq \rho(T_1x_1, T_2x_2)\\ &\leq \frac{1}{2}[\rho(x_1, T_2x_2) + \rho(x_2, T_1x_1) - 2\rho(\mathcal{A}, \mathcal{B})]\\ &\leq \frac{1}{2}[\rho(x_1, x_2) + \rho(x_2, T_2x_2) + \rho(x_2, x_1) + \rho(x_1, T_1x_1) - 2\rho(\mathcal{A}, \mathcal{B})]\\ &= \frac{1}{2}[2\rho(x_1, x_2)]\\ &\leq \rho(x_1, x_2)\end{aligned}$$

Thus $\rho(x_2, x_1) = \rho(T_2x_2, T_1x_1)$. Therefore, $x_2 = x_1$ and so that $\rho(x, T_1x) = \rho(x, T_2x) = \rho(\mathcal{A}, \mathcal{B})$. Hence proved.

Remark 1 If the considered non-self-mappings in Theorems 4, 5 are self-mappings then, the common best proximity becomes common fixed point.

4 Conclusion

In this paper, we proved results on **CBPP** for two non-self mappings on a closed bounded convex subset a Banach space. The main result of this paper is illustrated with suitable example. Proving the results obtained in this paper for other class of mappings is a problem of further research.

References

1. Eldred, A.A., Kirk, W.A., Veeraamani, P.: Proximal normal structure and relatively non expansive mappings. Stud. Math. **171**, 283–293 (2005)
2. Eldred, A.A., Veeraamani, P.: Existence and convergence of best proximity points. J. Math. Anal. Appl. **323**, 1001–1006 (2006)
3. Karapinar, E.: On best proximity point of -Geraghty contractions. Fixed Point Theory Appl. **200** (2013)
4. Prolla, J.B.: Fixed point theorems for set valued mappings and existence of best approximations. Numer. Funct. Anal. Optim. **5**, 449–455 (1983)
5. Reich, S.: Approximate selections, best proximations, fixed points and invariant sets. J. Math. Anal. Appl. **62**, 104–113 (1978)
6. Sehgal, V.M., Singh, S.P.: A generalization to multifunctions of Fans best approximations theorem. Proc. Am. Math. Soc. **102**, 534–537 (1988)
7. Sehgal, V.M., Singh, S.P.: A theorem on best approximations. Numer. Funct. Anal. Optim. **10**, 181–184 (1989)
8. Dey, L.K., Mondal, S.: Some common best proximity point theorems in a complete metric space. Afr. Mat. (2016). https://doi.org/10.1007/s13370-016-0432-1
9. Amini-Harandi, A.: Common best proximity points theorems in metric spaces. Optim. Lett. **8**, 581–589 (2014). https://doi.org/10.1007/s11590-012-0600-7
10. Mongkolkeha, C., Kumam, P.: Some common best proximity points for proximity commuting mappings. Optim. Lett. **7**(8), 1825–1836 (2013)

11. Mongkolkeha, C., Kumam, P.: Common best proximity points for proximity commuting mapping with Geraghtys functions. Carpath. J. Math. **31**(3), 359–364 (2015)
12. Basha Sadiq, S., Shahzad, N., Jeyaraj, R.: Common best proximity points, global optimization of multiobjective functions. Appl. Math. Lett. **24**(6), 883–886 (2011)
13. Sen, M.D., Agarwal, R.P.: Common fixed points and best proximity points of two cyclic self-mappings. Fixed Point Theory Appl. **136** (2012)
14. Zhang, J., Su, Y., Cheng, Q.: A note on A best proximity point theorem for Geraghty-contractions. Fixed Point Theory Appl. **99** (2013)
15. Petryshyn, W., V.: Construction of fixed points of demicompact mappings in Hilbert space. J. Math. Anal. Appl. **14**, 276–284 (1966)
16. Caballero, J., Harjani, J., Sadarangani, K.: A best proximity point theorem for Geraghty-contractions. Fixed Point Theory Appl. **231** (2012)
17. Dey, L.K., Mondal, S.: Best proximity point of F-contraction in a complete metric space. Bull. Allahabad Math. Soc. **30**(2), 173–189 (2015)
18. Kannan, R.: Some results on fixed points—II. Am. Math. Mon. **76**, 405–408 (1969)
19. Almeida, A., Karapinar, E., Sadarangani, K.: A note on best proximity theorems under weak-property. Abstr. Appl. Anal. 716825 (2014)
20. Chatterjea, S.K.: Fixed point theorems. C. R. Acad. Bulg. Sci. **25**, 727–730 (1972)

Dynamical Analysis of Conformable Fractional-Order Rosenzweig-MacArthur Prey–Predator System

P. Kowsalya, R. Kaviya, and P. Muthukumar

Abstract This study investigates the dynamical behavior of the conformable fractional-order Rosenzweig-MacArthur prey–predator system with the various fractional orders. The dynamical behaviors such as existence, uniqueness, non-negativity, and uniform boundedness of the system's solution have been achieved. This work verifies the parametric conditions for the existence of non-negative interior equilibrium points of the system. The local stability and global stability have been discussed using the Routh-Hurwitz stability test and choosing the suitable Lyapunov function. The numerical examples are illustrated to verify the achieved theoretical results. Finally, a conclusion is given with future directions in this study.

Keywords Conformable fractional calculus · Existence and uniqueness · Prey–predator system · Rosenzweig-MacArthur model · Stability analysis

1 Introduction

The Lotka–Volterra system is a pair of differential equations that describe the dynamics of ecological systems in which two species interact, one a predator and one its prey. It was developed independently by Alfred Lotka and Vita Volterra in the 1920s and is characterized by oscillations in the population size of both predator's oscillations and prey's oscillations (see [5]). Continuously, several studies on prey–predator interactions in mathematical ecology were carried out (see [7, 10]). There are several models known for studying the behavior of prey–predator systems, one of which is the Rosenzweig-MacArthur model (see [2, 4]). This model is related to the extinction of species influenced by the extended carrying capacity of prey in the ecosystem (see [9, 14]). The Rosenzweig-MacArthur prey–predator (RMPP) system is broadly used to describe prey–predator systems (see [6, 20]).

P. Kowsalya · P. Muthukumar (✉)
Department of Mathematics, The Gandhigram Rural Institute (Deemed to be University), Gandhigram 624302, Tamil Nadu, India

R. Kaviya
Department of Science and Humanities, Karpagam College of Engineering, Coimbatore 641 032, Tamil Nadu, India

P. Balasubramaniam et al. (eds.), *Mathematical Modelling and Computational Intelligence Techniques*, Springer Proceedings in Mathematics & Statistics 376,
https://doi.org/10.1007/978-981-16-6018-4_6

Integer-order calculus is a particular case of fractional-order calculus, and the fractional derivative is the generalization of the integer-order derivative (see [12]). Many researchers are interested in studying this fractional calculus in present times. The advantages of fractional calculus have brought a comprehensive study from researchers (see [13, 18]). Several fractional derivatives in literature, such as Riemann-Liouville fractional derivative and Caputo fractional derivative, exist with different properties. For instance, Riemann-Liouville fractional derivative does not satisfy the $\mathcal{D}^{\alpha}(1) = 0$, but the Caputo fractional derivative can (see [8]).

The ancient fractional derivatives have considerable difficulties and do not satisfy some significant properties satisfied by the integer-order calculus. For example, both the Riemann-Liouville definition and Caputo definition derivatives do not fulfill the product rule, quotient rule, chain rule and have some challenges in the calculation (see [12]). In this sense, the studies on the fractional-order system are complex than the integer-order system. In literature, the fractional-order RMPP system has been studied less than the studies on integer-order RMPP system. Khalil et al. introduced a new fractional calculus called "Conformable Fractional-order Derivative" (CFD) in [8] to overcome this difficulty in studying fractional calculus. This conformable fractional-order calculus is well suited to deal with problems in the usual fractional-order calculus. The conformable fractional-order calculus satisfies the integer-order calculus properties, such as the derivative of the product and the quotient of two functions and chain rule. We can easily obtain the results of various fractional-order systems via CFD because of these advantages. Thabet Abdeljawad (see [1]) developed some basic concepts for the CFD.

In literature, Abdourazek Souahi et al. (see [15]) studied the stability of conformable fractional-order nonlinear systems. The heat differential equation, Newton mechanics, and diffusive transport are studied with CFD (see [3, 17, 19]). We are motivated to combine the CFD and the RMPP system by these advantages and the prominent applications of CFD. From this method, we can study the fruitful results on fractional-order RMPP systems without any complexity. For this purpose, the present work combines the CFD and the RMPPS and then analyzes the dynamical behavior of the conformable fractional-order Rosenweig-MacArthur prey–predator system (CFRMPPS).

This paper is organized as follows: Sect. 2 gives the preliminary results based on the CFD and also constructed the (conformable fractional Rosenzweig-MacArthur prey–predator system) CFRMPPS. Section 3 analyzes the existence and uniqueness of the solution of the system, non-negativity, and uniform boundedness of the CFRMPPS. Section 4 has reached suitable conditions for the existence of local stability and the global stability of the proposed CFRMPPS at the non-negative interior equilibrium point. In Sect. 5, the numerical examples are illustrated to verify the obtained theoretical results.

2 Preliminaries and System Description

In this section, the required basic definitions and principles are defined, and a description of the proposed CFRMPPS is presented. To know more details in this section, the reader can refer to [8].

2.1 Preliminaries

Definition 1 *(see [8]) Given a differentiable function* $g : (t_0, \infty) \to \mathbb{R},\ t_0 \geq 0$. *Then the conformable fractional derivative of g of order* α *is defined by*

$$T^{\alpha} g(t) = \lim_{\varepsilon \to 0} \frac{g(t + \varepsilon t^{1-\alpha}) - g(t)}{\varepsilon}, \qquad t > t_0 \geq 0,\ \alpha \in (0, 1].$$

Theorem 2 *(see [8]) Let* $\alpha \in (0, 1]$, *g be differentiable at a point* $t \in (t_0, \infty),\ t_0 \geq 0$. *In addition, g is differentiable, then*

$$T^{\alpha} g(t) = t^{1-\alpha} \frac{dg(t)}{dt}.$$

Lemma 3 *(see [16]) Consider the system* $\frac{dx(t)}{dt} = g(t, x(t)),\ t > t_0 \geq 0$, *with initial condition* $x_0 > 0,\ g : (t_0, \infty) \times \Omega \to \mathbb{R}^n,\ \Omega \in \mathbb{R}^n$. *If* $g(t, x(t))$ *satisfies the locally Lipschitz condition with respect to* $x(t)$ *then there exists a unique solution of the above system on* $(t_0, \infty) \times \Omega$.

Lemma 4 *(see [11]) Let* $g(t)$ *be a continuous function on* $[a, b]$. *If* $\frac{dg(t)}{dt} \geq 0,\ t \in (a, b)$, *then* $g(t)$ *is a non-decreasing function for each* $t \in [a, b]$ *and if* $\frac{dg(t)}{dt} \leq 0,\ t \in (a, b)$, *then* $g(t)$ *is a non-increasing function for each* $t \in [a, b]$.

Lemma 5 *(see [11]) Let* $u(t)$ *be a continuous function on* $(t_0, \infty),\ t \geq 0$ *and satisfying*

$$\frac{du(t)}{dt} \leq -\lambda u(t) + \mu,$$

with the initial condition $u(t_0) > 0,\ (\lambda, \mu) \in \mathbb{R}^3,\ \lambda \neq 0$ *and* $t_0 \geq 0$ *is the initial time. Then, its solution has the form*

$$u(t) \leq \left(u(t_0) - \frac{\mu}{\lambda}\right) exp[-\lambda(t - t_0)] + \frac{\mu}{\lambda}.$$

2.2 System Description

In this type of three-dimensional RMPP model, it consists of three populations, namely prey, mid-level predator, and top-level predator with the various fractional orders $(\alpha_1, \alpha_2, \alpha_3) \in (0, 1]$. Let $x_1(t)$, $x_2(t)$, and $x_3(t)$ denote the densities of prey, mid-level predator, and top-level predator, respectively, at time $t \in (t_0, \infty)$, $t_0 \geq 0$. For the notation purpose $x_1(t)$, $x_2(t)$, and $x_3(t)$ are simply notated as x_1, x_2, and x_3, respectively.

Model Construction

- Assume that the species x_1 is a prey of the predators x_2 and x_3; x_2 is a prey of the predator x_3.
- The parameters r_1 and r_2 are intrinsic growth rates of the species x_1 and x_2, respectively. The species x_1 and x_2 grows logistically with the carrying capacity k_1 and k_2 respectively.
- d_1, d_2, and d_3 are the death rates of the species x_1, x_2, and x_3, respectively.
- $c_1 \in (0, 1)$ and $c_2 \in (0, 1)$ represent the densities of the species x_1 and x_2, respectively, where the attack is half-saturated.
- h_1 and h_2 represent the species handling time of the species x_1 and x_2, respectively.
- $\theta_1 \in (0, 1)$ and $\theta_2 \in (0, 1)$ are the maximum value of per capita reduction rate of the species x_1 and x_2 due to the species x_3, respectively.
- a_1 and a_2 are the habitat complexity of the species x_1 and x_2, respectively.

Conformable Rosenzweig-MacArthur Prey–Predator System

By the motivation of the study on the fractional-order two-species RMPP system in [13], this work analyzes the following three species CFRMPPS with the various fractional-order according to the assumptions mentioned above,

$$\left.\begin{aligned} \mathcal{T}^{\alpha_1} x_1 &= r_1 x_1 \left(1 - \frac{x_1}{k_1}\right) - \frac{x_1 x_2}{1 + x_1} - \frac{a_1(1 - c_1) x_1 x_3}{1 + a_1(1 - c_1) h_1 x_3} - d_1 x_1, \\ \mathcal{T}^{\alpha_2} x_2 &= r_2 x_2 \left(1 - \frac{x_2}{k_2}\right) + \frac{x_1 x_2}{1 + x_1} - \frac{a_2(1 - c_2) x_2 x_3}{1 + a_2(1 - c_2) h_2 x_3} - d_2 x_2, \\ \mathcal{T}^{\alpha_3} x_3 &= \frac{\theta_1 a_1(1 - c_1) x_1 x_3}{1 + a_1(1 - c_1) h_1 x_3} + \frac{\theta_2 a_2(1 - c_2) x_2 x_3}{1 + a_2(1 - c_2) h_2 x_3} - d_3 x_3, \end{aligned}\right\} \tag{1}$$

with the initial conditions $x_i(t_0) > 0$, $t_0 \geq 0$, $i = 1, 2, 3$. According to Theorem 2, the CFRMPPS (1) can be rewritten as,

$$\left.\begin{aligned} t^{1-\alpha_1} \frac{\mathrm{d}x_1}{\mathrm{d}t} &= r_1 x_1 \left(1 - \frac{x_1}{k_1}\right) - \frac{x_1 x_2}{1 + x_1} - \frac{a_1(1 - c_1) x_1 x_3}{1 + a_1(1 - c_1) h_1 x_3} - d_1 x_1, \\ t^{1-\alpha_2} \frac{\mathrm{d}x_2}{\mathrm{d}t} &= r_2 x_2 \left(1 - \frac{x_2}{k_2}\right) + \frac{x_1 x_2}{1 + x_1} - \frac{a_2(1 - c_2) x_2 x_3}{1 + a_2(1 - c_2) h_2 x_3} - d_2 x_2, \\ t^{1-\alpha_3} \frac{\mathrm{d}x_3}{\mathrm{d}t} &= \frac{\theta_1 a_1(1 - c_1) x_1 x_3}{1 + a_1(1 - c_1) h_1 x_3} + \frac{\theta_2 a_2(1 - c_2) x_2 x_3}{1 + a_2(1 - c_2) h_2 x_3} - d_3 x_3, \end{aligned}\right\} \tag{2}$$

For $t > t_0 \geq 0$, the system (2) can be written as

$$
\begin{aligned}
\frac{\mathrm{d}x_1}{\mathrm{d}t} =& K_1\left(r_1x_1\left(1-\frac{x_1}{k_1}\right)-\frac{x_1x_2}{1+x_1}-\frac{a_1(1-c_1)x_1x_3}{1+a_1(1-c_1)h_1x_3}-d_1x_1\right),\\
\frac{\mathrm{d}x_2}{\mathrm{d}t} =& K_1\left(r_2x_2\left(1-\frac{x_2}{k_2}\right)+\frac{x_1x_2}{1+x_1}-\frac{a_2(1-c_2)x_2x_3}{1+a_2(1-c_2)h_2x_3}-d_2x_2\right),\\
\frac{\mathrm{d}x_3}{\mathrm{d}t} =& K_3\left(\frac{\theta_1a_1(1-c_1)x_1x_3}{1+a_1(1-c_1)h_1x_3}+\frac{\theta_2a_2(1-c_2)x_2x_3}{1+a_2(1-c_2)h_2x_3}-d_3x_3\right).
\end{aligned}
\tag{3}
$$

Here $K_i = t^{\alpha_i-1}$, $i = 1, 2, 3$. For $t > t_0 \geq 0$, the system (3) can be written in the general form with the initial condition $X(t_0)$,

$$
\frac{\mathrm{d}X(t)}{\mathrm{d}t} = F(t, X(t)),\ t \in (t_0, \infty). \tag{4}
$$

Here, $X(t) \in \mathbb{R}^3_+$ and $F(t, X(t)) : (t_0, \infty) \times \mathbb{R}^3_+ \to \mathbb{R}^3_+$, and

$$
X(t) = \begin{bmatrix} x_1 \\ x_2 \\ x_3 \end{bmatrix},\ X(t_0) = \begin{bmatrix} x_1(t_0) \\ x_2(t_0) \\ x_3(t_0) \end{bmatrix} \text{ and } F(t, X(t)) = \begin{bmatrix} F_1(t, X(t)) \\ F_2(t, X(t)) \\ F_3(t, X(t)) \end{bmatrix}.
$$

Here, $F(t, X(t))$ is the respective right hand side value of the system (3).

Remark 1 Comparing the systems (1) and (4) and Theorem 2, one can conclude that the integer-order RMPP system (4) is the equivalent form of the fractional-order RMPPS (1).

3 Dynamical Behavior of the System

In this section, the dynamical behaviors such as existence and uniqueness, non-negativity, and uniform boundedness of solution $X(t)$ of the CFRMPPS (1) are analyzed.

3.1 Existence and Uniqueness of the Solution

The existence and uniqueness of solution $X(t)$ of the system (1) is obtained by using the Lipschitz method.

Theorem 6 *If $F(t, X(t))$ satisfies the locally Lipschitz condition with respect to $X(t)$, then there exists a unique solution $X(t)$ of the system (1) on the region $\Omega = \left\{X(t) \in \mathbb{R}^3_+ \mid max\{|x_1|, |x_2|, |x_3|\} \leq M > 0\right\}$.*

Proof For any $X(t)$ and $\overline{X(t)} \in \Omega$,

$$\begin{aligned}
&\|F(t,X(t)) - F(t,\overline{X(t)})\| \\
&= \sum_{i=1}^{3} \mid F_i(t,X(t)) - F_i(t,\overline{X(t)}) \mid \\
&\leq K_1(r_1-d_1) \mid x_1 - \overline{x_1} \mid + \frac{2K_1 r_1 M}{k_1} \mid x_1 - \overline{x_1} \mid + \left| \frac{2K_1 x_1 x_2}{1+x_1} \right| + \left| \frac{2K_2 \overline{x_1}\,\overline{x_2}}{1+\overline{x_1}} \right| \\
&\quad + K_1 a_1(1-c_1)(1+\theta_1) \left| \frac{x_1 x_3}{1+a_1(1-c_1)h_1 x_3} - \frac{\overline{x_1}\,\overline{x_3}}{1+a_1(1-c_1)h_1\overline{x_3}} \right| \\
&\quad + K_2(r_2-d_2) \mid x_2 - \overline{x_2} \mid + \frac{2K_2 r_2 M}{k_2} \mid x_2 - \overline{x_2} \mid + K_2 a_2(1-c_2)(1+\theta_2) \\
&\quad \times \left| \frac{x_2 x_3}{1+a_2(1-c_2)h_2 x_3} - \frac{\overline{x_2}\,\overline{x_3}}{1+a_2(1-c_2)h_2\overline{x_3}} \right| + K_3 d_3 \mid x_3 - \overline{x_3} \mid \\
&\leq \Bigg(K_1(r_1-d_1) + \frac{K_1 a_1(1-c_1)(1+\theta_1)M + \theta_1^2(1-c_1)^2(1+\theta_1)h_1^2 M^2}{1+2a_1(1-c_1)h_1 M + a_1^2(1-c_1)^2 h_1^2 M^2} \\
&\quad + \frac{2K_1 M r_1}{k_1} \Bigg) |x_1 - \overline{x_1}| + \Bigg(K_2(r_2-d_2) + \frac{2K_2 M r_2}{k_2} \\
&\quad + \frac{K_2 a_2(1-c_2)(1+\theta_2)M + a_2^2(1-c_2)^2(1+\theta_2)h_2^2 M^2}{1+2a_2(1-c_2)h_2 M + a_2^2(1-c_2)^2 h_2^2 M^2} \Bigg) |x_2 - \overline{x_2}| \\
&\quad + \frac{4K_1 K_2 M^2}{1+M} + \Bigg(\frac{K_1 a_1(1-c_1)(1+\theta_1)M}{1+2a_1(1-c_1)h_1 M + a_1^2(1-c_1)^2 h_1^2 m^2} \\
&\quad + \frac{K_2 a_2(1-c_2)(1+\theta_2)M}{1+2a_2(1-c_2)h_2 M + a_2^2(1-c_2)^2 h_2^2 M^2} + K_3 d_3 \Bigg) |x_3 - \overline{x_3}| \\
&\leq L_1 \|(x_1, x_2, x_3) - (\overline{x_1}, \overline{x_2}, \overline{x_3})\|.
\end{aligned}$$

Here,

$$\begin{aligned}
L_1 &= \left(L + \frac{4K_1 K_2 M^2}{1+M} \right) \geq 0, \;\; L = \max\{l_i, l_3\}, \;\; \text{for } i = 1, 2, \\
l_i &= \Bigg(K_i(r_i - d_i) + \frac{K_i a_i(1-c_i)(1+\theta_i)M + a_i^2(1-c_i)^2(1+\theta_i)h_i M^2}{1+2a_i(1-c_i)h_i M + a_i^2(1-c_i)^2 h_i^2 M^2} \\
&\quad + \frac{2K_i M r_i}{k_i} \Bigg), \; l_3 = \sum_{i=1}^{2} \frac{K_i a_i(1-c_i)(1+\theta_i)M}{1+2a_i(1-c_i)h_i M + a_i^2(1-c_i)^2 h_i^2 M^2} + K_3 d_3.
\end{aligned}$$

If the Lipschitz constant $L_1 \geq 0$, then the function $F(t, X(t))$ satisfies Lipschitz condition with respect to $X(t)$. Using Lemma 3, there exists a unique solution $X(t)$ of the system (1) on the region Ω with initial condition $X(t_0)$. This completes proof of the theorem. □

3.2 Non-negativity and Boundedness

Considering the biological significance of the model, the solution $X(t)$ of the system (1) is non-negative and uniformly bounded in the region $\mathbb{R}^3_+$. The non-negativity and boundedness of the system (1) are given by the following theorem.

Theorem 7 *All the solutions of the proposed system* (1) *which start in* $\mathbb{R}^3_+$ *are non-negative and uniformly bounded.*

Proof First we show that the solution $X(t)$ of the system (1) is non-negative if it starts with the initial values $X(t_0)$. If not, then there exists a $t_1 > 0$ such that,

$$\left.\begin{array}{l} X(t) > 0,\ 0 \leq t < t_1, \\ X(t) = 0,\ t = t_1, \\ X(t_1^+) < 0 \end{array}\right\}. \tag{5}$$

Using (5) in the first equation of the system (4),

$$\left.\frac{\mathrm{d}X(t)}{\mathrm{d}t}\right|_{t=t_1} = 0.$$

According to Lemma 4, we have $X(t_1) = 0$, which contradicts the fact $X(t_1^+) < 0$. Therefore, $X(t) \geq 0,\ t > t_0 \geq 0$.

Next we show that all solution $X(t)$ of system (1) which initiate in $\mathbb{R}^3_+$ are uniformly bounded. Define a function,

$$V_1(t) = x_1(t) + x_2(t) + \frac{1}{\theta_1} x_3(t).$$

Taking time derivative,

$$\begin{aligned} \mathrm{d}V_1(t) =& \mathrm{d}x_1(t) + \mathrm{d}x_2(t) + \frac{1}{\theta_1}\mathrm{d}x_3(t) \\ =& K_1(r_1 - d_1)x_1\left(1 - \frac{x_1}{k_1}\right) + K_2(r_2 - d_2)x_2\left(1 - \frac{x_2}{k_2}\right) \\ & - \frac{K_2 a_2(1 - c_2)x_2 x_3}{1 + a_2(1 - c_2)h_2 x_3} + \frac{K_2\theta_2 a_2(1 - c_2)x_2 x_3}{\theta_1(1 + a_2(1 - c_2)h_2 x_3)} - \frac{K_3 d_3 x_3}{\theta_1}. \end{aligned}$$

Now, for each $\eta > 0$,

$$
\begin{aligned}
dV_1(t)+\eta V_1(t)=&K_1(r_1-d_1)x_1\left(1-\frac{x_1}{k_1}\right)+K_2(r_2-d_2)x_2\left(1-\frac{x_2}{k_2}\right)-\frac{K_3d_3x_3}{\theta_1}\\
&-\frac{K_2a_2(1-c_2)x_2x_3}{1+a_2(1-c_2)h_2x_3}+\frac{K_2\theta_2a_2(1-c_2)x_2x_3}{\theta_1(1+a_2(1-c_2)h_2x_3)}\\
&+\eta\left(x_1+x_2+\frac{x_3}{\theta_1}\right)\\
=&(K_1(r_1-d_1)+\eta)x_1-\frac{K_1r_1}{k_1}x_1^2+(K_2(r_2-d_2)+\eta)x_2\\
&-\frac{K_2r_2}{k_2}x_2^2+\left(\left(\frac{\eta-K_3d_3}{\theta_1}\right)+\left(\frac{\theta_2-\theta_1}{\theta_1}\right)\frac{K_2a_2(1-c_2)x_2}{1+a_2(1-c_2)h_2}\right)x_3\\
\leq&\frac{k_1}{4K_1r_1}(K_1(r_1-d_1)+\eta)^2+\frac{k_2}{4K_2r_2}(K_2(r_2-d_2)+\eta)^2\\
&+\left(\left(\frac{\eta-K_3d_3}{\theta_1}\right)+\left(\frac{\theta_2-\theta_1}{\theta_1}\right)\frac{K_2a_2(1-c_2)x_2}{1+a_2(1-c_2)h_2}\right)x_3. \qquad (6)
\end{aligned}
$$

If we take $\eta < K_3d_3$ and $\theta_2 < \theta_1$, then right-hand side of (6) is bounded in $\mathbb{R}^3_+$ and there exist a constant $p > 0$ such that, $(dV_1(t) + \eta V_1(t)) \leq p$, here $p = \sum_{i=1}^{2} \frac{k_i}{4Kr_i}(K(r_i - d_i) + \eta)^2$, $K = \max\{K_1, K_2, K_3\}$. Applying the Lemma 5,

$$
V_1(t) \leq \left(V_1(t_0) - \frac{p}{\eta}\right) e^{[-\eta t]} + \frac{p}{\eta} = V_1(t_0)e^{[-\eta t]} + \frac{p}{\eta}(1 - e^{[-\eta t]}),
$$

for $t \to \infty$, thus $V_1(t) \to \frac{p}{\eta}$. Therefore, $0 < V_1(t) \leq \frac{p}{\eta}$. Hence, the solution $X(t)$ of the system (1) that starts from $\mathbb{R}^3_+$ is confined in the region $B = \{X(t) \in \mathbb{R}^3_+ | 0 < V_1(t) \leq \frac{p}{\eta} + \varepsilon$, for any $\varepsilon > 0$, $0 < \eta < K_3d_3$ and $p = \sum_{i=1}^{2} \frac{k_i}{4Kr_i}(K(r_i - d_i) + \eta)^2\}$. This completes the proof. □

4 Equilibrium Points and Stability Analysis

Here, the existence of non-negative equilibrium points of the system (1) is obtained by finding the suitable non-negative parameters.

4.1 Equilibrium Points

In order to obtain the equilibrium points of the system (1), we set $\frac{dX(t)}{dt} = 0$. Then, the system (1) has following equilibrium points:

1. The equilibrium point $E_1(x_1^*, x_2^*, 0)$ is

$$x_1^* = \frac{(r_2/k_2)x_2^* - (r_2 - d_2)}{1 + (r_2 - d_2) - (r_2/k_2)x_2^*}, \quad x_2^* = \frac{A_1 \pm \sqrt{B_1}}{2((r_2/k_2) - (r_2/k_2)^2(r_1 - d_1))}.$$

Here,

$$\begin{aligned} A_1 =& -(2(r_1 - d_2)(r_2 - d_2)(r_2/k_2) - (r_1/k_1)(r_2/k_2) - (r_2 - d_2) - 1), \\ B_1 =& A_1^2 - 4((r_2/k_2) - (r_2/k_2)^2(r_1 - d_1))((r_2 - d_2)r_1/k_1 \\ & - ((r_1 - d_1)(r_2 - d_2)^2) + (r_1 - d_1)). \end{aligned}$$

2. The equilibrium point $E_2(0, x_2^*, x_3^*)$ is

$$x_2^* = \frac{d_3 + \theta_2 h_2 d_3 x_3^*}{\theta_2(1 - c_2)}, \quad x_3^* = \frac{A_2 \pm \sqrt{B_2}}{2((r_2/k_2)d_3\theta_2(1 - c_2)h_2^2)}.$$

Here,

$$\begin{aligned} A_2 =& -(((r_2 - d_2)\theta_2(1 - c_2)^2 h_2) - (r_2/k_2)d_3(1 - c_2)h_2 - (r_2/k_2)d_3\theta_2 h_2 \\ & - (1 - c_2)^2\theta_2) \\ B_2 =& A_2^2 - 4((-r_2/k_2)d_3\theta_2 h_2^2(1 - c_2))((r_2 - d_2)\theta_2(1 - c_2) - (r_2/k_2)d_3). \end{aligned}$$

3. The equilibrium point $E_3(x_1^*, 0, x_3^*)$ is

$$x_1^* = \frac{d_3}{(1 - c_1)\theta_1}\left(1 + \frac{(1 - c_1)h_1(A_3 \pm \sqrt{B_3})}{2(r_1/k_1)d_3(1 - c_1)h_1^2}\right) \quad x_3^* = \left(\frac{A_3 \pm \sqrt{B_3}}{2(-(r_1/k_1)d_3(1 - c_1)^2 h_1^2)}\right)$$

Here,

$$\begin{aligned} A_3 =& -((r_1 - d_1)(1 - c_1)^2\theta_1 h_1 - 2(r_1/k_1)d_3(1 - c_1)h_1 - (1 - c_1)^2\theta_1) \\ B_3 =& A_3^2 - 4(-(r_1/k_1)d_3(1 - c_1)^2 h_1^2)((r_1 - d_1)(1 - c_1)\theta_1 - (r_1/k_1)d_3). \end{aligned}$$

4.2 *Stability Analysis*

In this part, the local stability analysis of the system (1) obtained via the Routh–Hurwitz (R–H) stability criterion. Also, the global stability of the system (1) is obtained by choosing a suitable Lyapunov function.

Jacobian matrix at $E(x_1^*, x_2^*, x_3^*)$
To construct the Jacobian matrix $B = (b'_{ij})$ for all $b'_{ij} \neq 0$ of system (1) at the non-negative interior equilibrium point $E(x_1^*, x_2^*, x_3^*)$ for investigating the local stability analysis.

$$B = \begin{bmatrix} b'_{11} & b'_{12} & b'_{13} \\ b'_{21} & b'_{22} & b'_{23} \\ b'_{31} & b'_{32} & b'_{33} \end{bmatrix}.$$

Here,

$$\begin{aligned}
b'_{11} =& \frac{\partial f_1}{\partial x_1} = K_1\left((r_1 - d_1) - \frac{2r_1x_1}{k_1} - \frac{x_2}{(1+x_1)^2} - \frac{a_1(1-c_1)x_3}{1+a_1(1-c_1)h_1x_3}\right),\\
b'_{22} =& \frac{\partial f_2}{\partial x_2} = K_2\left((r_2 - d_2) - \frac{2r_2x_2}{k_2} + \frac{x_1}{1+x_1} - \frac{a_2(1-c_2)x_3}{1+a_2(1-c_2)h_2x_3}\right),\\
b'_{33} =& \frac{\partial f_3}{\partial x_3} = \frac{K_1\theta_1a_1(1-c_1)x_1}{(1+a_1(1-c_1)h_1x_3)^2} + \frac{K_2\theta_2a_2(1-c_2)x_2}{(1+a_2(1-c_2)h_2x_3)^2} - K_3d_3,\\
b'_{12} =& \frac{\partial f_1}{\partial x_2} = \frac{-K_1x_1}{1+x_1}, \qquad b'_{13} = \frac{\partial f_1}{\partial x_3} = \frac{-K_1a_1(1-c_1)x_1}{(1+a_1(1-c_1)h_1x_3)^2},\\
b'_{21} =& \frac{\partial f_2}{\partial x_1} = \frac{K_2x_2}{(1+x_1)^2}, \qquad b'_{23} = \frac{\partial f_2}{\partial x_3} = \frac{K_2a_2(1-c_2)x_2}{(1+a_2(1-c_2)h_2x_3)^2},\\
b'_{31} =& \frac{\partial f_3}{\partial x_1} = \frac{K_3\theta_1a_1(1-c_1)x_3}{1+a_1(1-c_1)h_1x_3}, \quad b'_{32} = \frac{\partial f_3}{\partial x_2} = \frac{K_3\theta_2a_2(1-c_2)x_3}{1+a_2(1-c_2)h_2x_3}.
\end{aligned}$$

Local Stability Analysis
The characteristic equation of the Jacobian matrix B at the non-negative interior equilibrium point $E(x_1^*, x_2^*, x_3^*)$ of the system (1) is

$$\lambda^3 + B_1\lambda^2 + B_2\lambda + B_3 = 0. \tag{7}$$

Here,

$$\begin{aligned}
B_1 =& b'_{11} + b'_{22} + b'_{33},\\
B_2 =& b'_{11}(b'_{22} + b'_{33}) + b'_{22}b'_{33} - (b'_{23}b'_{32} + b'_{12}b'_{21} + b'_{13}b'_{31}),\\
B_3 =& b'_{11}(b'_{23}b'_{32} - b'_{22}b'_{33}) + b'_{12}(b'_{21}b'_{33} - b'_{23}b'_{31}) + b'_{13}(b'_{22}b'_{31} - b'_{32}b'_{31}).
\end{aligned}$$

Thus, the above discussion gives the following local stability theorem.

Theorem 8 *The equilibrium point $E(x_1^*, x_2^*, x_3^*)$ is locally asymptotic stable for the system* (1) *if it satisfies $B_j > 0,\ j = 1, 2, 3$ and $U_1 = B_1B_2 - B_3 > 0$ with values are defined in equation* (7). □

Global Stability Analysis

Theorem 9 *If the following conditions are hold* (i) $x_1^* < x_1$ *or* $x_1 < \frac{x_2 x_1^*}{x_2^*}$, (ii) $x_1 < x_1^*$ *and* $x_3 < x_3^*$ *or* $x_1 > x_1^*$ *and* $x_3 < x_3^*$, (iii) $x_1 x_3^* < x_1^* x_3$, (iv) $x_2 < x_2^*$ *and* $x_3 < x_3^*$ *or* $x_2 > x_2^*$ *and* $x_3 < x_3^*$, *then the interior equilibrium point* $E(x_1^*, x_2^*, x_3^*)$ *of the system* (1) *is globally asymptotic stable.*

Proof Let us consider the following Lyapunov function,

$$V_2 = \sum_{i=1}^{3} \left(x_i - x_i^* - x_i^* ln \left(\frac{x_i}{x_i^*} \right) \right).$$

The time derivative of V_2,

$$\frac{\mathrm{d}V_2}{\mathrm{d}t} = \sum_{i=1}^{3} \left(\frac{x_i - x_i^*}{x_i} \right) \frac{\mathrm{d}x_i}{\mathrm{d}t}.$$

After the simple algebraic manipulation, we get

$$\begin{aligned}
\frac{\mathrm{d}V_2}{\mathrm{d}t} =& -\frac{K_1 r_1}{k_1}(x_1 - x_1^*)^2 - \frac{K_2 r_2}{k_2}(x_2 - x_2^*)^2 \\
&+ K_1(x_1 - x_1^*)\left(\frac{x_2}{1 + x_1} - \frac{x_2^*}{1 + x_1^*} \right) + K_2(x_2 - x_2^*)\left(\frac{x_1^*}{1 + x_1^*} - \frac{x_1}{1 + x_1} \right) \\
&+ K_1(x_1 - x_1^*)a_1(1 - c_1)\left(\frac{x_3}{1 + a_1(1 - c_1)h_1 x_3} - \frac{x_3^*}{1 + a_1(1 - c_1)h_1 x_3^*} \right) \\
&+ K_2(x_2 - x_2^*)a_2(1 - c_2)\left(\frac{x_3}{1 + a_2(1 - c_2)h_2 x_3} - \frac{x_3^*}{1 + a_2(1 - c_2)h_2 x_3^*} \right) \\
&+ K_3(x_3 - x_3^*)\theta_1 a_1(1 - c_1)\left(\frac{x_1}{1 + a_1(1 - c_1)h_1 x_3} - \frac{x_1^*}{1 + a_1(1 - c_1)h_1 x_3^*} \right) \\
&+ K_3(x_3 - x_3^*)\theta_2 a_2(1 - c_2)\left(\frac{x_2}{1 + a_2(1 - c_2)h_2 x_3} - \frac{x_2^*}{1 + a_2(1 - c_2)h_2 x_3^*} \right), \\
=& -\frac{K_1 r_1}{k_1}(x_1 - x_1^*)^2 - \frac{K_2 r_2}{k_2}(x_2 - x_2^*)^2 \\
&- \frac{K_1(x_1 - x_1^*)(x_2 - x_2^*) + K_1(x_1 - x_1^*)(x_2 x_1^* - x_1 x_2^*)}{(1 + x_1)(1 + x_1^*)} \\
&- \frac{K_2(x_2 - x_2^*)(x_1 - x_1^*)}{(1 + x_1^*)(1 + x_1)} - \frac{K_1(1 - \theta_1)a_1(1 - c_1)(x_1 - x_1^*)(x_3 - x_3^*)}{(1 + a_1(1 - c_1)h_1 x_3)(1 + a_1(1 - c_1)h_1 x_3^*)} \\
&+ \frac{K_1 \theta_1 (a_1(1 - c_1))^2 h_1 (x_3 - x_3^*)(x_1 x_3^* - x_1^* x_3)}{(1 + a_1(1 - c_1)h_1 x_3)(1 + a_1(1 - c_1)h_1 x_3^*)}
\end{aligned}$$

$$- \frac{K_2(1-\theta_2)a_2(1-c_2)(x_2-x_2^*)(x_3-x_3^*)}{(1+a_2(1-c_2)h_2x_3)(1+a_2(1-c_2)h_2x_3^*)}$$
$$+ \frac{K_2\theta_2(a_2(1-c_2))^2h_2(x_2x_3^*-x_2^*x_3)}{(1+a_2(1-c_2)h_2x_3)(1+a_2(1-c_2)h_2x_3^*)}.$$

By the assumption $(i)-(iv)$, we get

$$\frac{dV_2}{dt} < 0.$$

Therefore, the system (1) is globally asymptotic stable at the interior equilibrium point $E(x_1^*, x_2^*, x_3^*)$.

5 Numerical Example

In this section, the obtained theoretical results are illustrated numerically with the graphical representation.

Example 1 *Choose the suitable non-negative parameter set with the various fractional orders to investigate the local stability of the system* (1). *Let* $r_1 = 2.65;\ r_2 = 2.1;\ k_1 = 8.9;\ k_2 = 1.9;\ c_1 = 0.1;\ c_2 = 0.09;\ h_1 = 0.08;\ h_2 = 0.09;\ a_1 = 0.085;\ a_2 = 0.05;\ \theta_1 = 0.01;\ \theta_2 = 0.04$ *with the initial condition* $x_1(t_0) = 4.91,\ x_2(t_0) = 8.2,\ x_3(t_0) = 0.53$.

Remark 2 The parameter set with the fractional orders satisfies the R–H stability test in Theorem 8 (see Table 1). But the equilibrium points of the system and the stability nature differed when the fractional orders are varied. From this analysis, the small changes in the fractional orders of the system can change the density of the species level. Hence, the fractional orders having much influence on the dynamical behavior of the system (see Table 1). The stable dynamical and phase-space representation of the system (1) are graphically presented in Figs. 1, 2, 3, 4, 5, and 6 with their respective equilibrium points.

Conclusion In this work, the new problem of three species CFRMPPS was constructed. The existence-uniqueness, non-negativity, and uniform boundedness of the

Table 1 R-H stability test

S. No.	$(\alpha_1, \alpha_2, \alpha_3)$	B_1	B_2	B_3	U_1	Stability
1.	(0.25,0.5,0.6)	1.4292	2.7880	0.0119	0.3985	Stable
2.	(0.75,0.25,0.8)	1.4525	0.1077	0.0019	0.1565	Stable
3.	(0.5,0.45,0.3)	1.4531	0.1094	0.0020	0.1590	Stable

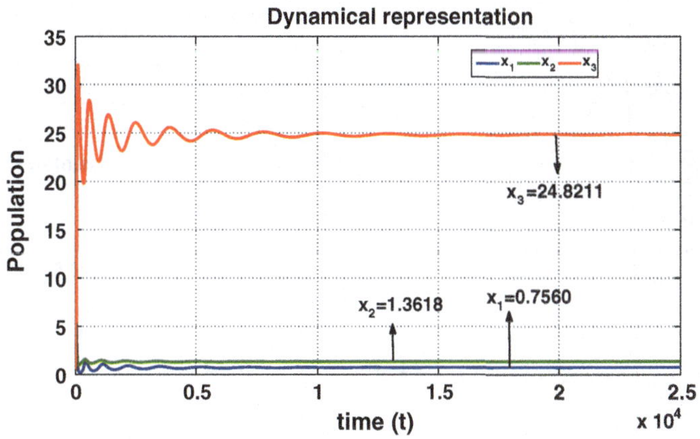

Fig. 1 Stable dynamical representation of the system (1) with the fractional orders $(\alpha_1, \alpha_2, \alpha_3) = (0.25, 0.5, 0.6)$

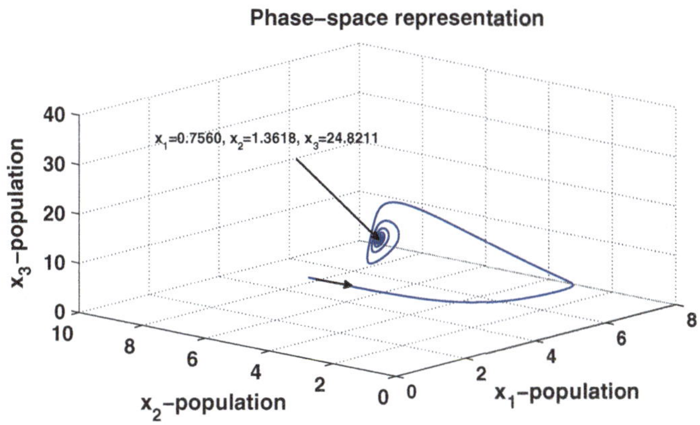

Fig. 2 Stable phase-space representation of the system (1) with the fractional orders $(\alpha_1, \alpha_2, \alpha_3) = (0.25, 0.5, 0.6)$

solutions of the system are verified. The non-negative equilibrium points of the system are obtained by finding the appropriate parameter set. The stability behavior of the system is analyzed in two ways: First, the local stability of the system was examined by the R–H stability criterion. Second, the global stability of the system was obtained by choosing a suitable Lyapunov function. Finally, the obtained theoretical results are justified in the numerical example with the graphical representation. In the future, it is believed that essential research is needed in fractional-order prey–predator models and Susceptible-Exposed-Infectious-Recovered-Vaccinated (SEIRV) epidemic models.

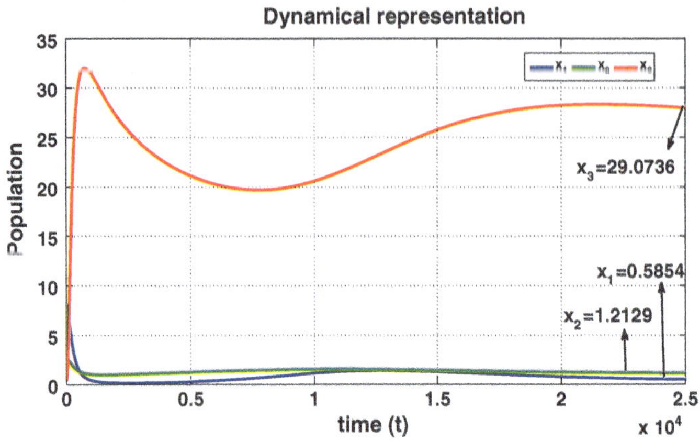

Fig. 3 Stable dynamical representation of the system (1) with the fractional orders $(\alpha_1, \alpha_2, \alpha_3) = (0.75, 0.25, 0.8)$.

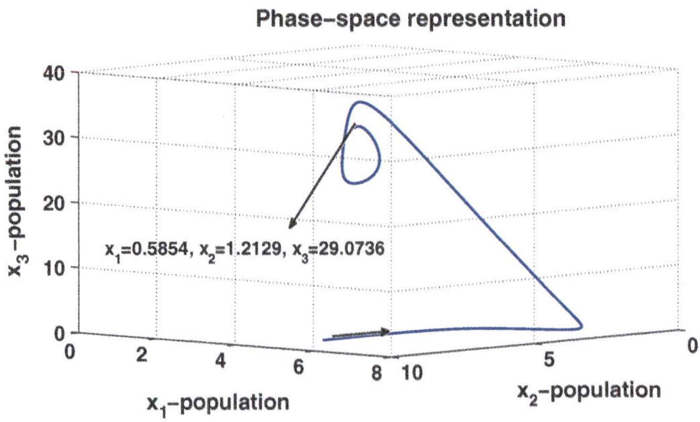

Fig. 4 Stable phase-space representation of the system (1) with the fractional orders $(\alpha_1, \alpha_2, \alpha_3) = (0.75, 0.25, 0.8)$.

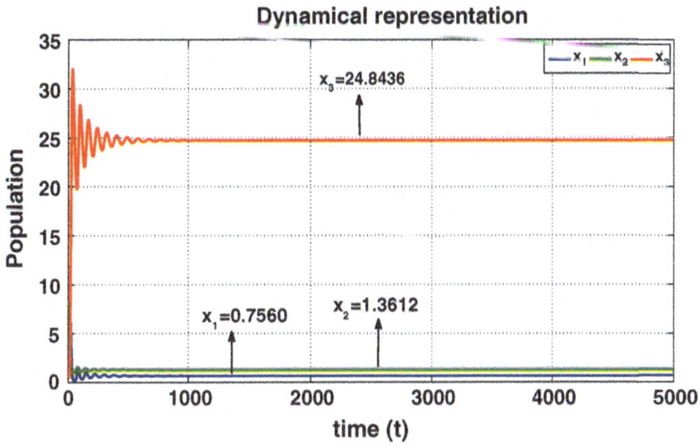

Fig. 5 Stable dynamical representation of the system (1) with the fractional orders $(\alpha_1, \alpha_2, \alpha_3) = (0.5, 0.45, 0.3)$.

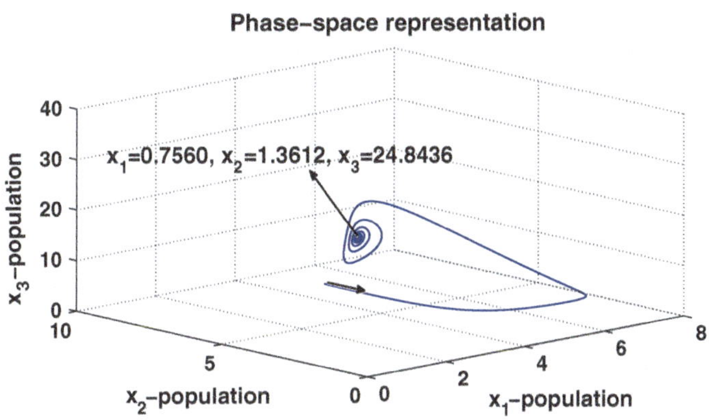

Fig. 6 Stable phase-space representation of the system (1) with the fractional orders $(\alpha_1, \alpha_2, \alpha_3) = (0.5, 0.45, 0.3)$.

References

1. Abdeljawad, T.: On conformable fractional calculus. J. Comput. Appl. Math. **279**, 57–66 (2015). https://doi.org/10.1016/j.cam.2014.10.016
2. Bakhanova, Y.V., Kazakov, A.O., Korotkov, A.G., Levanova, T.A., Osipov, G.V.: Spiral attractors as the root of a new type of "bursting activity" in the Rosenzweig–MacArthur model. Eur. Phys. J. Special Top. **227**(No. 7–9), 959–970. Springer, Verlag (2018). https://doi.org/10.1140/epjst/e2018-800025-6
3. Cenesiz, Y., Kurt, A., Nane, E.: Stochastic solutions of conformable fractional Cauchy problems. Statistics Probab. Lett. **124**, 126–131 (2017) https://doi.org/10.1016/j.spl.2017.01.012
4. Ducrot, A., Liu, Z., Magal, P.: Large speed traveling waves for the Rosenzweig–MacArthur predator–prey model with spatial diffusion. Physica D: Nonlinear Phenomena **415**, 132730 (2020). https://doi.org/10.1016/j.physd.2020.132730

5. Goel, N. S., Maitra, S. C., Montroll, E. W.: On the Volterra and other nonlinear models of interacting populations. Revi. Mod. Phys. **43**(2), 231 (1971). https://doi.org/10.1103/RevModPhys.43.231
6. Joshua, E.E., Akpan, E.T.: Global stability and Hopf-bifurcation analysis of biological systems using delayed extended Rosenzweig–MacArthur model. Mod. Appl. Sci. **12**(2), 171–181 (2018). https://doi.org/10.5539/mas.v12n2p171
7. Kaviya, R., Muthukumar, P.: The impact of immigration on a stability analysis of Lotka-Volterra system. IFAC-Papers OnLine **53**(1), 214–219 (2020). https://doi.org/10.1016/j.ifacol.2020.06.037
8. Khalil, R., Al Horani, M., Yousef, A., Sababheh, M.: A new definition of fractional derivative. J. Comput. Appl. Math. **264**, 65–70 (2014). https://doi.org/10.1016/j.cam.2014.01.002
9. Kruff, N., Lax, C., Liebscher, V., Walcher, S.: The Rosenzweig–MacArthur system via reduction of an individual based model. J. Math. Bio. **78**(1–2), 413–439 (2019). https://doi.org/10.1007/s00285-018-1278-y
10. Kumar, S., Kumar, R., Cattani, C., Samet, B.: Chaotic behaviour of fractional predator-prey dynamical system. Chaos Solitons Fractals 135, 109811 (2020). https://doi.org/10.1016/j.chaos.2020.109811
11. Li, H. L., Zhang, L., Hu, C., Jiang, Y. L., Teng, Z.: Dynamical analysis of a fractional-order predator-prey model incorporating a prey refuge. J. Appl. Math. Comput. **54**(1–2), 435–449 (2017). https://doi.org/10.1007/s12190-016-1017-8
12. Miller, K. S., Ross, B.: An Introduction to the Fractional Calculus and Fractional Differential Equations. Wiley (1993)
13. Mondal, S., Biswas, M., Bairagi, N.: Local and global dynamics of a fractional-order predator–prey system with habitat complexity and the corresponding discretized fractional-order system. J. Appl. Math. Comput. **63**, 1–30 (2020). https://doi.org/10.1007/s12190-020-01319-6
14. Moustafa, M., Mohd, M. H., Ismail, A. I., Abdullah, F. A.: Dynamical analysis of a fractional-order Rosenzweig–MacArthur model incorporating a prey refuge. Chaos Solitons Fractals **109**, 1–13 (2018). https://doi.org/10.1016/j.chaos.2018.02.008
15. Souahi, A., Makhlouf, A. B., Hammami, M. A.: Stability analysis of conformable fractional-order nonlinear systems. Indagationes Mathematicae **28**(6), 1265–1274 (2017). https://doi.org/10.1016/j.indag.2017.09.009
16. Odibat, Z. M., Shawagfeh, N. T.: Generalized Taylor's formula. Appl. Math. Comput. **186**(1), 286–293 (2007). https://doi.org/10.1016/j.amc.2006.07.102
17. Wang, Y.: Dynamic analysis and synchronization of conformable fractional-order chaotic systems. Eur. Phys. J. Plus **133**(11), 481 (2018). https://doi.org/10.1140/epjp/i2018-12300-y
18. Wang, Z., Xie, Y., Lu, J., Li, Y.: Stability and bifurcation of a delayed generalized fractional-order prey–predator model with interspecific competition. Appl. Math. Comput. 3**47**, 360–369 (2019). https://doi.org/10.1016/j.amc.2018.11.016
19. Yang, S., Wang, L., Zhang, S.: Conformable derivative: application to non-Darcian flow in low-permeability porous media. Appl. Math. Lett. **79**, 105–110 (2018). https://doi.org/10.1016/j.aml.2017.12.006
20. Zhou, Z., Van Gorder, R. A.: Turing instability and colony formation in spatially extended Rosenzweig-MacArthur predator–prey models with allochthonous resources. Bull. Math. Biol. **81**(12), 5009–5053 (2019). https://doi.org/10.1007/s11538-019-00667-0

Computation of Probabilities of Mixed Poisson–Weibull Distribution

Yeh Ching Low and Seng Huat Ong

Abstract Mixed Poisson distributions are well-researched statistical models in the count data literature. They are able to model over dispersion and have applications in various fields such as in modeling automobile claims, citation counts, and species abundance. However, only a few mixed Poisson distributions have probability mass functions in closed form, thus limiting their uses due to computational issues. This paper considers a two-parameter mixed Poisson–Weibull distribution and proposes two general methods for the computation of its probabilities: (a) employing an alternating series formula and (b) Monte Carlo simulation technique by exploiting the invertibility of the Weibull distribution function. An EM-type algorithm is considered for maximum likelihood estimation of the Poisson–Weibull parameters. Application of this model is illustrated through comparison of model fit for two real data sets with some popular two-parameter mixed Poisson distributions: negative binomial, Poisson–lognormal and Poisson–inverse Gaussian distributions. It is shown that the Poisson–Weibull distribution gives a much better fit than these distributions.

Keywords Weibull distribution · Over dispersion · Alternating series · Monte Carlo integration · EM algorithm · Poisson–inverse Gaussian · Poisson–lognormal · Goodness-of-fit

---

Y. C. Low
Department of Computing and Information Systems, Sunway University, 47500 Petaling Jaya, Selangor, Malaysia
e-mail: yehchingl@sunway.edu.my

S. H. Ong (✉)
Department of Actuarial Science and Applied Statistics, UCSI University, 56000 Kuala Lumpur, Malaysia
e-mail: ongsh@ucsiuniversity.edu.my

P. Balasubramaniam et al. (eds.), *Mathematical Modelling and Computational Intelligence Techniques*, Springer Proceedings in Mathematics & Statistics 376,
https://doi.org/10.1007/978-981-16-6018-4_7

1 Introduction

Data in the form of counts arise in many applications such as in the modeling of car insurance claims, citation counts for publications and species abundance. The Poisson distribution is a standard distribution for modeling count data. However, the phenomenon of over dispersion (variance is larger than the mean) is often present in real-life count data. Over dispersion may be caused by clustering, unobserved heterogeneity or presence of excess zeroes in the data. The Poisson distribution has mean equal to the variance, and thus, it is not able to model over dispersed data. This prompted researchers to extend and generalize the Poisson distribution in different ways. One method is through the mixed Poisson distributions where the Poisson mean θ is taken to be a random variable. This leads to a model exhibiting over dispersion. The distribution of the random Poisson mean θ is known as the mixing distribution. For an overview on mixed Poisson distributions, their properties and applications, see Gupta and Ong [10], Karlis and Xekalaki [16] and Nikoloulopoulos and Karlis [17]. Recently, Chee [3] considered modeling of count data by using nonparametric Poisson mixtures.

The probability mass function (pmf) of a mixed Poisson distribution is given by

$$\Pr(X = k) = \int_0^\infty \frac{e^{-\theta}\theta^k}{k!} f(\theta)\mathrm{d}\theta, \quad k = 0, 1, 2, \ldots \tag{1}$$

where $f(\theta)$ denotes the probability density function (pdf) of the mixing distribution. The variance of a mixed Poisson distribution is always greater than that of the simple Poisson model [16] with the same mean. Therefore, a mixed Poisson distribution is always over dispersed. Shaked's [28] Two Crossings Theorem implies that a mixed Poisson distribution, compared to its simple Poisson counterpart, will always have a higher zero count probability and a longer right tail.

In the existing literature, several distributions have been proposed as the mixing distribution, such as the gamma distribution which resulted in the negative binomial (NB) [8] or Poisson–gamma (P-G) distribution, inverse Gaussian distribution which gives rise to the Poisson–inverse Gaussian (P-iG) distribution [11, 26] and the lognormal distribution leading to the Poisson–lognormal (P-LN) distribution [1]. The NB distribution is a popular model for count data exhibiting over dispersion. The P-iG distribution with pmf in terms of the modified Bessel special function has a simple three-term recurrence relation to facilitate computation of probabilities for most of the k counts (see [20]), and this distribution has been proposed as an alternative to the NB distribution [33]. Apart from the NB and P-iG distributions, only a few mixed Poisson distributions are investigated in detail because of the intractability of the pmfs. For instance, due to lack of a closed form, the numerical evaluation of the P-LN pmf is more involved [1, 12] especially for large counts k. This paper examines the mixed Poisson–Weibull (P-W) distribution which is obtained when the mixing distribution is the Weibull distribution. The Weibull distribution has a closed form cumulative distribution function (cdf) and is very popular model for applications, particularly in reliability.

There is only one published study [4] for the P-W distribution, in which a regression model is developed for modeling motor vehicle crash data, and this is compared with the NB regression model. Since the likelihood function of the P-W distribution is complicated, Cheng et al. [4] developed Bayesian inference, where the Markov chain Monte Carlo (MCMC) simulation is based upon the WinBUGS software [30]. They showed that their P-W generalized linear model (GLM) performed as well as the P-G GLM.

We consider the P-W distribution in a non-Bayesian framework. This obviates the need to use the computer-intensive MCMC simulation. Since the P-W pmf does not have a closed form, we propose and examine two general approaches to compute its probabilities: (a) an alternating series formula and (b) Monte Carlo simulation technique. The rapid computer sampling from the Weibull distribution by the method of inverse cdf is facilitated by the invertibility of the Weibull cdf. Thus, it is easy to implement the Monte Carlo simulation technique which is shown to be highly accurate. Maximum likelihood estimation is conducted through a much simpler expectation-maximization (EM) algorithm [15] that only requires estimation of the Weibull distribution. Overcoming the computational bottleneck allows the Poisson–Weibull distribution to be a practical alternative to other mixed Poisson models.

The paper is arranged as follows. In Sect. 2, to facilitate the ensuing discussion, we briefly describe three popular mixed Poisson distributions, namely the NB, P-iG and P-LN distributions. The proposed methods for the computation of the mixed Poisson probabilities are discussed in Sect. 3. In Sect. 4, we present an EM type algorithm for maximum likelihood estimation. To illustrate its application, we perform model fitting on two real data sets in Sect. 5 with comparison to the NB, P-iG and P-LN distributions. Finally, Sect. 6 gives the conclusion.

2 Mixed Poisson Distributions

2.1 Negative Binomial, Poisson–Inverse Gaussian, and Poisson–Lognormal Distributions

The NB distribution is introduced by Greenwood and Yule [8] as a mixed Poisson distribution with the gamma distribution as mixing distribution. Since there are various representations of its pmf in the literature, we use the representation which is given as

$$\Pr(X = k) = \frac{\Gamma(k+\alpha)}{k!\Gamma(\alpha)}\left(\frac{\beta}{1+\beta}\right)^{\alpha}\left(\frac{1}{1+\beta}\right)^{k} \quad (2)$$

for $k = 0, 1, 2, \ldots$, and $\alpha, \beta > 0$ [15]. It has been frequently applied to model over dispersed count data due to its relative simplicity since the pmf has a simple form with only two parameters.

If the inverse Gaussian distribution is used as the mixing distribution, the mixed Poisson is known as the P-iG distribution [11, 26]. The pmf of the P-iG distribution is given as

$$\Pr(X = k) = \sqrt{\frac{2\alpha}{\pi}} \frac{\exp(\alpha\sqrt{1-\theta})(\alpha\theta/2)^k}{k!} K_{k+\frac{1}{2}}(\alpha) \tag{3}$$

for $k = 0, 1, 2, \ldots$ and $\alpha > 0$, $0 < \theta < 1$. The P-iG distribution is useful in modeling heavy-tailed data. To address the computational issues of its probabilities, Shaban [27] provided a series representation as well as a probability recurrence relation and examined approximations and limiting cases. Ong [20] expressed the P-iG distribution as a mixed Poisson distribution with the inverted gamma as mixing distribution. This result is used to improve the approximation for large counts and, hence, provide a method to compute the tail probabilities. Based on the results in Ong [19], a Taylor expansion for the P-iG pmf is also derived.

Bulmer [1] introduced the Poisson–lognormal distribution in the study of species abundance. The pmf is given by

$$\Pr(X = k) = \frac{1}{\sqrt{2\pi}\sigma k!} \int_0^\infty e^{-\theta}\theta^{k-1} \exp\left(-\frac{(\log\theta - \mu)^2}{2\sigma^2}\right) d\theta \tag{4}$$

for $k = 0, 1, 2, \ldots$, and $-\infty < \mu < \infty$, $\sigma > 0$, where the lognormal distribution is the mixing distribution with pdf $f(\theta) = \frac{1}{\sigma\theta\sqrt{2\pi}}\exp\left(-\frac{(\log\theta-\mu)^2}{2\sigma^2}\right)$. Even though the Poisson–lognormal distribution is suited to fit long-tailed data found in diverse areas, for example, in species abundance [1], bibliometry [31] and food safety risk assessments [32], its application is limited by an intractable pmf. An approximation of the Poisson–lognormal pmf has been derived by Bulmer [1] for large values of k. This is achieved by applying Taylor series expansion to the pmf which has been expressed as an expectation with respect to a gamma pdf. The probabilities for small k are calculated by numerical integration. Grundy [9] given the probabilities for $k = 0, 1$ by employing an infinite series to approximate the integral. Cassie [2] examined the application of the lognormal distribution to approximate the mixed Poisson probabilities. In a more recent discussion on maximum likelihood estimation of Poisson–lognormal parameters, Izsák [12] applied numerical integration for small k after making prior modifications to the integral, and the saddle point method is used for large k.

2.2 The Poisson–Weibull Distribution

Suppose Y is a Weibull random variable. The Weibull distribution has pdf given by

$$f(y) = \alpha\beta^{-\alpha}y^{\alpha-1}\exp\left(-\frac{y}{\beta}\right)^{\alpha} \tag{5}$$

for $y > 0$ and $\alpha, \beta > 0$. The parameters α and β are the shape and scale parameters, respectively. In reliability studies, the Weibull distribution is of special importance because it can model lifetimes with increasing, decreasing or constant failure rates. Special cases of the Weibull distribution are the exponential distribution ($\alpha = 1$) and the Rayleigh distribution ($\alpha = 2$ and $\beta = \sigma\sqrt{2}$). More information on the Weibull distribution can be found, for instance, in Johnson et al. [13].

Definition 2.1 (Poisson–Weibull distribution) Let X be a discrete random variable and $X|\Theta \sim \text{Poisson}(\Theta)$, where Θ is a real-valued and nonnegative Weibull random variable with pdf $f(\theta)$ given by (5). The unconditional X has pmf given by

$$\Pr(X = k) = \frac{\alpha\beta^{-\alpha}}{k!}\int_0^{\infty} e^{-\theta}\theta^{k+\alpha-1}\exp(-(\theta/\beta)^{\alpha})\mathrm{d}\theta, \quad k = 0, 1, 2, \dots \tag{6}$$

The P-W distribution has probability generating function (pgf) which is defined as

$$G(z) = \frac{\alpha\beta^{-\alpha}}{k!}\int_0^{\infty} e^{\theta(z-1)}\theta^{k+\alpha-1}\exp(-(\theta/\beta)^{\alpha})\mathrm{d}\theta$$

This can be rewritten as an infinite series by applying the connection between the moment generating function (mgf) $M(t)$ of the mixing distribution and the pgf $G(z)$ of the mixed Poisson distribution [16, p. 38], which is $G(z) = M(z-1)$. This gives

$$G(z) = \sum_{n=0}^{\infty}\frac{(z-1)^n\beta^n}{n!}\Gamma\left(1+\frac{n}{\alpha}\right) \tag{7}$$

where the Weibull mgf is $M(t) = \sum_{n=0}^{\infty}\frac{t^n\beta^n}{n!}\Gamma\left(1+\frac{n}{\alpha}\right)$.

Theorem 2.1 *Let X be a P-W random variable. Then, its mean μ, variance σ^2, and index of dispersion d are:*

$$\mu = \beta\Gamma\left(\frac{1}{\alpha}+1\right) \tag{8}$$

$$\sigma^2 = \beta\Gamma\left(\frac{1}{\alpha}+1\right) + \beta^2\Gamma\left(\frac{2}{\alpha}+1\right) - \beta^2\left[\Gamma\left(\frac{1}{\alpha}+1\right)^2\right], \text{ and} \tag{9}$$

$$d = 1 + \beta\frac{\Gamma\left(\frac{2}{\alpha}+1\right)}{\Gamma\left(\frac{1}{\alpha}+1\right)} - \mu \tag{10}$$

Proof The P-W mean and variance can directly be determined from the Weibull moments since factorial moments of the mixed Poisson distribution are the moments about the origin of the mixing distribution [16]. The Weibull distribution has mean $\beta\Gamma\left(\frac{1}{\alpha}+1\right)$ and variance $\beta^2\left\{\Gamma\left(\frac{2}{\alpha}+1\right) - \left[\Gamma\left(\frac{1}{\alpha}+1\right)\right]^2\right\}$, and the formulae for μ, σ^2 and d are easily obtained.

When $\alpha = 1$, this reduces to the geometric distribution [14].

3 Computation of Poisson–Weibull Probabilities

The computation of mixed Poisson probabilities has been studied by various authors. Willmot [34] obtained a general formula to approximate these probabilities and demonstrated that the behavior of the mixed Poisson right tail is similar to the mixing distribution. A similar result has been derived by Perline [24]. Karlis and Xekalaki [16] discussed numerical integration using Gauss–Laguerre polynomials. They also proposed a method in which the mixed Poisson probability function is expressed as an infinite series in terms of the moments of the mixing distribution. Willmot [35] proposed recursive relations for the mixed Poisson probability functions. Ong [19] proposed a three-term recurrence formula to facilitate computation the full beta distribution with pmf

$$\Pr(X = k) = \frac{1}{B(p,q)}\int_0^\infty \frac{e^{-\theta}\theta^k}{k!}\frac{b^p\theta^{p-1}}{(1+b\theta)^{p+q}}d\theta$$

for $k = 0, 1, \ldots, b, p, q > 0$ where $B(p, q)$ is the beta function. This distribution has been derived with the beta distribution of the second kind, scale parameter $1/b$, as the mixing distribution. However, in general, three-term recurrence relations are subjected to issues of numerical stability.

In the following sections, we propose two general methods to evaluate the P-W probability mass function, namely a method which uses an alternating series formula and a Monte Carlo simulation technique.

3.1 *Alternating Series Formula for Poisson–Weibull Probabilities*

A formula for the P-W probabilities $\Pr(X = k) = P(k)$ is obtained by differentiating the pgf (7) and setting $z = 0$. For $k \geq 1$, we have

$$P(k) = \frac{1}{k!}\frac{\partial^k G}{\partial z^k}\bigg|_{z=0} = \frac{(-1)^k}{k!}\sum_{n=k}^{\infty}(-1)^n n(n-1)\dots(n-(k-1))\frac{\beta^n}{n!}\Gamma\left(\frac{n}{\alpha}+1\right)$$
$$= \frac{(-1)^k}{k!}\sum_{n=k}^{\infty}(-1)^n c_n \tag{11}$$

where $c_n = n(n-1)\dots(n-(k-1))\frac{\beta^n}{n!}\Gamma\left(\frac{n}{\alpha}+1\right), k \geq 1$.

To compute the alternating series (11), we derive a recurrence relation for the coefficients c_n.

Consider c_{n+1}/c_n. We get

$$\frac{c_{n+1}}{c_n} = \frac{\beta}{(n-k+1)}\frac{\Gamma\left(\frac{(n+1)}{\alpha}+1\right)}{\Gamma\left(\frac{n}{\alpha}+1\right)} \tag{12}$$

Due to the quick growth of the gamma function $\Gamma(x)$ with x, we use the following asymptotic formula from Johnson et al. [14, Eq. (1.32)] to calculate the ratio of gamma functions in (12):

$$\frac{\Gamma(x+a)}{\Gamma(x+b)} \sim x^{a-b}\left(1 + \frac{(a-b)(a+b-1)}{2x} + \cdots\right).$$

The recurrence Formula (12) for c_n becomes

$$c_{n+1} = \frac{\beta}{(n-k+1)}\left(\frac{n}{\alpha}\right)^{1/\alpha}\left(1 + \frac{\left(\frac{1}{\alpha}\right)\left(\frac{1}{\alpha}+1\right)}{2(n/\alpha)}\right)c_n, \quad k \geq 1$$

with $c_k = \beta^k\Gamma\left(\frac{k}{\alpha}+1\right)$. This two-term recurrence relation is computationally stable, and the asymptotic approximation for the ratio of gamma functions is rather accurate even for small values of the argument. Table 1 illustrates the accuracy of the alternating series Formula (11) with the values from numerical integration as benchmark.

For $\Pr(X = 0) = P(0)$, we have

$$P(0) = G(0) = \sum_{n=0}^{\infty}(-1)^n d_n \tag{13}$$

Table 1 Poisson–Weibull probabilities computed by (a) numerical integration, (b) Monte Carlo estimator (15) and (c) alternating series formula for $\alpha = 1$ and $\beta = 0.5$

k	$\Pr(X = k)$ Method (a)	$\Pr(X = k)$ Method (b)	$\Pr(X = k)$ Method (c)
2	0.0740740704	0.0740740742	0.0740740741
3	0.0246913569	0.0246913584	0.0246913580
4	0.0082304518	0.0082304534	0.0082304527
5	0.0027434840	0.0027434852	0.0027434842
10	0.0000112901	0.0000112895	0.0000112901
15	0.0000000465	0.0000000461	0.0000000465
20	0.0000000002	0.0000000002	0.0000000002

where $d_n = \frac{\beta^n}{n!}\Gamma\left(\frac{n}{\alpha}+1\right)$, $d_0 = 1$. The recurrence formula for d_n is

$$d_{n+1} = \frac{\beta}{(n+1)}\left(\frac{n}{\alpha}\right)^{1/\alpha}\left(1 + \frac{(\frac{1}{\alpha})(\frac{1}{\alpha}+1)}{2(n/\alpha)}\right)d_n.$$

To speed up the convergence of the alternating series (11) and (13), the method of Cohen et al. [6] may be used.

3.2 *Monte Carlo Simulation Technique*

Let Θ be a random variable having pdf $f(\theta)$, and it is required to compute

$$E[g(\Theta)] = \int g(\theta) f(\theta) \mathrm{d}\theta$$

for a function $g(\theta)$. A crude Monte Carlo estimator to approximate this integral uses N simulated realizations $\theta_1, \theta_2, \ldots, \theta_N$ of Θ and is given by

$$E[g(\Theta)] \approx \frac{1}{N}\sum_{i=1}^{N} g(\theta_i) \tag{14}$$

The mixed Poisson probabilities (1) may be approximated by (14):

$$\Pr(X = k) = \int_0^\infty \frac{e^{-\theta}\theta^k}{k!} f(\theta)\mathrm{d}\theta \approx \frac{1}{N}\sum_{i=1}^{N} g(\theta_i; k) \tag{15}$$

where $g(\theta_i; k) = \frac{e^{-\theta_i}\theta_i^k}{k!}$, and $\{\theta_1, \theta_2, \ldots, \theta_N\}$ is a random sample from $f(\theta)$.

In Monte Carlo simulation studies, several methods are available to generate random samples from a prescribed probability distribution. A general technique is the inverse transform method that uses the inverse of the cumulative distribution function (cdf). This method is very fast when the cdf is available in closed form. Otherwise, the acceptance rejection or envelope rejection method is an alternative. Ong and Lee [21] have applied the mixed Poisson formulation of the NB distribution to formulate an envelope rejection method for generating NB samples. The mixing formulation has also been used by Ong [18] to generate samples from bivariate distributions with given marginal and correlation. As such, the method for simulation of the random numbers θ_N, $i = 1, 2, \ldots, N$ from the mixing distribution $f(\theta)$ is selected according to its suitability.

A general drawback of Monte Carlo simulation is the long computation time. In practice, to attain convergence, the number of points N required is unrealistically high. To mitigate this problem, variance reduction methods are used to speed up convergence. A simple technique to reduce the sampling size in order to speed up the convergence of estimator (15) is as follows. In the generation of θ_i in the simulation, the (pseudo) random numbers are substituted with low F-discrepancy sequence of numbers. This method is also called the quasi-Monte Carlo method since the low F-discrepancy sequence of numbers is a deterministic sequence.

In the one-dimensional case, the set of N points with the lowest F-discrepancy is given by

$$u_i = \frac{2i-1}{2N}, \quad i = 1, 2, 3, \ldots N \tag{16}$$

Therefore, an algorithm for computing mixed Poisson probabilities is as follows.

1. Generate lowest-F discrepancy numbers (16).
2. Compute $\theta_i = F^{-1}\left(\frac{2i-1}{2N}\right)$, where $F(\theta)$ is the cdf of the mixing distribution.
3. Compute the estimator (15).

With modification, the algorithm is applicable to arbitrary mixing distributions. In particular, for P-W probabilities, $F(\theta)$ is the Weibull cdf with inverse

$$F^{-1}(\theta) = \beta[-\log(1-\theta)]^{\frac{1}{\alpha}}$$

Remark Recently, Ong et al. [23] gave a general method for the Monte Carlo computation of mixed Poisson probabilities. This method does not require random samples from the mixing distribution which is very helpful if sampling from the mixing distribution is difficult. The proposed Monte Carlo method here differs by exploiting the invertibility of the Weibull cdf and, thus, samples from the Weibull mixing distribution.

3.3 Application of the Computational Approaches

This section illustrates the application of the computational approaches discussed. Table 1 lists the Poisson–Weibull probabilities evaluated using numerical integration with (6), Monte Carlo simulation and the alternating series Formulas (11) and (12) for selected values of k. The parameters selected are $\alpha = 1$and $\beta = 0.5$. It is shown that the values obtained agrees to at least eight decimal places.

Some Poisson–Weibull probability plots are displayed in Fig. 1 to show distributional features like mode, skewness, and tail length. It is seen that the distribution is able to represent data with high zero counts, right skewness, and symmetrical shapes.

Numerical computation shows that when $\alpha < 1$, there is a high mode at $k = 0$ for small β. As β increases, the probabilities and mode shift away from $k = 0$. But when $\alpha > 1$, the mode shifts away from $k = 0$ as β increases, and the distribution changes shape from being right skewed to almost symmetrical. As the value of α becomes larger and larger, the mode moves further away from $k = 0$ when β increases. It is well known that unimodality of the mixing distribution implies unimodality of the mixed Poisson distribution [16, p. 41]. Since the Weibull distribution is unimodal, the mixed Poisson–Weibull distribution is also unimodal.

4 Parameter Estimation

When the mixed Poisson pmf is not in a closed form, statistical inference procedures such as parameter estimation will require numerical methods. Such procedures may be unstable and give biased estimates. For example, in maximum likelihood (ML) estimation, numerical methods are used in the evaluation of the Poisson–lognormal likelihood function [1, 32]. To overcome this, Karlis [15] considered an expectation-maximization type algorithm in the ML estimation of mixed Poisson distributions. Dempster et al. [7] were the first to propose the EM algorithm for ML estimation for data containing missing values. The EM algorithm converts the maximization of the mixed Poisson likelihood function to maximizing a usually simpler function involving the mixing distribution.

In the formulation of the EM algorithm for mixed Poisson distributions, the random sample of size n, $Y_i = (X_i, \theta_i)$, $i = 1, 2, \ldots, n$, is regarded as the complete data. This consists of observed X_i and unobserved realizations of θ_i at each X_i. As stated in Sect. 1, random variable θ has the mixing pdf $f(\theta; \boldsymbol{\phi})$ with parameter vector $\boldsymbol{\phi}$. In the E-step for the $(k+1)$-th iteration, the estimates $\phi^{(k)}$ are used to compute pseudo-values $E(h_j(\theta)|X_i, \phi^{(k)})$, for $i = 1, 2, \ldots, n, j = 1, 2, \ldots, m$, where $h_j(.)$ are some functions of the mixing distribution. The expectation is calculated based on the conditional distribution $f(Y; X, \phi^{(k)})$. At the M-step, based on the pseudo-values from the E-step, $Q(\phi; \phi^{(k)}) = E(\log p(Y|\phi)|X, \phi^{(k)})$ is maximized over ϕ.

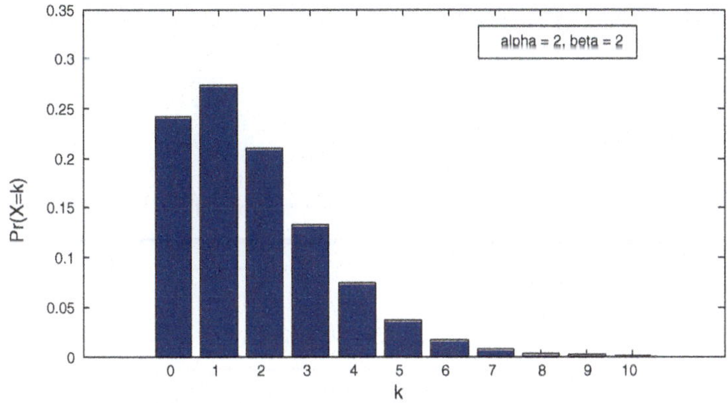

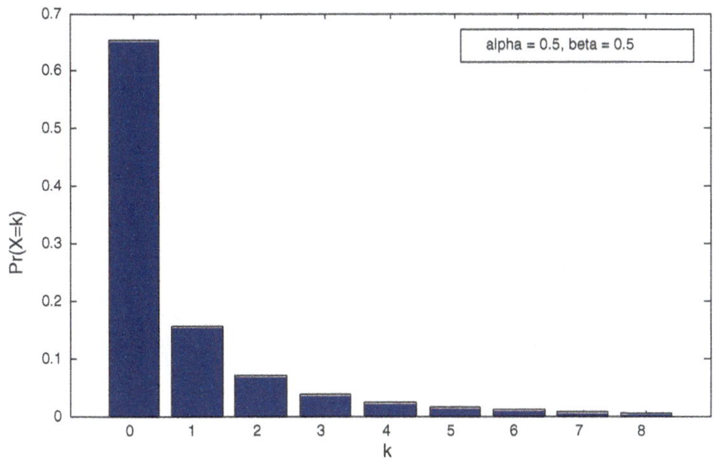

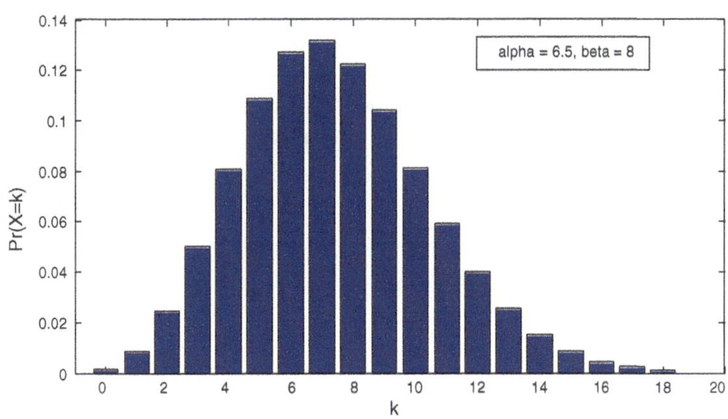

Fig. 1 Probability plots of the Poisson–Weibull distribution

For details of the EM algorithms for the NB, P-iG, and P-LN distributions, see Karlis [15]. The same approach for the P-W distribution is followed, and the algorithm is:

E-step. Compute the pseudo-values

$$t_i = E(\theta_i{}^{\alpha}|x_i) = \frac{\int e^{-\theta_i}\theta_i{}^{x_i+\alpha_{\text{old}}-1}(\theta_i{}^{\alpha})\exp\left(-\left(\frac{\theta_i}{\beta_{\text{old}}}\right)^{\alpha_{\text{old}}}\right)d\theta}{\int e^{-\theta_i}\theta_i{}^{x_i+\alpha_{\text{old}}-1}\exp\left(-\left(\frac{\theta_i}{\beta_{\text{old}}}\right)^{\alpha_{\text{old}}}\right)d\theta}$$

$$s_i = E(\log\theta_i|x_i) = \frac{\int e^{-\theta_i}\theta_i{}^{x_i+\alpha_{\text{old}}-1}(\log\theta_i)\exp\left(-\left(\frac{\theta_i}{\beta_{\text{old}}}\right)^{\alpha_{\text{old}}}\right)d\theta}{\int e^{-\theta_i}\theta_i{}^{x_i+\alpha_{\text{old}}-1}\exp\left(-\left(\frac{\theta_i}{\beta_{\text{old}}}\right)^{\alpha_{\text{old}}}\right)d\theta}$$

We use numerical integration to evaluate the integrals.

M-step. The ML estimate of the Weibull parameter β is given by

$$\widehat{\beta} = \left(\frac{1}{n}\sum_{i=1}^{n} x_i{}^{\alpha}\right)^{\frac{1}{\alpha}}$$

and a simplified likelihood equation for α is given as

$$\frac{n}{\alpha} - \frac{n}{\sum_{i=1}^{n} x_i{}^{\alpha}}\sum_{i=1}^{n} x_i{}^{\alpha}\log(x_i) + \sum_{i=1}^{n}\log(x_i) = 0$$

First compute $\beta_{\text{new}} = \left(\frac{1}{n}\sum_{i=1}^{n} t_i\right)^{\frac{1}{\alpha_{\text{old}}}}$ and then obtain α_{new} by numerically maximizing $Q = n\log\alpha - n\alpha\log\beta_{\text{new}} + (\alpha - 1)\sum_{i=1}^{n} s_i - \left(\frac{1}{\beta_{\text{new}}}\right)^{\alpha}\sum_{i=1}^{n} t_i$ with respect to α. The iterations are terminated when the relative change in the log-likelihood function values is less than 10^{-10}.

Although the E-M algorithm simplifies ML estimation, a drawback is that it may converge to the local maximum of the likelihood function (see, for instance, Chung and Lindsay [5]). The standard practice is to use multiple initial values for the algorithm.

5 Applications

To exemplify the application of the P-W distribution, and to compare the goodness-of-fit with the two-parameter mixed Poisson distributions discussed in Sect. 2, we use two real data sets found in the literature. We perform ML estimation by the EM

Table 2 Systemic adverse events after vaccination [25]

Number of adverse events	Observed frequency	Negative binomial (Poisson–gamma)	Poisson–inverse Gaussian	Poisson–Weibull	Poisson–lognormal
0	1437	1409.08	1350.43	1414.65	1345.99
1	1010	1068.66	1154.38	1052.44	1152.32
2	660	670.66	689.13	670.79	694.42
3	428	391.63	374.28	397.89	377.16
4	236	220.16	200.62	225.43	200.20
5	122	120.88	108.92	123.49	107.27
6	62	65.32	60.29	65.88	58.79
7	34	34.89	34.02	34.38	33.12
8	14	18.47	19.54	17.61	19.19
9	8	9.71	11.39	8.87	11.43
10	4	5.08	6.72	4.41	6.99
11	4	2.64	4.01	2.16	4.38
12	1	1.37	2.42	1.05	2.80
13 or more	0	1.46	3.85	0.95	5.93
Total	4020	4020	4020	4020	4020
Number of classes		14	14	14	14
Chi-square		12.5283	48.7646	8.6431	51.7539
d.f		11	11	8	11
p-value		0.3253	<0.0001	0.3846	<0.0001
AIC		20,364.00	13,524.53	13,480.53	13,529.75

algorithm outlined previously. Several sets of randomly generated initial values are used to check whether the algorithm converges to the global maximum.

Table 2 displayed the data and goodness-of-fit results for clinical trial data on systemic adverse events for first four anthrax vaccine absorbed (AVA) study injections [25]. The sample mean is 1.5070 with standard deviation 1.7040, and the dispersion index of 1.93 implies over dispersion. It is seen that the P-W distribution gives the best fit with much smaller values of AIC and chi-square goodness-of-fit statistic.

Table 3 gave the observed and expected count frequencies and fitting results for the number of quarterly sales [29]. The mean is 3.5502 with standard deviation 3.3636, giving a dispersion index of 3.19. The P-W distribution gives the best fit when compared with the other three distributions.

The ML estimates and log-likelihood function values are given in Table 4.

Table 3 Quarterly sales data [29]

Number of sales	Observed frequency	Negative binomial (Poisson–gamma)	Poisson–inverse Gaussian	Poisson–Weibull	Poisson–lognormal
0	514	490.94	413.28	506.53	418.77
1	503	537.63	589.53	523.43	579.09
2	457	480.49	544.18	467.03	536.90
3	423	397.18	428.32	391.23	428.46
4	326	314.98	316.99	315.43	320.58
5	233	243.45	229.83	247.54	233.47
6	195	184.90	166.19	190.26	168.54
7	139	138.65	120.75	143.78	121.74
8	101	102.98	88.39	107.12	88.41
9	77	75.91	65.24	78.83	64.72
10	56	55.61	48.55	57.39	47.80
11	40	40.54	36.40	41.37	35.64
12	37	29.43	27.49	29.56	26.83
13	22	21.29	20.88	20.95	20.39
14	9	15.35	15.96	14.74	15.63
15	7	11.04	12.26	10.30	12.08
16	10	7.92	9.46	7.15	9.42
17	9	5.67	7.33	4.93	7.40
18	3	4.05	5.71	3.38	5.85
19	2	2.89	4.45	2.31	4.66
20	2	2.06	3.49	1.57	3.74
21	2	1.47	2.74	1.06	3.02
22 or more	1	3.57	10.57	2.12	14.86
Total	3168	3168	3168	3168	3168
Number of classes		23	23	23	23
Chi-square		18.7663	85.9350	17.0039	82.6772
d.f		20	20	20	20
p-value		0.5371	<0.0001	0.6527	<0.0001
AIC		15,057.22	15,128.43	15,054.80	15,128.24

6 Concluding Remarks

In general, difficulties in computation have hampered the application of many mixed Poisson distributions. In this paper, the Poisson–Weibull distribution is considered for modeling over dispersed data, and the probabilities are computed by using a

Table 4 Parameter estimates and log-likelihood function value

Data	Parameter estimates	Log-likelihood value
Systemic adverse effects after vaccination	$\widehat{\alpha} = 1.5268$; $\widehat{\beta} = 0.9870$ (NB)	−6740.60 (NB)
	$\widehat{\alpha} = 2.5208$; $\widehat{\theta} = 0.6782$ (P-iG)	−6760.27 (P-iG)
	$\widehat{\alpha} = 1.2669$; $\widehat{\beta} = 1.6226$ (P-W)	−6738.27 (P-W)
	$\widehat{\mu} = 0.1372$; $\widehat{\sigma} = 0.7469$ (P-LN)	−6762.87 (P-LN)
Shmueli et al.'s data on quarterly sales	$\widehat{\alpha} = 1.5817$; $\widehat{\beta} = 2.2505$ (NB)	−7526.61 (NB)
	$\widehat{\alpha} = 3.3987$; $\widehat{\theta} = 0.8394$ (P-iG)	−7562.22 (P-iG)
	$\widehat{\alpha} = 1.2984$; $\widehat{\beta} = 3.8505$ (P-W)	−7525.40 (P-W)
	$\widehat{\mu} = 0.9853$; $\widehat{\sigma} = 0.7709$ (P-LN)	−7562.27 (P-LN)

Legend: *NB* Negative binomial; *P-iG* Poisson–inverse Gaussian; *P-W* Poisson–Weibull; *P-LN* Poisson–lognormal

series formula and a Monte Carlo simulation technique. The Monte Carlo integration exploits the fact that the Weibull cdf is easily inverted, and thus, generating samples from the Weibull distribution is very fast and straightforward. This also avoids the need to do separate analysis for small and very large counts as in the Poisson–lognormal distribution. Maximum likelihood estimation is done with an EM type algorithm which is based upon the mixing distribution, that is, the Weibull distribution. This leads to much simplification even though the Poisson–Weibull pmf is not in a closed form. Other variants of the EM-type algorithm like the Monte Carlo EM can be considered to further improve the computations. Two data sets have been considered for empirical modeling, and the Poisson–Weibull distribution shows a better fit in comparison with some well-known mixed Poisson models. In providing viable methods for computing Poisson–Weibull probabilities and a simple EM algorithm for parameter estimation, it is envisaged that the Poisson–Weibull distribution will be an attractive model for data analysts. Since the choice of the mixing distribution in the mixed Poisson is wide open, a generalization of the Poisson–Weibull distribution may be obtained by using a generalized Weibull distribution (see [22]) as mixing distribution. This will be considered elsewhere.

Acknowledgements The authors wish to thank the reviewers for their useful comments which have greatly improved the paper. The authors are supported by Ministry of Higher Education grant FRGS/1/2020/STG06/SYUC/02/1.

References

1. Bulmer, M.G.: On fitting the Poisson lognormal distribution to species-abundance data. Biometrics **30**(1), 101–110 (1974)
2. Cassie, R.M.: Frequency distribution models in the ecology of plankton and other organisms. J. Anim. Ecol. **31**(1), 65–92 (1964)
3. Chee, C.S.: Modelling of count data using nonparametric mixtures. AStA Adv. Stat. Anal. **100**, 239–257 (2015)
4. Cheng, L., Geedipally, S.R., Lord, D.: The Poisson–Weibull generalized linear model for analyzing motor vehicle crash data. Saf. Sci. **54**, 38–42 (2013)
5. Chung, Y., Lindsay, B.G.: Convergence of the EM algorithm for continuous mixing distributions. Statist. Probab. Lett. **96**, 190–195 (2015)
6. Cohen, H., Villegas, F.R., Zagier, D.: Convergence acceleration of alternating series. Exp. Math. **9**(1), 3–11 (2000)
7. Dempster, A.P., Laird, N.M., Rubin, D.B.: Maximum likelihood from incomplete data via the EM algorithm. J. Roy. Stat. Soc. B **39**(1), 1–38 (1977)
8. Greenwood, M., Yule, G.U.: An inquiry into the nature of frequency distributions representative of multiple happenings with particular reference to the occurrence of multiple attacks of disease or of repeated accidents. J. R. Stat. Soc. Ser. A **83**, 255–279 (1920)
9. Grundy, P.M.: The expected frequencies in a sample of an animal population in which the abundances of species are log-normally distributed, Part 1. Biometrika **38**(3), 427–434 (1951)
10. Gupta, R.C., Ong, S.H.: Analysis of long-tailed count data by Poisson mixtures. Commun. Stat. Theor. Methods **34**(3), 557–573 (2005)
11. Holla, M.S.: On a Poisson-inverse Gaussian distribution. Metrika **11**, 115–121 (1967)
12. Izsák, R.: Maximum likelihood fitting of the Poisson lognormal distribution. Environ. Ecol. Stat. **15**(2), 143–156 (2008)
13. Johnson, N.L., Kotz, S., Balakrishnan, N.: Continuous Univariate Distributions, vol. 1. Wiley, New York (1994)
14. Johnson, N.L., Kemp, A.W., Kotz, S.: Univariate Discrete Distributions, 3rd edn. Wiley, New York (2005)
15. Karlis, D.: EM algorithm for mixed Poisson and other discrete distributions. ASTIN Bull. **35**(1), 3–24 (2005)
16. Karlis, D., Xekalaki, E.: Mixed Poisson distributions. Int. Stat. Rev. **73**(1), 35–58 (2005)
17. Nikoloulopoulos, A.K., Karlis, D.: On modeling count data: a comparison of some well-known discrete distributions. J. Stat. Comput. Simul. **78**(3), 437–457 (2008)
18. Ong, S.H.: The computer generation of bivariate binomial variables with given marginals and correlation. Commun. Stat. Comput. Simul. **21**, 285–299 (1992)
19. Ong, S.H.: Computation of probabilities of a generalized log-series and related distributions. Commun. Stat. Theor. Methods **24**(1), 253–271 (1995)
20. Ong, S.H.: A note on the mixed Poisson formulation of the Poisson-inverse Gaussian distribution. Commun. Stat. Simul. Comput. **27**(1), 67–78 (1998)
21. Ong, S.H., Lee, W.J.: Computer generation of negative binomial variates by envelope rejection. Comput. Stat. Data Anal. **52**, 4175–4183 (2008)
22. Ong, S.H., Biswas, A., Peiris, S., Low, Y.C.: Count distribution for generalized Weibull duration with applications. Commun. Stat. Theor. Methods **44**(19), 4203–4216 (2015)
23. Ong, S.H., Lee, W.J., Low, Y.C.: A general method of computing mixed Poisson probabilities by Monte Carlo sampling. Math. Comput. Simul. **170**, 98–106 (2020)
24. Perline, R.: Mixed Poisson distributions tail equivalent to their mixing distributions. Statist. Probab. Lett. **3**, 229–233 (1998)
25. Rose, C.E., Martin, S.W., Wannemuehler, K.A., Plikaytis, B.: On the use of zero-inflated and hurdle models for modeling vaccine adverse event count data. J. Biopharm. Stat. **16**(4), 463–481 (2006)
26. Sankaran, M.: Mixtures by the inverse Gaussian distribution. Sankhya Ser. B **30**, 455–458 (1968)

27. Shaban, S.A.: Computation of the Poisson-inverse Gaussian distribution. Commun. Stat. Theor. Methods **10**(14), 1389–1399 (1981)
28. Shaked, M.: On mixtures from exponential families. J. Roy. Stat. Soc. B **42**, 192–198 (1980)
29. Shmueli, G., Minka, T.P., Kadane, J.B., Borle, S., Boatwright, P.: A useful distribution for fitting discrete data: revival of the Conway-Maxwell-Poisson distribution. J. Roy. Stat. Soc. Ser. C (Appl. Stat.) **54**(1), 127–142 (2005)
30. Spiegelhalter, D.J., Thomas, A., Best, N.G., Lun, D.: WinBUGS version 1.4.1 user manual. MRC Biostatistics Unit, Cambridge. Available from: http://www.mrcbsu.cam.ac.uk/bugs/welcome.shtml (2003)
31. Stewart, J.A.: The Poisson-lognormal model for bibliometric/scientometric distributions. Inf. Process. Manage. **30**(2), 239–251 (1994)
32. Williams, M.S., Ebel, E.D.: Methods for fitting the Poisson-lognormal distribution to microbial testing data. Food Control **27**, 73–80 (2012)
33. Willmot, G.E.: The Poisson-inverse Gaussian distribution as an alternative to the negative binomial. Scand. Actuar. J. **1987**(3–4), 113–127 (1987)
34. Willmot, G.E.: Asymptotic tail behaviour of Poisson mixtures with applications. Adv. Appl. Probab. **22**, 147–159 (1990)
35. Willmot, G.E.: On recursive evaluation of mixed Poisson probabilities and related quantities. Scand. Actuar. J. **18**, 114–133 (1993)

Cost of Energy for Distributed Energy Resources-Based Power Generation in a Rural Microgrid: Impact of Controlling Parameters

Md Mustafa Kamal and Imtiaz Ashraf

Abstract This paper examines the variation in the cost of energy with changes in the parameters of four different types of renewable energy-based power generation systems: solar photovoltaic system, (SPV), wind energy system (WES), micro hydro plant (MHP), and biomass energy system (BES). The factors controlling the cost of energy (COE) are the annual capacity factor (K), the rate of discount (r), the installed capital cost (P), and the operation and maintenance costs (O&M). For this study, the impact of different parameters affecting the cost of energy in a group of villages in the Indian state of Uttarakhand was analyzed. The findings contribute to insights into the relative importance of each of the control variables for each type of renewable energy system. The optimal parameters for rural microgrid planning are suggested.

Keywords Cost of energy · Parameter variation · Microgrid · Solar · Hydro · Wind · Biomass

1 Introduction

Energy is the main contributor to the overall growth of a country. Currently a large proportion of energy production comes from conventional energy sources. The major drawback of these fuels is that they are being very rapidly depleted and have adverse environmental effects due to the release of greenhouse gases during combustion. Renewable energy sources are environmentally friendly and freely available alternatives to conventional fuels [1]. For the past few years, renewable energy resources, such as solar, wind, hydro, biomass, and biogas, have been widely used in the generation of electricity. Renewable energy-based power generation has a greater potential in developing nations where large populations live in rural areas where the development of grid infrastructure is costly. Locally available energy resources can be integrated in the electrification of remote rural communities. A microgrid formed of mixed renewable resources can work both in grid-connected and standalone modes.

M. M. Kamal (✉) · I. Ashraf
Department of Electrical Engineering, Aligarh Muslim University, Aligarh 202002, India

P. Balasubramaniam et al. (eds.), *Mathematical Modelling and Computational Intelligence Techniques*, Springer Proceedings in Mathematics & Statistics 376,
https://doi.org/10.1007/978-981-16-6018-4_8

Most energy models used for the electrification of rural communities work in standalone mode [2]. Isolated communities in mountain areas can access electricity via renewable sources.

Microgrid planning is a very difficult task due to a large number of constraints. For the electrification of rural communities, efficient planning that takes the specialized needs of the area into account is important. One model proposes an isolated microgrid consisting of various AC/DC distributed energy resources [3]. The planning aims to provide electrification at minimum cost with optimal-sized components. A microgrid model for electrification of a rural community in Uttarakhand (India) was studied to minimize the cost of electricity in standalone and grid-connected modes of operation [4]. A remote microgrid for an isolated community with a mix of different energy resources was planned to provide reliable electrification with minimum operating cost [5]. A novel microgrid planning methodology proposed in [6] incorporated efficient conditions for smooth operation in grid-connected and standalone modes. Authors in [7] planned a rural microgrid for an isolated community in Malaysia, with a technoeconomic analysis and component sizing targeting the lowest energy cost for consumers.

Since the cost of energy (COE) is determined by various factors, including the translational cost of components, variations in the diesel price, incident radiation and wind speed, sensitive analysis is necessary in order to understand the optimal cost of energy. In [8], a microgrid model for off-grid operation in a residential location in Nigeria was developed. The suggested model optimized the lowest energy cost using the grasshopper optimization algorithm. The effect of varying different parameters on the COE was analyzed. A microgrid model for the grid-connected operation was proposed and the effect of different parameters on the COE carried out in [9]. In [10], a hybrid model for the electrification of a rural community is proposed, consisting of various local renewable energy resources. Optimization techniques were used to minimize COE and component size, and the variation of COE according to different parameters was studied. A microgrid model for a mountainous area of Uttarakhand (India) was proposed and optimized using biography-based optimization, and the proposed model was subjected to sensitivity analysis [11]. Challenges associated with the integration of renewable energy resources in the system are discussed in [12]. A novel model for the electrification of an Indian rural community was proposed in [13]. The suggested model was analyzed for its social, economic, and technical impacts on the electrification programme, and verified by the MATLAB and HOMER systems models. [14] suggests a microgrid model for a small island in Thailand, and system performance models are evaluated using HOMER software. In [15], a microgrid model that uses locally available resources for the electrification of a rural community in Uttarakhand (India) is proposed. Different methods for resource allocation of integrated resources are compared.

The above literature survey suggests the importance of modelling and the effect of various components on the planning of a microgrid for a rural community. The prime objective of every electrification programme is to supply energy at the lowest cost to customers. So, it is important to study the various factors affecting the determination of the cost of energy. In the present study we will investigate the effects on the COE of varying different parameters.

2 Study Area

Uttarakhand is a state in north-western India located at 30.04 N/78.14 E. In the year 2000 Uttarakhand became India's 27th state. Its capital city is Dehradun. Its total area is 53,483 km^2, of which 65% is forest cover and 86% is mountainous. It has 13 districts, of which Theri Garhwal (headquarters in Theri) was chosen for the case study. A large area of the district is covered by the Himalayan region. The district's population is 618,913 (2011 census), and its area is 4080 km^2. Due to the low population density and the difficulty of developing grid infrastructure, the area has limited access to electricity. The total number of villages in the area is 259, of which 55 are unelectrified. 24 villages have been electrified, using renewable energy sources available in the locality, by Uttaranchal Renewable Development Agency (UREDA). The Jaunpur block is taken as the study area. The block is divided into four regions, of which Region 4 has been chosen for the case study, which examines social and economic aspects of 12 unelectrified villages. Table 1 gives basic facts on the region.

Due to its remote and mountainous nature, all the regions in this block have low connectivity for roads, water and electricity, and very few basic amenities are available. Table 2 provides information on the economic and social status of the area, taken from the social and economic report of the district [22].

The renewable energy development agency of Uttarakhand is now promoting renewable-based electrification of the area. The proportion of unelectrified villages in the block is 7.5%. Access to electricity in these areas is difficult and extension of the central grid is not a feasible option. Electrification is only possible using renewable energy resources and diesel generators, as most of the villages are located in remote mountainous terrain. Providing electricity by diesel generator is not feasible

Table 1 Basic information on the study region

Item	Total
Country	India
State	Uttarakhand
Name of block	Jaunpur
Total number of *panchayats*	111
Number of villages	259
Number of installed water mills (Gharat)	53
Number of available springs	33
Number of available waterfalls	9
Number of reserve forest	23

Table 2 Social and economic status of the Jaunpur block

Static and features	Region 1	Region 2	Region 3	Region 4
Total population	13,867	7580	19,378	9856
Income status	High	Medium	Medium	Low
Unelectrified villages	2	7	4	12
Economic status	Weak	Weaker	Very poor and weak infrastructure	Weakest and most backward region
Social status	Poor	Very poor	Very poor	Very poor

Table 3 Total available capacity in the area

Village	Available resources				
	MHP (kWh/yr)	SPV (kWh/m^2/yr)	WES (kWh/m^2/yr)	BES (kWh/yr)	Total (kWh/yr)
Silla	2857	1864	1271	13,509	19,493
Nihaldanda	N/A	1864	1271	1807	4939
Dandagaon	N/A	1864	1271	2119	5252
Gawalidanda	N/A	1864	1271	1528	4663
Chifalti	1374	1864	1271	2625	7126
Jaintwari	N/A	1864	1271	3140	6273
Talyakatal	N/A	1864	1271	3959	7092
Sandna laga	15,266	1864	1271	23,143	4154
Kund	73,718	1864	1271	508,808	585,653
Ragargaon	N/A	1864	1271	5884	9017
Airralgaon	8677	1864	1271	51,827	63,633
Sera	26,298	1864	1271	23,044	52,472

in the long run because of the increasing fuel cost and greenhouse gas emissions. The only solution to provide reliable electrification of the area is to combine the available energy resource. Table 3 shows green energy resources available in the area (data taken from the district energy sources report published by the government of Uttarakhand).

3 Load Assessment

The area under study has huge potential for renewable energy resource-based power production (see Table 3). The 12 selected villages have abundant hydro, solar, wind and biomass resources that more than suffice to meet the area's energy requirements. The area has low population density and most people depend on agriculture for their living. There are only a few agriculture-based small-scale industries. For the load

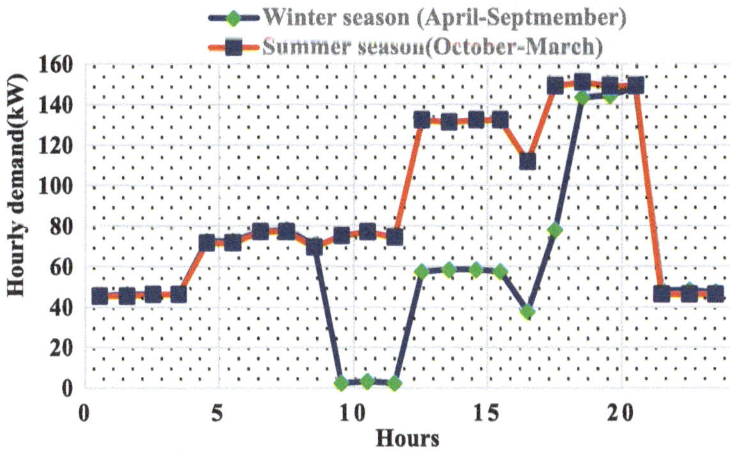

Fig. 1 Load profile of the area

estimation, a door-to-door data collection survey was conducted. The estimated load of the area was divided into two seasons, summer and winter. The peak load in the winter season is 77 kW and in the summer season 142 kW. The peak energy demand of the area is 152.65 kW and 148.67 kW in summer and winter respectively. Figure 1 shows the load profile of the area.

Table 4 shows the different types of loads: commercial, domestic, and community. The data are taken from the area survey and the questionnaire.

4 Resource Estimation

The microgrid as currently planned is formed by the integration of solar, wind, biomass, hydro, and biogas energy systems. It is necessary to develop a mathematical model for the implementation of the energy model.

The solar photovoltaic system (SPV) consists of several PV modules. The module power is represented by Eqs. (1)–(5).

$$S_p = SP_r y_s\left(\frac{I_t}{I_{\mathrm{st}}}\right)\left[1 + \partial_p(t_{\mathrm{cel}} - t_s)\right] \tag{1}$$

The cell temperature t_{cel} °C can be formulated as

$$t\partial I_t = \eta_s I_t + g_l(t_{\mathrm{cel}} - t_\partial) \tag{2}$$

$$t_{\mathrm{cel}} = t_\partial + I_t\left(\frac{t\partial}{g_l}\right)\left(1 - \frac{\eta_s}{t\partial}\right) \tag{3}$$

Table 4 Load distribution of the area

Total load									
Domestics loads					Community loads				
Apparatus	Types	Rating (W)	No	Total (W)	Apparatus	Types	Rating (W)	No	Total (W)
Homes	CFL	15	1120	16,800	Post office	Fans	65	9	585
	Television	75	600	45,000		CFL	35	9	540
	Fans	65	600	38,000		Computers	100	5	500
	Radios	15	600	9000		Printer	40	3	120
					Provenance	Lights	35	5	150
						Heaters	1500	5	7500
						Fans	65	5	325
						coolers	250	5	1250
Commercial loads					Hospitals	Lights	35	15	525
Apparatus	Types	Ratings (W)	No	Total (W)		Fans	65	15	975
Shops	Lights	30	30	900		Heater	1500	5	7500
	Fans	35	25	875		Refrigerator	475	5	2375
Flour mills		5000	6	3000		Cooler	250	5	1250
Industrial loads					Schools	Lights	30	35	1050
Sawmill		5000	5	25,000		Fans	60	70	4200
					Street lights	Lights	70	30	2100

Table 5 Specification of solar photovoltaic system

Parameters	Values
Nominal power output	300 W
Power tolerance	0–5 W
Module efficiency	15.4%
Open-circuit voltage (V_{oc})	36.6 V
Short-circuit current (I_{sc})	14.8 A
Maximum voltage (V_{mp})	44.8 V
Maximum current (I_{mp})	8.71 A
Temperature co-efficient of V_{oc}	−0.41%/°C
Temperature co-efficient of I_{sc}	−0.32%/°C
(O&M)	0.70 ₹/kWh
Useful life	25 years

The value of $\left(\frac{t\partial}{g_l}\right)$ is difficult to measure. Based on the report provided by the manufacturer, with the nominal cell temperature at 800 W/m^2, and temperature at 20 °C under no-load conditions, the equation is rewritten as

$$\frac{t\alpha}{g_l} = \frac{t_{\text{cel,nom}} - t_{\partial,\text{nom}}}{I_{t,\text{nom}}} \tag{4}$$

The final PV cell temperature can be obtained for the value of $(\partial t) = 0.90$

$$t_{\text{cel}} = t_\alpha + I_t \frac{t_{\text{cel,nom}} - t_{\alpha,\text{nom}}}{I_{t,\text{nom}}} \left(1 - \frac{\eta_s}{0.90}\right) \tag{5}$$

The solar panel used in the modelling of the system is of 300 W. The specification of the solar panel used is shown in Table 5, and is based on the data sheet supplied by the manufacturer [16].

Solar radiation data was collected from the terrain located between 28.47 latitude to 78.14 longitude. The area has annual average solar radiation of 5.37 kWh/m^2 and in May the maximum radiation recorded was 7.420 kWh/m^2. The study area has enormous solar potential, with total annual solar energy potential recorded as 1946 kWh/m^2/yr. Monthly average temperature and solar radiation of the study area is shown in Figs. 2 and 3.

The state of Uttarakhand has enormous hydro-power potential because of the Himalayas and its rivers and waterfalls. The installed hydro-power capacity is 15 kW. The total annual micro-hydro-power capacity in the study area is 256,331 kWh/yr. The stream can be calculated using Eq. (6).

$$P_{\text{MHP}}(t) = \rho \times Q(t) \times g \times H \times \eta_{\text{MHG}} \tag{6}$$

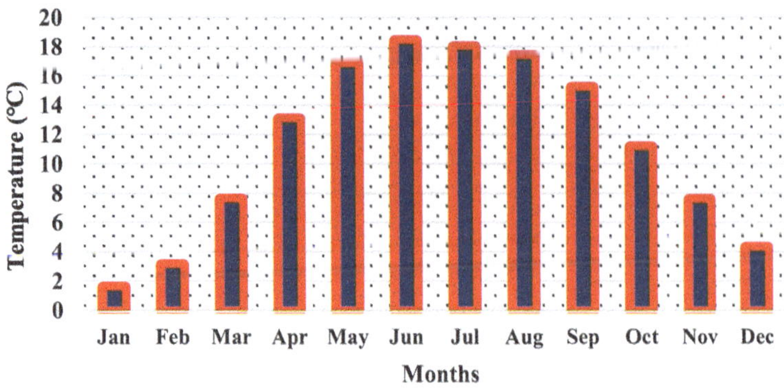

Fig. 2 Average temperature of the area

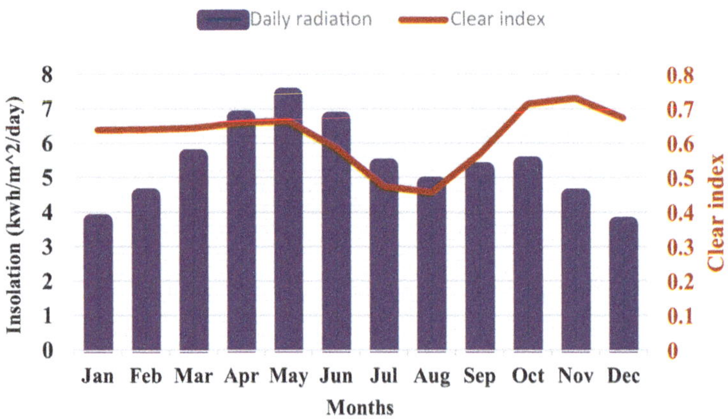

Fig. 3 Average annual insolation and clear index of the area

Hydro-turbines are used to harness the potential of hydro energy. Various hydro-turbine ranges are available, with choice depending on the size of the plant. In the present work two types of hydro-turbine, 25 kW and 1 kW, are used. Table 6 shows the different parameter specifications of the 1 kW and 25 kW micro-hydro-turbines, taken from the data sheet supplied by the manufacturer [16].

A biomass generator converts biomass to potential electricity. The main content of the biomass energy system in the study area is forest foliage and agriculture waste. The area is surrounded by Himalayan forest, so the overall biomass potential of Region 4 in Jaunpur block is 92.21 tons/yr from agriculture waste and 36.36 tons/yr from forest foliage. To convert biomass into electricity, a biomass gasifier-based generating system is used. The biomass gasifier operates for 10 h/day, and its power capacity is 30 kW. Table 7 shows the parameter specifications of a 40 kW biomass system, taken from the data sheet supplied by the biomass generator manufacturer [17]. The mathematical power of biomass is represented by Eq. (7).

Table 6 Specification of hydro-turbine

Parameters	Values	Values
Capacity (kW)	25	1
Q (m^3/s)	0.145	0.102–0.041
H_{NET} (m)	25	2–5
Efficiency (η) (%)	70	50
$C_{CW\text{-}MHP}$	₹1,200,000	₹48,000
C_{pen}	₹200,000	₹8000
$C_{T\&V}$	₹300,000	₹12,000
$C_{G\&C}$	₹300,000	₹12,000
$C_{O\&M}$	0.60₹/kWh	0.60₹/kWh
Useful life	25 years	25 years

Table 7 Biomass generator specification

Parameters	Values
Rating	40 kW
$C_{CW\text{-}BMGS}$	₹200,000
(O&M)	0.80₹/kWh
$C_{EG\text{-}BMGS}$	₹1,100,000
C_{BM}	2₹/kg
S_{BM}	1.5 kg/kWh
η	20%
C_G	₹940,000
Useful life of gasifier	10,000 h
Useful life of civil works	25 years
Useful life	15,000 h

$$P_{bms} = {(c_t \times CV_{bms} \times \eta_{bms})}/{(365 \times 860)} \tag{7}$$

where P_{bms} is total biomass generator output, CV_{bms} cost of biomass generator, and η_{bms} is the efficiency of biomass engine.

The biogas generator converts dung, mainly from cattle, into electricity. The total potential of biogas in the study area was estimated by counting the total cattle population in the 12 villages of Region 4, which is 2223. The availability of dung is 18,104 kg/day.

Biogas produced from the dung amounts to 650.50 m^3/day, of which 401.58 m^3/day is used for cooking and the remainder for power generation. The biogas generator works nine h/day and total output power is 35 kW. It is assumed that the quantity of dung is 1.5 kg/day from goats, 11 kg/day dung from mules/cows/horses, and 16 kg/day from buffalo. Table 8 shows the specifications of the 50 kW biogas system, taken from the data sheet supplied by the manufacturer.

Table 8 Specifications of biogas generator

Parameters	Values
Rating	50 kW
$C_{\text{OW-BGGS}}$	₹400,000
$C_{\text{EG-BGGS}}$	₹1,600,000
C_{CD}	0.20₹/kg
Q_{CD}	25 kg/m^3
S_{BGGS}	0.5 m^3/kWh
η	27%
(O&M)	0.80₹/kWh
Civil work lifetime	25 years
Life of engine	20,000 h

Biogas availability is calculated by Eq. (8) [18, 19].

$$P_{\text{BCG}}(t) = \left[\left(g_{\text{total}}\left(\frac{m^3}{d}\right) \times CV_b \times \eta_{\text{Bio}}\right)/860 \times \left(\frac{t}{d}\right)\right] \tag{8}$$

Wind energy is a greener from of energy. The area has variable wind speed of 3–13 m/s. Low-speed wind turbines are used to harness wind energy. The power output of the wind energy system depends on power co-efficient (Cp) and wind speed (V). The wind power output in t hours is calculated using Eq. (9).

$$P_w = \begin{cases} P_r\left(\frac{v^3(t)-v_c^3(t)}{v_r^3(t)-v_c^3(t)}\right) & \text{for } v_c(t) \le v(t) \le v_r(t) \\ p_r & \text{for } v_r(t) \le v(t) \le v_f \\ 0 & \text{for } v(t)\langle v_c \text{ and } v(t)\rangle v_f \end{cases} \tag{9}$$

The energy produced by energy conversion of E_{WES} in kWh is estimated by the formula

$$E_{\text{WES}}(t) = P_{\text{WES}}(t) \times \Delta t \tag{10}$$

The monthly average wind speed of the selected site is shown in Fig. 4. Various ranges of wind turbine are available for the harnessing of the potential of wind. The study area has wind speed in the range 5–7 m/s, so the 3.2 kW model wind turbine was selected. Table 9 shows the specifications of the 3.2 kW wind turbine, taken from the selected model manufacturer's data sheet. V_{CI} is the cut-in speed of the wind turbine, V_{CO} is the cut-out speed, V_f the rated speed of wind turbine, p_rated is the rated power of wind turbine and Δt is the step time [20, 21].

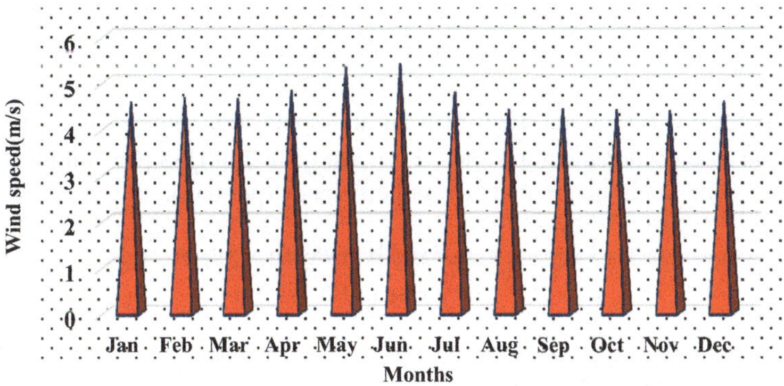

Fig. 4 Yearly average wind speed of the area

Table 9 Specification of wind turbine

Parameters	Values
Rated power of wind turbine	3.2 kW
(O&M) cost	0.61₹/kWh
Speed of rotor	12 m/s
Capital cost (C_{WT})	₹200,000
V_{CO}	24 m/s
$C_{IN/WT}$	₹100,000
A_{WT}	15.89 m^2
V_{CI}	3.1 m/s
Useful life	25 years

5 Formulating the Problem

The microgrid planned for the electrification of the rural community will be formed by an appropriate combination of resources. The energy resources available in the study area are estimated. The basic structure of the renewable system framework has zero fuel cost (apart from purchase of biomass), high capital costs, and low operation and maintenance cost. The simplest form of energy cost calculation for a renewable energy source is obtained by adding the O&M cost and capital recovery cost per unit cost (Eq. 11).

$$c = \left[\frac{r(1+r)^n}{(1+r)^n - 1}\right]\left\lfloor\frac{P}{87.6K}\right\rfloor + [\text{O\&M}] \tag{11}$$

The amortization period of a plant is equal to the lifetime of the system. The annual capacity factor of a system is defined as the total generation of energy on a yearly basis:

$$K = \frac{\text{energy generated in kWh per year}}{(\text{system rating in kWh})(8760\text{ h per year})}$$

If biomass is used as the fuel, then an additional term $\frac{0.3413F}{\eta_{\text{bio}}}$ is added to the right-hand side of the equation, where F = cost of fuel in US$ per million BTU at the site of generation, η_{bio} = plant overall efficiency (per unit):

$$\eta_{\text{bio}} = \frac{3143\text{ BTU/kWh}}{\text{heat rate in BTU/kWh}} \tag{12}$$

To calculate the cost of energy and analyze the effect of different parameters on per-unit cost, we made the following assumptions:

1. The system is an energy displacer and no firm capacity is expected from it.
2. No further increase in O&M will be incurred during the said period.
3. The effect of tax structure and insurance costs is not included.
4. The depreciation value at the end of the said period (amortization period) is assumed to be zero.

Power plant equipment deteriorates on a regular basis due to wear and tear. Replacement is therefore necessary after a fixed lifetime. Depreciation means that during the said period, all equipment is working well and providing output at its rated value. In the present planning for the amortization period the solar, wind and other equipment is working well and providing output at its best. Hence it is assumed that depreciation is zero. If non-zero value is considered, the reduction in production each year should be included in the planning.

6 Results and Discussion

Electricity is the main driving force that can change a person's socioeconomic standing. The cost of energy, though important, should be kept to a minimum when planning a rural microgrid. To minimize the COE, an appropriate combination of resources is needed. Table 3 shows the status of different energy resources in the study area. Per-unit cost of energy is calculated using Eq. (10). The parameters that can affect the cost of energy are interest rate (r), installed capital cost (P), annual capacity factor (K), and operation and maintenance costs (O&M). The effect on the cost of energy of an incremental change of each parameter by a specific amount was included in the study, and results are discussed in this section.

With Uttarakhand state's enormous solar energy potential, solar-based projects are widely used for the electrification of remote areas. The current installed capacity of solar photovoltaic energy in the study area is 5 kW, and the potential per annum is 1863 kWh. The installed capital cost per kW varies between ₹ 300,000 and 350,000. The estimated annual capacity factor is 0.15; operation and maintenance cost for each kWh is 1.125. For calculation purposes, useful life is taken to be 25 years. Figures 5,

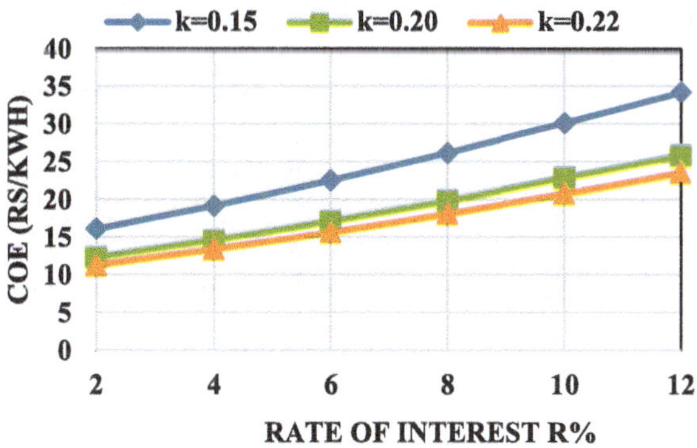

Fig. 5 Variation of COE w.r.t *r* for different value of *K*

6, 7 and 8 show the variation of COE with respect to different parameters for a solar photovoltaic system, suggesting that as the annual capacity factor increases for a given rate of interest, the cost of energy falls sharply.

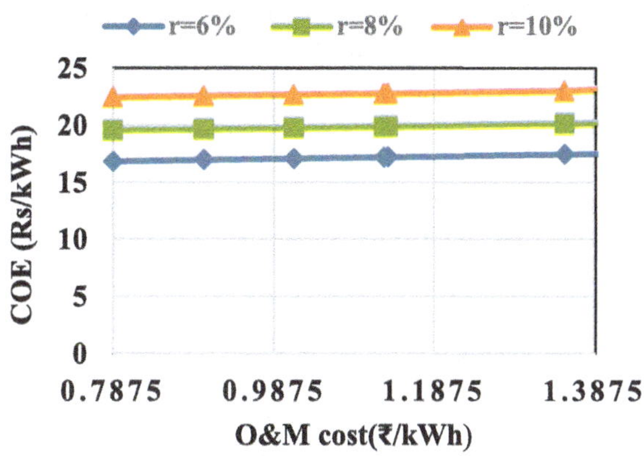

Fig. 6 Variation of COE w.r.t O&M for different value of *r*

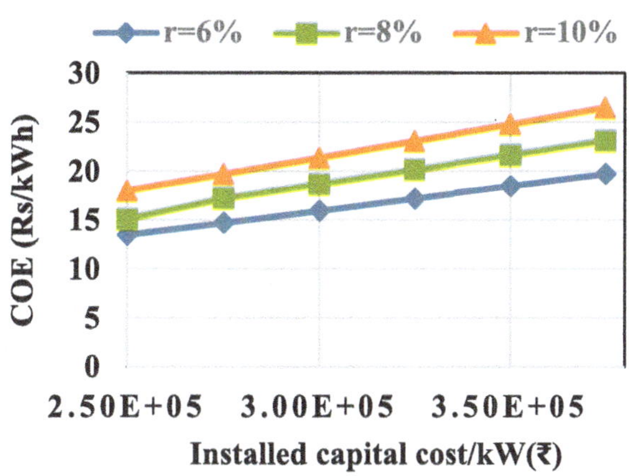

Fig. 7 Variation of COE w.r.t *P* for different value of *r*

Fig. 8 Variation of COE w.r.t *K* for different value of *r*

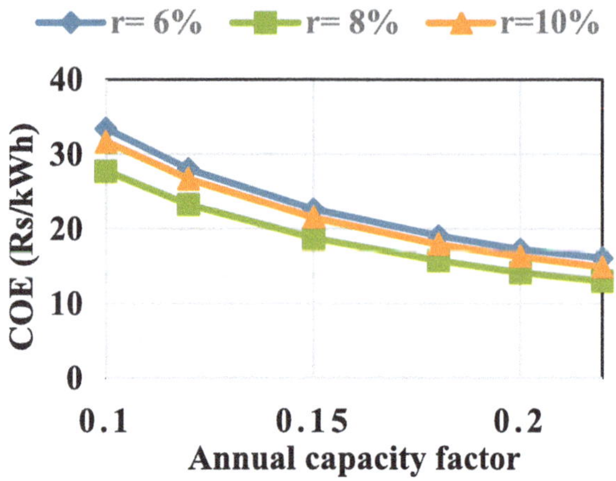

The total installed capacity of the wind energy system is 3 kW, and potential per annum is 1270 kWh. The installed capital cost is ₹ 45,000–70,000; estimated annual capacity factor is 024–0.4; and operation and maintenance cost per kWh is ₹0.0959. The estimated useful life for calculation purposes is taken as 20 years. Each parameter was varied within certain permissible limits and the per-unit cost of energy calculated using three different interest rates: 6%, 8%, and 10%. Figures 9, 10, 11 and 12 show the resulting variation in the COE for the wind energy system (WES). Results suggest that the installed capacity cost and O&M do not impact much, but with increase in the interest rate the COE increases and as the annual capacity factor increases, the COE decreases.

Due to its many waterfalls, Uttarakhand has enormous micro hydro-power potential. The installed capacity is 19.26 kW, and potential per annum is 128,166 kWh. The installed capital cost per kW is ₹ 125,000. The estimated annual capacity factor is 0.76, and operation and maintenance cost is 0.3425 ₹/kW. For the calculation, the estimated useful life is taken to be 25 years. Figures 13, 14, 15 and 16 show the variation of COE for micro hydro-power system [14–17].

Fig.9 Variation of COE w.r.t *k* for values of *r*

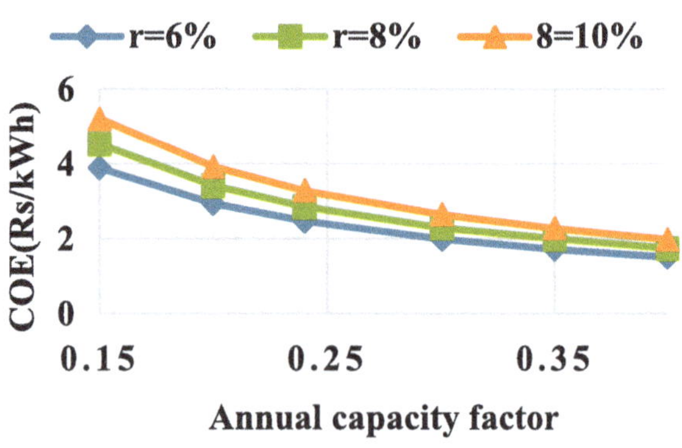

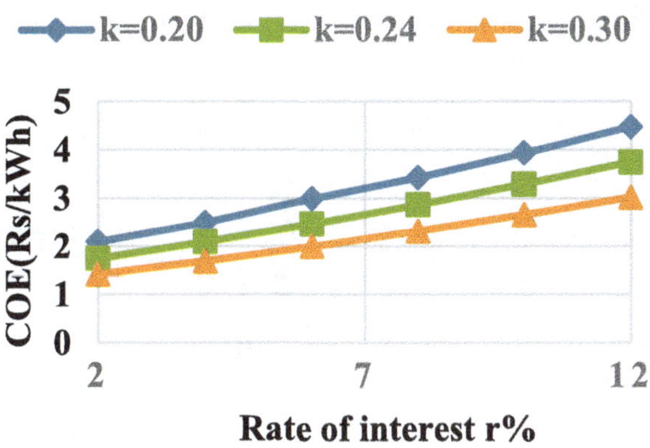

Fig.10 Variation of COE w.r.t r for different value of K

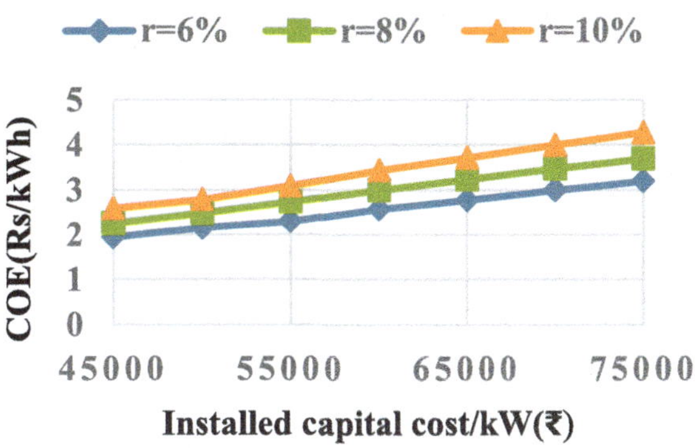

Fig. 11 Variation of COE w.r.t P for values of r

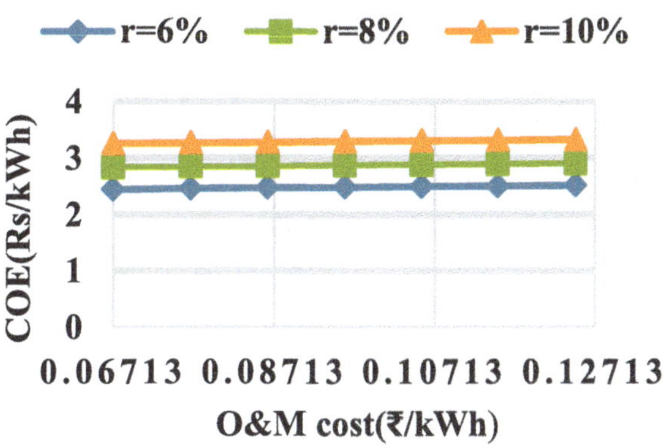

Fig.12 Variation of COE w.r.t O&M for values of r

A large area of Uttarakhand is covered by Himalayan forest, providing high capacity of the biomass energy system to meet the demand for energy. The installed capacity of the biomass energy system is 70 kW, and potential per annum is 641,384 kWh. The installed capital cost per KW is ₹ 230,000–275,000. The estimated annual capacity factor is 0.9, and operation and maintenance cost is 1.1458 ₹/kW.

Fig. 13 Variation of COE w.r.t *r* for values of *r*

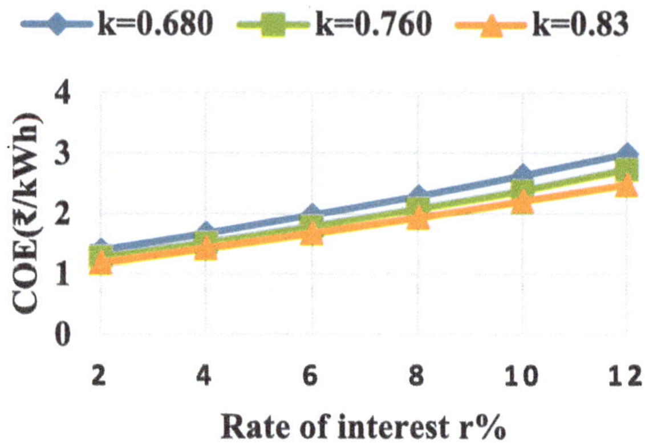

Fig.14 Variation of COE w.r.t O&M for values of *r*

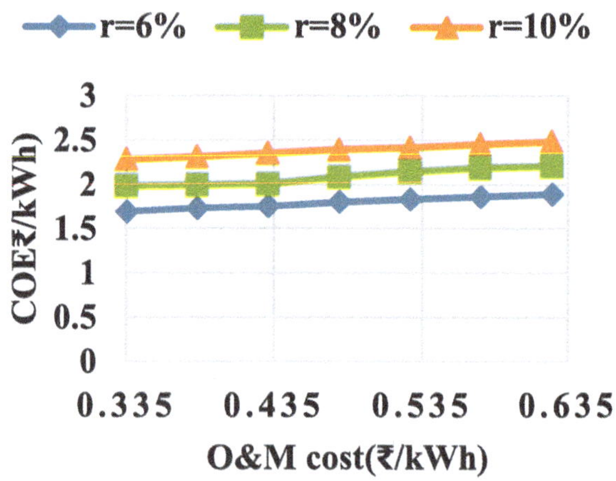

Fig.15 Variation of COE w.r.t *K* for values of *r*

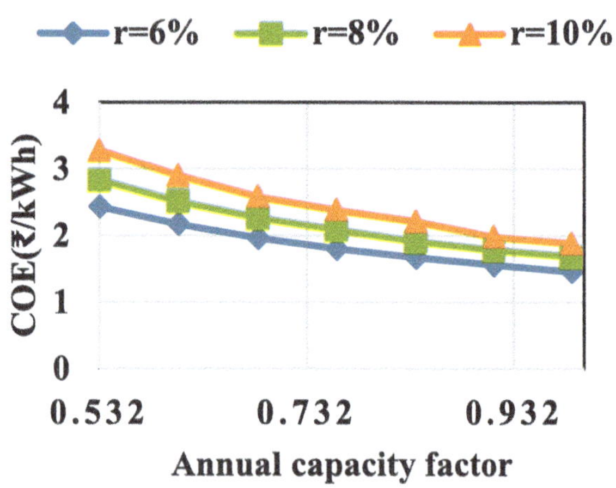

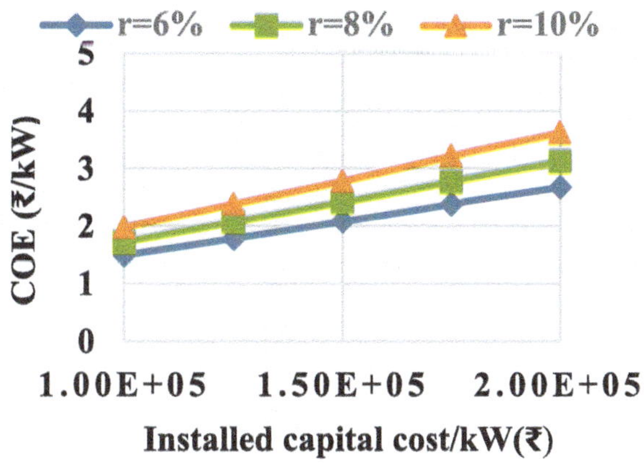

Fig.16 Variation of COE w.r.t *P* for values of *r*

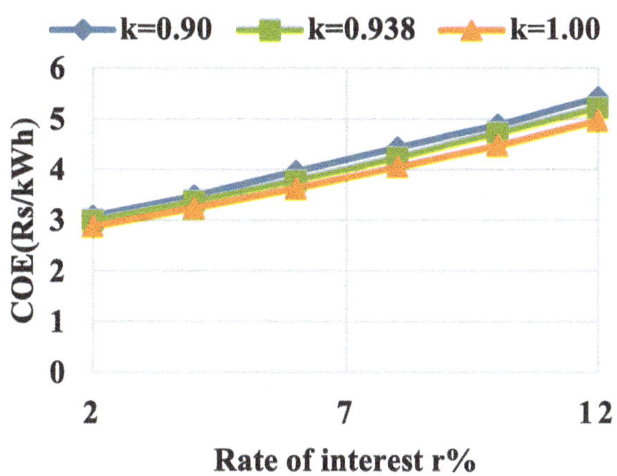

Fig. 17 Variation of COE w.r.t *r* for values of *K*

For the calculation, the estimated useful life is taken to be 20 years. Figures 17, 18, 19 and 20 show the variation in COE for the biomass energy system (BES).

Planning the cost of energy for a rural area includes a detailed assessment of the relative impact of different parameters. The parameters affecting the per-unit cost of energy are annual capacity factor (*K*), installed capacity cost (*P*), rate of interest (*r*), and operation and maintenance cost (O&M). These parameters vary according to the geographical structure of the area and market conditions. For each energy resource the authors considered the three interest rates, 6%, 8%, and 10%. Similarly, for the other parameters—annual capacity factor, installed capacity costs, and O&M charges—the parameters were varied up to a permissible limit beyond which variation is not regarded as feasible and which will incur higher energy cost.

This study of the Himalayan areas of India incorporates solar, wind, biomass, and biogas energy systems in standalone operation mode. For modelling purposes it has been assumed that production is constant throughout the given lifetime, and the highly intermittent nature of renewable resources that depend on environmental

Fig. 18 Variation of COE w.r.t O&M values of *r*

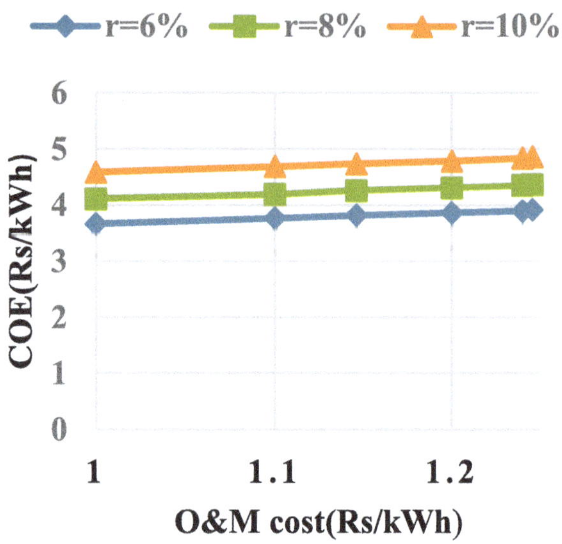

Fig. 19 Variation of COE w.r.t *K* for values of *r*

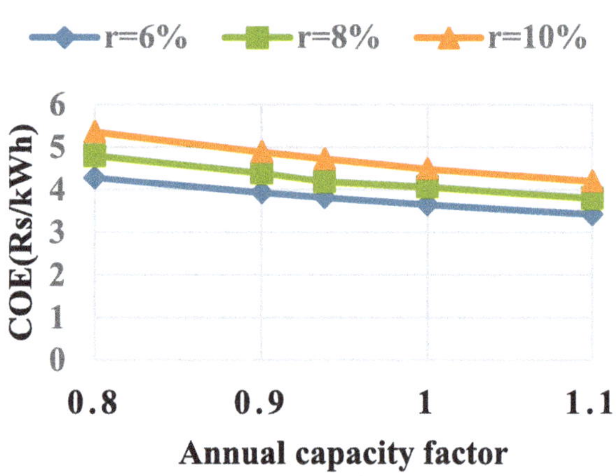

conditions and geographical location has not been taken into account. Total system output as well as COE value is directly affected by the intermittency of renewable sources in both grid-connected and standalone mode. Another limitation of these types of studies is that they can only compare similar types of planning. So this is not a generalized model. The inclusion of storage devices affects the COE, and the planning model discussed here does not include battery systems, which are always needed for reliable planning.

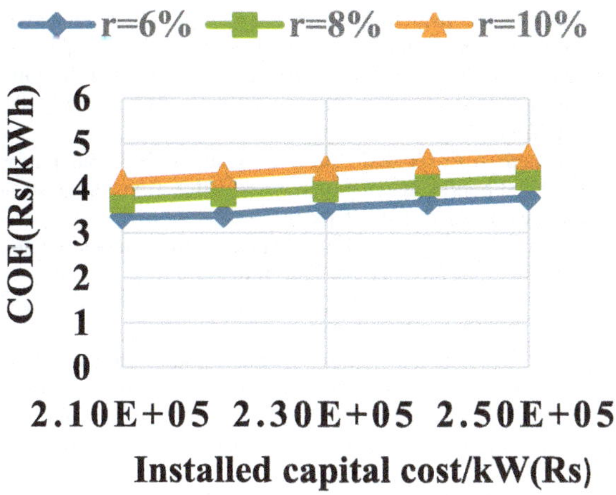

Fig. 20 Variation of COE w.r.t *P* for values of *r*

7 Conclusion

Rural electrification has numerous benefits for the overall development of both region and country. The implementation of the hybridized energy system is not easy. Challenges of a technical, economic, political, and socio-cultural nature need to be overcome for successful implementation of the project. The market infrastructure should provide support through proper policy planning and regulation to achieve the optimum cost of electrification. Economic considerations are the main factor in the implementation of rural hybrid microgrids, because installation costs are generally high in developing countries. The present study has investigated the potential for integration of available energy resources to provide the optimum cost of energy for a remote rural community. A group of 12 villages were selected for the study, in which the energy demand is fulfilled by different resources available in the area. The cost of energy (COE) was calculated for the model under consideration. The COE depends on rate of interest (r), annual capacity factor (K), installed capital cost (P), O&M cost, and amortization period (n). The per-unit cost of energy was calculated by varying each parameter up to a specified limit. The interest rate varies from 6 to 8%. Results suggest that variation in the annual capacity factor has a greater effect on the COE than the other parameters. It was found that for a fixed interest rate of 6%, the COE is optimum for the model. The optimum values of the per-unit cost of energy for SPV, BES, MHP, and WES are ₹ 15.27, ₹ 3.81, ₹ 1.80, and ₹ 3.00 respectively.

References

1. Kiptoo, M.K., Lotfy, M.E., Adewuyi, O.B., Conteh, A., Howlader, A.M., Senjyu, T.: Integrated approach for optimal techno-economic planning for high renewable energy-based isolated

microgrid considering cost of energy storage and demand response strategies. Energy Convers. Manag. **215**, 112917 (2020). http://doi.org/10.1016/j.enconman.2020.112917
2. Adefarati, T., Bansal, R.C., Bettayeb, M., Naidoo, R.: Optimal energy management of a PV-WTG-BSS-DG microgrid system. Energy **217**, 119358 (2021). http://doi.org/10.1016/j.energy.2020.119358
3. Hamad, A.A., Nassar, M.E., El-Saadany, E.F., Salama, M.M.A.: Optimal configuration of isolated hybrid AC/DC microgrids. IEEE Trans. Smart Grid **10**(3), 2789–2798 (2019). http://doi.org/10.1109/TSG.2018.2810310
4. Kamal, M., Ashraf, I.: Environmental effects modeling and assessment of economic viability of grid-connected photovoltaic system for rural electrification. Energy Sources Part A Recover. Util. Environ. Eff. 1–18 (2021). https://doi.org/10.1080/15567036.2021.1905108
5. Che, L., Zhang, X., Shahidehpour, M., Alabdulwahab, A., Abusorrah, A.: Optimal interconnection planning of community microgrids with renewable energy sources. IEEE Trans. Smart Grid **8**(3), 1054–1063 (2017). https://doi.org/10.1109/TSG.2015.2456834
6. Osama, R.A., Zobaa, A.F., Abdelaziz, A.Y.: A planning framework for optimal partitioning of distribution networks into microgrids. IEEE Syst. J. **14**(1), 916–926 (2020). https://doi.org/10.1109/JSYST.2019.2904319
7. Haidar, A.M.A., Fakhar, A., Helwig, A.: Sustainable energy planning for cost minimization of autonomous hybrid microgrid using combined multi-objective optimization algorithm. Sustain. Cities Soc. **62**(December 2019), 102391 (2020). http://doi.org/10.1016/j.scs.2020.102391
8. Bukar, A.L., Tan, C.W., Yiew, L.K., Ayop, R., Tan, W.S.: A rule-based energy management scheme for long-term optimal capacity planning of grid-independent microgrid optimized by multi-objective grasshopper optimization algorithm. Energy Convers. Manag. **221**, 113161 (2020). http://doi.org/10.1016/j.enconman.2020.113161
9. Chakir, A., et al.: Optimal energy management for a grid connected PV-battery system. Energy Rep. (6), 218–231 (2019). http://doi.org/10.1016/j.egyr.2019.10.040
10. Han, Y., Zhang, G., Li, Q., You, Z., Chen, W., Liu, H.: Hierarchical energy management for PV/hydrogen/battery island DC microgrid. Int. J. Hydrogen Energy **4**, 5507–5516 (2019). https://doi.org/10.1016/j.ijhydene.2018.08.135
11. Upadhyay, S., Sharma, M.P.: Selection of a suitable energy management strategy for a hybrid energy system in a remote rural area of India. Energy **94**, 352–366 (2016). https://doi.org/10.1016/j.energy.2015.10.134
12. Hossain, M.A., Pota, H.R., Squartini, S., Abdou, A.F.: Modified PSO algorithm for real-time energy management in grid-connected microgrids. Renew. Energy 746–757 (2019). http://doi.org/10.1016/j.renene.2019.01.005
13. Kumar, A., Member, G.S., Singh, A.R., Deng, Y.A.N.: A novel methodological framework for the design of sustainable rural microgrid for developing nations. IEEE Access **6**, 24925–24951 (2018). https://doi.org/10.1109/ACCESS.2018.2832460
14. Veilleux, G., et al.: Techno-economic analysis of microgrid projects for rural electrification: a systematic approach to the redesign of Koh Jik off-grid case study. Energy Sustain. Dev. **54**, 1–13 (2020). https://doi.org/10.1016/j.esd.2019.09.007
15. Kamal, M.M., Ashraf, I., Fernandez, E.: Energy resource allocation for distributed energy resource based power generation in a rural microgrid. In: IEEE international conference on electrical electronics, communication, computer and optimization techniques (ICEECCOT), pp. 571–575 (2020). http://doi.org/10.1109/iceeccot43722.2018.9001504
16. Chauhan, A., Saini, R.P.: Discrete harmony search based size optimization of integrated renewable energy system for remote rural areas of Uttarakhand state in India. Renew. Energy **94**, 587–604 (2016). https://doi.org/10.1016/j.renene.2016.03.079
17. Sundaramoorthy, K., Sankar, A.: Performance evaluation of a control strategy developed for a hybrid energy system integrated in DC-AC microgrids. Electr. Power Compon. Syst. **42**(15), 1762–1770 (2014). https://doi.org/10.1080/15325008.2014.949914
18. Bhatt, A., Sharma, M.P., Saini, R.P.: Feasibility and sensitivity analysis of an off-grid micro hydro—photovoltaic—biomass and biogas—diesel—battery hybrid energy system for a remote area in Uttarakhand state, India. Renew. Sustain. Energy Rev. **61**, 53–69 (2016). https://doi.org/10.1016/j.rser.2016.03.030

19. Ismail, M.S., Moghavvemi, M., Mahlia, T.M.I.: Genetic algorithm based optimization on modeling and design of hybrid renewable energy systems. **85**, 120–130 (2014). http://doi.org/10.1016/j.enconman.2014.05.064
20. Fathima, A.H., Palanisamy, K.: Optimization in microgrids with hybrid energy systems—a review. Renew. Sustain. Energy **45**, 431–446 (2015). http://doi.org/10.1016/j.rser.2015.01.059
21. Tan, Y., Meegahapola, L., Muttaqi, K.M.: A review of technical challenges in planning and operation of remote area power supply systems. **38**, 876–889 (2014). http://doi.org/10.1016/j.rser.2014.07.034
22. Uttarakhand Economic Survey.: https://des.uk.gov.in/pages/display/163-economic-survey--2020-21. Accessed 04 May 2021

Image Processing

Adaptive Learning Rate-Based Convolutional Neural Network Models for Brain Tumor Images Classification

Thiyagarajan Padmapriya, Thiruvenkatam Kalaiselvi, and Karuppanagounder Somasundaram

Abstract In the proposed work, we have introduced three CNN models based on fixed, scheduled, and adaptive learning rates for classifying three types of fatal brain tumors such as glioma, meningioma, and pituitary tumors using a publicly available dataset generated by Cheng et al. Learning rates have a significant impact on increasing or decreasing the accuracy of CNN models. Initially, we have performed some experiments for fixing the model hyperparameters. Then, three kinds of CNN models have been developed by fixing the learning rates in model 1 and named as fixed learning rate network (FLRNET), by scheduling the learning rates in model 2 and named as learning rate schedule network (LRSNET), and by adaptively setting the learning rates in model 3 and named as adaptive learning rate schedule network (ALRSNET). Finally, we have evaluated the three proposed models on a standard MRI dataset that contains 3064 T1 weighted MRI images of brain tumor types. We have attained the highest accuracy of 99.2% with adaptive model ALRSNET, which outperformed other state-of-the-art methods available in the literature.

Keywords Artificial intelligence · Deep learning · Convolutional neural networks · Learning rates · Brain tumor · Glioma tumor · Meningioma tumor · Pituitary tumor

1 Introduction

Cancer is a fatal disease in the world. A survey concluded that every sixth death in the world is due to cancer [1]. It is ranked second after cardiovascular diseases in causing death. Brain tumors can be categorized into different types based on their shape, location, and texture. There are various types of tumors, such as acoustic neuroma tumors, meningioma tumors, pituitary tumors, glioma tumors, and CNS lymphoma tumors [2]. Among all the brain tumors, the incidence rates in glioma tumors are approximately forty-five, whereas for meningioma and pituitary tumors,

---

T. Padmapriya · T. Kalaiselvi (✉) · K. Somasundaram
Department of Computer Science and Applications, The Gandhigram Rural Institute-Deemed to be University, Dindigul, India
e-mail: kalaiselvi.gri@gmail.com

P. Balasubramaniam et al. (eds.), *Mathematical Modelling and Computational Intelligence Techniques*, Springer Proceedings in Mathematics & Statistics 376,
https://doi.org/10.1007/978-981-16-6018-4_9

it is fifteen [3]. Based on the type of tumors, the doctors can decide the specific treatment such as surgery, chemotherapy, and radiotherapy [4].

One of the most popular medical imaging modalities is magnetic resonance imaging (MRI). MRI is used for producing apparent images of internal organs of humans. These images are generated by MRI in the format of two dimension and three dimension. Since MRI declines conspicuous pictures on the tissues of brain, it is considered as an accurate technique for detecting cancer. Cancer types are recognized and identified by the experienced radiologists, which is a time-consuming procedure. Diagnosis of a tumor by humans is unreliable because of the tumor's shapes and visibility in the MRI images [5].

Artificial intelligence (AI) plays a vital role in computer vision. One of the popularly used image classification techniques in AI is convolutional neural networks (CNN). CNN models play a more significant role in several fields of computer vision, such as medical imaging [6], plant disease identification, handwritten digits classification, and general image classification.

Several state-of-the-art methods were carried out on brain tumor-type classification. Kalaiselvi et al. [7] had proposed a CNN model with five layers including batch normalization and stopping criteria (FLSCBN), which produced the best accuracy when compared with the state-of-the-art methods. Ismael et al. [8] had proposed an enhanced approach using residual networks for classifying brain tumor types from 3064 MRI brain images that consist three brain tumors such as glioma, meningioma, and pituitary. They had achieved an accuracy of 99%. Ali Mohammed et al. [9] had proposed four-layered convolutional neural network architecture for classifying brain tumors into three classes, and they have recorded their accuracy results for cropped lesions as 98.52%, and for uncropped lesions, they had achieved 98.52%, and for segmented lesion images, they had attained 97.62%.

Nyoman et al. [10] had developed five convolutional neural network models with a maximum of two convolutional layers. Their overall validation accuracy is 84.19%. They have not used any hyperparameters for training their model. And thus, their model did not provide any satisfactory results. Deepak et al. [11] had proposed a brain tumor classification model using transfer learning. They had used GoogleNet as the pre-trained model for extracting features from brain MRI images. They had achieved a classification accuracy of 98 percent. Mustafa et al. [12] had developed a method for classifying brain tumor types using statistical features. These features are obtained from the combination of discrete wavelet transform (DWT) and Gabor filter techniques. They had used a back propagation neural network classifier to test the classification accuracy. They had obtained 91.9% accuracy. Rehman et al. [13] had used AlexNet, GoogleNet, and VGGNet to classify brain tumors. Among them, VGG16 architecture attained an accuracy of 98.69% in the classification of brain tumors.

In all the works mentioned above, the authors have chosen a neural network model approximately and trained for brain tumor classification. Most of the works are based on transfer learning models. Some of the model architectures used in the literature for brain tumor classification does not contain any hyperparameter for training the model.

In the proposed work, we have developed CNN-based models for classifying the three types of fatal brain tumors such as glioma, meningioma, and pituitary tumors based on the learning rates. Initially, we have used various hyperparameters for training the model from scratch without using transfer learning. We have tried fixing the number of epochs, batch normalization, and appropriate optimizers for training the models. We have also experimented by varying the image size of input data. We have realized that the learning rate plays a vital role in training the models for producing good accuracy. Then, three kinds of CNN models have been developed by fixing the learning rates in model 1 and named as fixed learning rate network (FLRNET), by scheduling the learning rates in model 2 and named as learning rate schedule network (LRSNET), and by adaptively setting the learning rates in model 3 and named as adaptive learning rate schedule network (ALRSNET). After performing various experiments for training the models, we have chosen the best model for classifying brain tumor types.

The rest of the article is organized as follows. Section 2 elaborates on the importance and explanation of learning rate in training the neural network models. Section 3 describes the materials and metrics used for the proposed work. The methodology and experimental results are discussed in Sect. 4. In Sect. 5, conclusions and future enhancements are discussed.

2 Learning Rate

Learning rate (LR) acts as a parameter for tuning in an optimization algorithm. It plays a vital role in determining each iteration's step size, and it moves toward the minimum loss function. The learning rate also shows at what speed the algorithm learns. During the learning rate fixation, we should be aware of the convergence rate and outreach. The gradient of the loss function determines the descent direction, while the learning rate determines the larger step taken toward that direction. There is a possibility of overshooting, and a decrease in convergence rate if the learning rate is too high or too low. The learning leaps over minima if the learning rate is too high. If the learning rate is too low, it will take more time to converge, or else it gets fixed in unacceptable local minimum. To attain faster convergence and preventing oscillations, the LR can be assorted during training. This LR variation is implemented by using the learning rate scheduler or an adaptive learning rate.

2.1 Scheduling Learning Rate

The learning rate is modified during learning using the learning rate schedule, and it is mostly modified between epochs/iterations. It is mainly carried out by the parameters of decay and momentum. Decay enables the learning process to avoid oscillations during a situation of a high LR, which makes the learning leap commutation

Table 1 Learning schedules

S. No.	Types	Formula
1	Time-based learning	$\eta(n+1) = \eta n/((1+dn))$
2	Step-based learning	$\eta n = \eta o$ dfloor$\{(1+n)/r\}$
3	Exponential learning	$\eta n = \eta o\ e^{(-dn)}$

where η is the learning rate
d is a decay parameter
n is the step of the iteration
r denotes drop rate or how often the rate should be dropped

over minima. These oscillations are managed by the hyperparameters. Momentum increases the learning rate and speeds up learning, while the cost of gradient error is advancing toward the same direction in a prolonging manner. It also avoids the local minima by overwhelming small bounce.

The learning rate schedule can be categorized into three types such as time-based learning, step-based learning, and exponential learning as given in Table 1. In time-based learning, the learning rate is altered based on the learning rate of the previous iteration. In step-based learning, the learning rate is changed by steps based on some predefined steps. In exponential learning schedules, a decreasing exponential function is used instead of steps.

2.2 *Adaptive Learning Rate*

Adaptive learning rate overcomes the issue caused by the learning rate scheduler. The problem with the learning rate scheduler is that they all rely on the hyperparameters which are generally set by manual before starting the experiments. But, in the adaptive learning rate, the LR is reduced automatically if there is no improvement in accuracy or loss during model training. Thus, the adaptive learning rate increases the efficiency of the neural network models.

3 Materials and Metrics

We have used a brain tumor dataset that is publicly available. The dataset consists of 3064 T1-weighted brain tumor MRI images. This brain tumor dataset consists of three brain tumors (glioma, meningioma, and pituitary tumors), which have high incident rates. It contains images collected from two-dimensional MRI scans from two hundred and thirty-three cancer patients. The dataset was initially generated by Cheng et al. [14]. Some of the sample images of brain tumors in axial, coronal, and sagittal orientations are shown in Fig. 1. The brain tumor dataset contains 708 slices for meningioma, 1426 slices for glioma, and 930 slices of pituitary tumor.

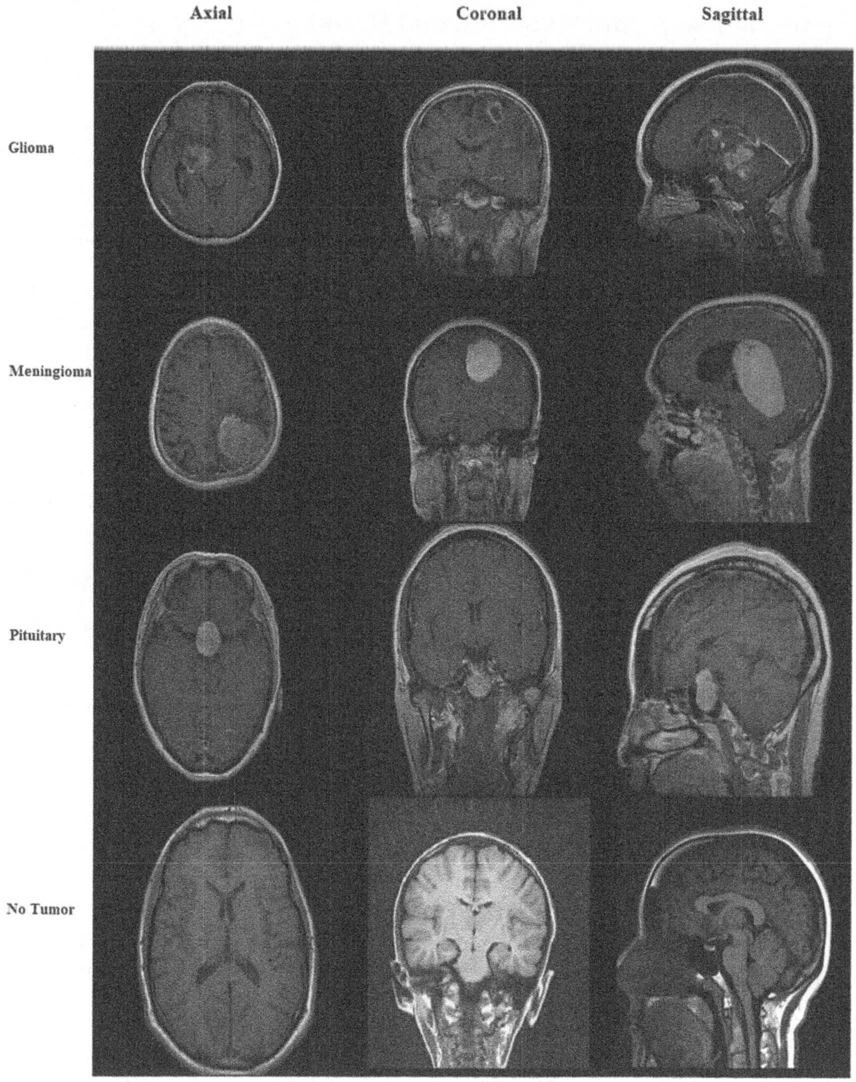

Fig. 1 Sample MRI brain images for glioma, pituitary, meningioma, and non-tumor slices in different orientations

We have used the accuracy metric for evaluating our model's performance. Accuracy can be defined as follows.

$$\text{Accuracy} = \frac{\text{number of correct predictions}}{\text{total number of predictions}} \tag{1}$$

4 Methodology and Experimental Results

The workflow of the proposed work is shown in Fig. 2.

4.1 Data Preprocessing

The model performance is based on the amount of data used for training. Preprocessing of images is done to improve the performance of the model. We have resized the images to 100 * 100 and converted them into the grayscale mode. The dataset is split into training and testing sets in 80:20 ratios. The dataset used in our proposed

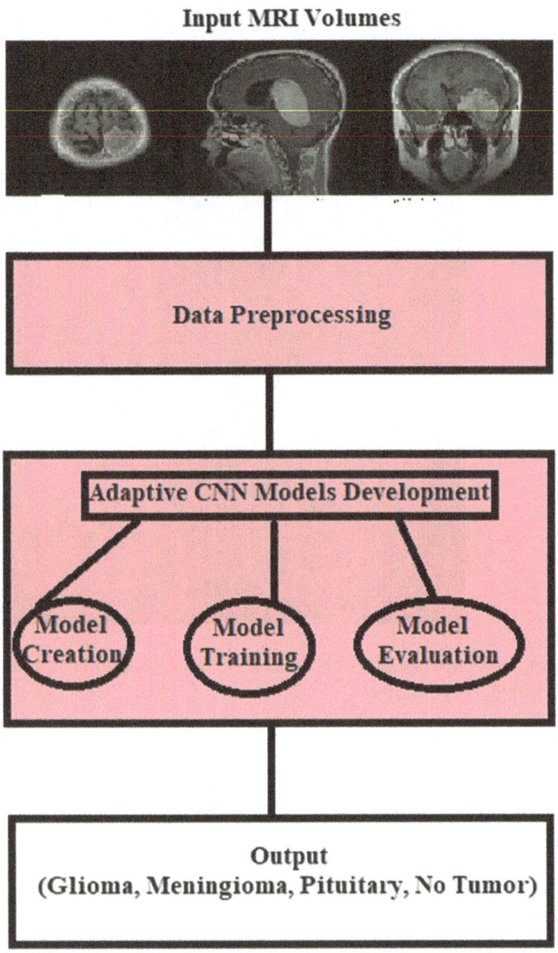

Fig. 2 Workflow of the proposed work

work is small to train deep neural networks. So, we have implemented data augmentation techniques such as horizontal and vertical flips, rotation, shifting and zooming, correction of brightness, which enhanced the accuracy of the proposed models.

4.2 Adaptive CNN Models Development

Experiments for fixing hyperparameters have been done and followed by the development of learning rate CNN models. The process of fixing the hyperparameters for the model is described below.

4.2.1 Fixing Model Hyperparameters

The general architecture of the proposed model is shown in Fig. 3. After preprocessing of MRI brain images, we concentrated on the image size. When the image size is larger, the model did not perform well. After conducting experiments by varying the image size, we finally fixed 100 * 100, which showed higher results. Then, we have selected the number of layers to five and epochs to 30 based on the work we have performed earlier [6, 7]. Further, we performed experiments using batch normalization and selecting suitable optimizers such as Adam and RMSprop, which also increased our models' test accuracy. Table 2 shows the models which we have developed for fixing the hyperparameters with 90% and above test accuracy.

From Table 2, it is concluded that the parameters used in row 1 are chosen for fixing hyperparameters for the general architecture of proposed CNN models.

4.2.2 Learning Rate-Based CNN Models

We have developed three CNN models based on learning rates. In the model 1, we have fixed learning rate to 0.0001 for all the iterations and named as fixed learning rate network (FLRNET). The accuracy obtained by the FLRNET model is not satisfactory. Hence, we have developed a model 2 by varying the learning rate for every ten epochs. For the first ten epochs, the learning rate was fixed to 0.001. From epoch 10 to 20, the learning rate was set to 0.0001. After 20 epochs, the learning rate was reduced to 0.00001. We named this model 2 as learning rate schedule network (LRSNET). LRSNET produced an accuracy of 96 percent, which is better than FLRNET.

For model 3 the learning rates are not scheduled based on epochs as in above model 2, LRSNET. In model 3, the learning rates are changed automatically when the loss starts to increase for continuous five iterations. If there is no improvement in validation loss for five epochs, then the learning rate will be reduced. The lower bound of the learning rate is fixed to 0.0001, and the learning rate is reduced by 0.1 after no improvement in the loss function. We named this model 3 as an adaptive learning rate schedule network (ALRSNET).

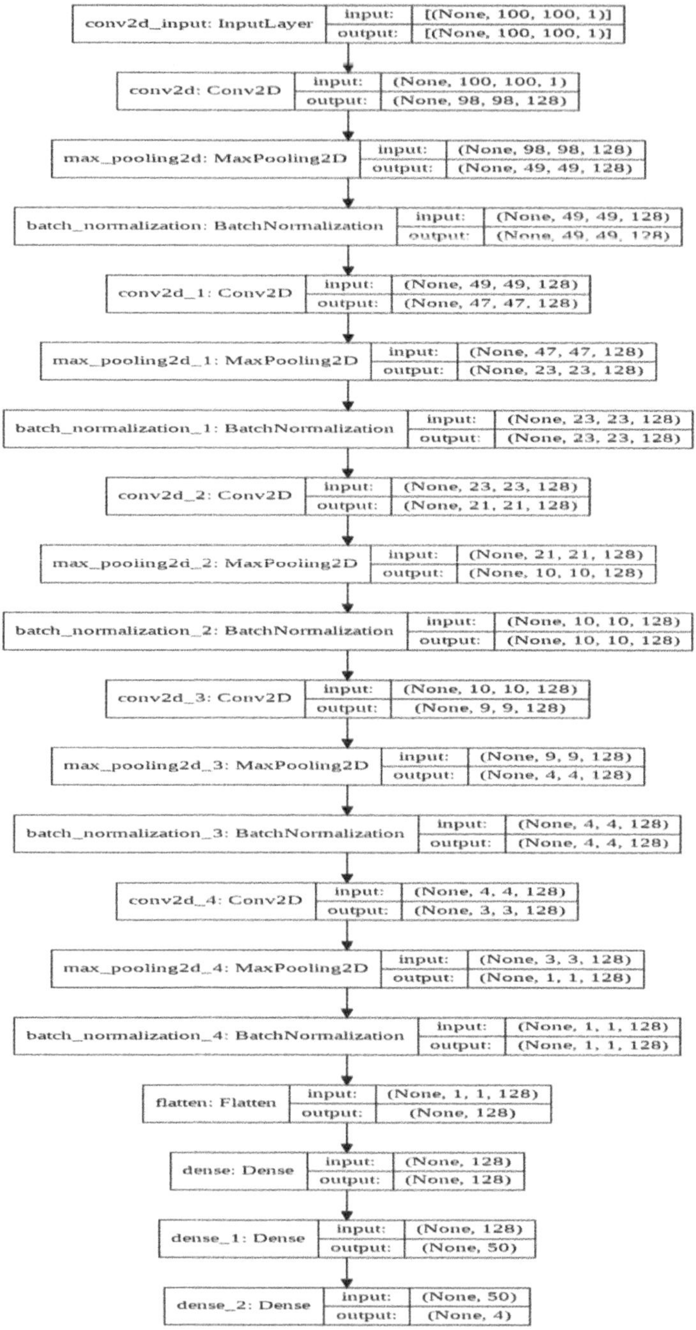

Fig. 3 CNN architecture used for the proposed work

Table 2 List of experiments conducted for fixing hyperparameters in proposed CNN models

S. No.	Model hyperparameters	Accuracy
1	**RMSprop with batch normalization, image size (100, 100)**	**95**
2	RMSprop without batch normalization, image size (100, 100)	93
3	RMSprop with batch normalization, image size (224, 224)	93
4	Adam with batch normalization, image size (100, 100)	94
5	Adam without Batch Normalization, image size (100, 100)	93
6	Adam with batch normalization, image size (224, 224)	92.7

Table 3 Accuracy of proposed CNN models

S. No.	Model hyperparameters	Accuracy
1	FLRNET	95
2	LRSNET	96
3	ALRSNET	99.2

4.2.3 Model Evaluation

Experiments are conducted on FLRNET, LRSNET, and ALRSNET with the dataset mentioned in Sect. 3, and the results were compared for performance analysis. We have recorded the accuracy of these models, and it is shown in Table 3. Among these three models, the adaptive model ALRSNET performed well, and it attained highest accuracy of 99.2%. We have also compared the results of our models with the state-of-the-art methods. The comparison of FLRNET, LRSNET, and ALRSNET with state-of-the-art methods for brain tumor type's classification using the same dataset is inferred in Table 4. The proposed adaptive model ALRSNET outperformed all the other state-of-the-art methods taken for performance analysis (Figs. 4, 5, and 6).

Table 4 State-of-the-art methods for brain tumor types classification

S.NO	Authors	Method	Accuracy
1	Ismael et al. [8]	Residual networks	99
2	Ali Mohammed et al. [9]	Four-layered CNN model	98.52
3	Nyoman et al. [10]	CNN models	84.19
4	Deepak et al. [11]	GoogleNet	98
5	Mustafa et al. [12]	Backpropagated neural networks	91.9
6	Rehman et al. [13]	VGG16 networks	98.69
7	FLRNET (proposed)	CNN	95
8	LRSNET (proposed)	CNN	96
9	ALRSNET (proposed)	CNN	99.2

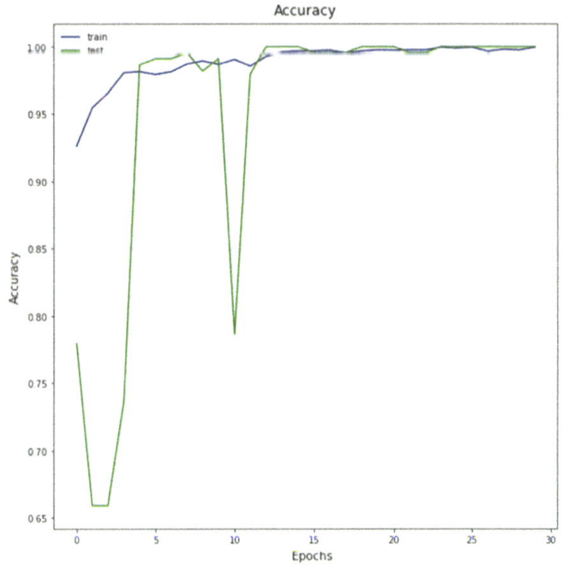

Fig. 4 Training and testing accuracy plot of ALRSNET model

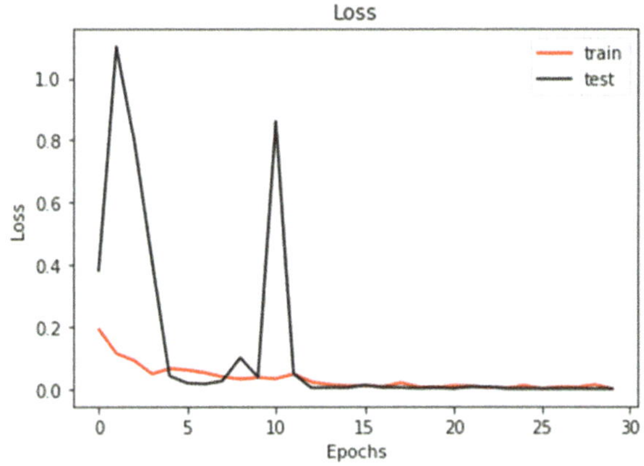

Fig. 5 Plot of training and testing loss in ALRSNET model

5 Conclusion and Future Enhancements

In the proposed work, we have developed three CNN models by varying the learning rates for brain tumor type's classification from MRI images. The dataset used for training our model is small to train deep neural networks. We have also employed data augmentation methods, which enhanced the accuracy of the proposed models. When compared with the state-of-the-art methods, the proposed ALRSNET model outperformed well. The computation time of proposed ALRSNET model is 136 s. In the future, this work can be extended by increasing the amount of data used for

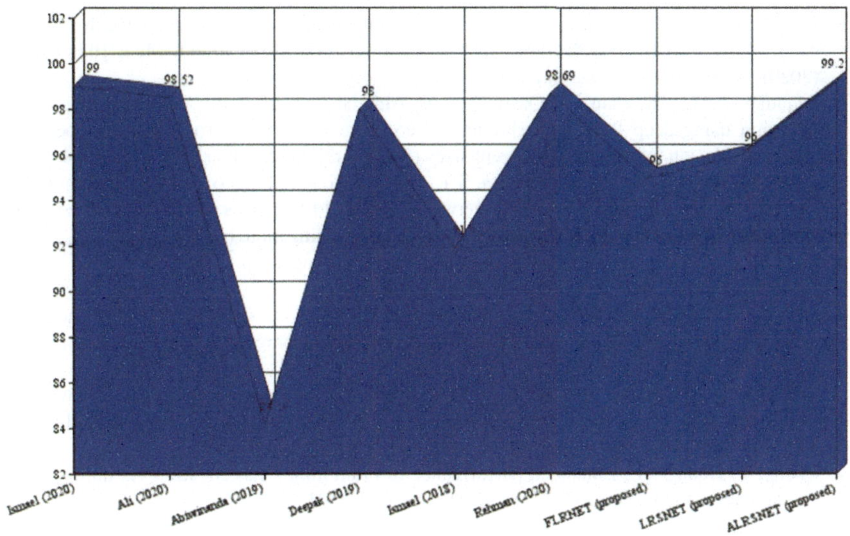

Fig. 6 Accuracies of the state-of-the-art methods and the proposed models

training. Also, a model can be developed for identifying more number of brain tumor types.

1. References

1. Roser, M., Ritchie, H.: Cancer (2019). [Online]. Available: https://ourworldindata.org/cancer. Last accessed 25 Dec 2020
2. Louis, D.N., Perry, A., Reifenberger, G., Von Deimling, A., Figarella-Branger, D., Cavenee, W.K., Ohgaki, W., Wiestler, O.D., Kleihues, P., Ellison, D.W.: The 2016 World Health Organization classification of tumors of the central nervous system: a summary. Acta Neuropathologica **131**(6), 803–820 (2016)
3. Nawab, Z.A.R, Swati, K., Zhao, Q., Kabir, M., Ali, Z., Ahmed, S., Lu, J., Ali, F.: Content-based brain tumor retrieval for MR images using transfer learning. IEEE Access **7**, 17809–17822 (2019)
4. Pereira, S., Meier, R., Alves, V., Reyes, M., Carlos, A.: Automatic brain tumor grading from MRI data using convolutional neural networks and quality assessment. In: iMIMIC—Workshop on Interpretability of Machine Intelligence in Medical Image Computing (2018)
5. Kumar, S., Dabas, C., Godara, S.: Classification of brain MRI tumor images: a hybrid approach. Procedia Comput. Sci. **122**, 510–517 (2017). [Online]
6. Kalaiselvi, T., Padmapriya, T., Sriramakrishnan, P., Priyadharshini, V.: Development of automatic glioma brain tumor detection system using deep convolutional neural networks. Int. J. Imaging Syst. Technol. **30**(4), 926–938 (2020)
7. Kalaiselvi, T., Padmapriya, S.T., Sriramakrishnan, P., Somasundaram, K.: Deriving tumor detection models using convolutional neural networks from MRI of human brain scans. Int. J. Inf. Technol., 1–6 (2020)

8. Ismael, S.A.A., Mohammed, A., Hefny, H.: An enhanced deep learning approach for brain cancer MRI images classification using residual networks. Artif. Intell. Med. **102**, 101779 (2020)
9. Alqudah, A.M., Alquraan, H., Qasmieh, I.A., Alqudah, A., Al-Sharu, W.: Brain tumor classification using deep learning technique—a comparison between cropped, uncropped, and segmented lesion images with different sizes (2020). arXiv preprint arXiv:2001.08844.
10. Abiwinanda, N., Hanif, M., Hesaputra, S.T., Handayani, A., Mengko, T.R.: Brain tumor classification using convolutional neural network. In: World Congress on Medical Physics and Biomedical Engineering 2018 (pp. 183–189). Springer, Singapore (2019)
11. Deepak, S., Ameer, P.M.: Brain tumor classification using deep CNN features via transfer learning. Comput. Biol. Med. **111**, 103345 (2019)
12. Ismael, M.R., Abdel-Qader, I.: Brain tumor classification via statistical features and back-propagation neural network. In: 2018 IEEE International Conference on Electro/Information Technology (EIT), pp. 0252–0257. IEEE (2018)
13. Rehman, A., Naz, S., Razzak, M.I., Akram, F., Imran, M.: A deep learning-based framework for automatic brain tumors classification using transfer learning. Circuits Systems Signal Process. **39**(2), 757–775 (2020)
14. Cheng, J., Huang, W.: Enhanced performance of brain tumor classification via tumor region augmentation and partition. PLoS ONE **10**(10), 1–13 (2015)

Extended Discrete Cosine Transform

S. Praveenkumar, K. Somasundaram, S. Magesh, and T. Kalaiselvi

Abstract Lossy compression is a valuable technique in reducing the file size of an image for storing and transmitting across computer networks. Discrete cosine transform (DCT) has been one of the efficient transforms that is being used in popular compressor like JPEG. For fast compression, transforms with integer coefficients were developed. Few such transforms are integer discrete cosine transform (IntDCT) and integer-based discrete Tchebichef transform (DTT). In this paper, we extend the type II 8×8 DCT and produce a transform with integer coefficients. Simulated results show that the proposed method performed better than the other integer-based transforms in terms of the PSNR values and is also competitive with DCT.

Keywords DCT · Image compression · Integer DCT · Transform coding · Transforms with integer coefficients

1 Introduction

An image conveys the most information than a plain text. Images are used in all fields, such as passport photo, X-ray, magnetic resonance images (MRI), satellite images, sports, in cinema, TV, personal identification biometrics, to quote a few. Storing and transmitting images cost more than a plain text of the same size. Large organizations collect and store biometric identification such as face, iris, fingerprint, and voice

S. Praveenkumar
Qualcomm Atheros, 1700 Technology Drive, San Jose, CA 95110, USA

K. Somasundaram (✉) · T. Kalaiselvi
Department of Computer Science and Applications, Gandhigram Rural Institute, Gandhigram, Dindigul, Tamil Nadu 624302, India

K. Somasundaram
Department of BCA, PSGR Krishnammal College for Women, Coimbatore 641004, India

S. Magesh
Quicken Loans, LLC, 1050 Woodward Ave, Detroit, MI 48226, USA

P. Balasubramaniam et al. (eds.), *Mathematical Modelling and Computational Intelligence Techniques*, Springer Proceedings in Mathematics & Statistics 376,
https://doi.org/10.1007/978-981-16-6018-4_10

for authentication and access. In hospitals, patient-related personal ID and clinical-related images, like MRI and X ray, are stored for fast retrieval of information for patient monitoring and diagnosis. Government of India, for example, collects facial images, fingerprint, and iris images of its about 138 billion people. Storage of such images requires a huge amount of storage space. Therefore, methods are required to reduce the storage and transmission cost.

Usually, a compressor system contains two parts. First part, compressor, takes the image and produces a compressed image, which has lesser file size than the original image, using some technique. The second part, decompressor, takes the compressed image file, decompresses it, and reconstructs the image. The image compression methods broadly fall under two categories, namely lossy and lossless compression. In a lossless compression, the reconstructed image will be exactly the same image as that of the original image. There will not be any loss in the pixel values in each of the pixels in that image. Huffman coding is one such method [1]. The compression achieved by such methods is very small. In lossy compression, the reconstructed image will not be the same as that of the original but an approximate one. There will be some loss in the values of the individual pixels that make up the reconstructed image. The compression achieved by this type of compressors is high. Most applications make use of lossy compressors.

Compressors make use of the intrinsic properties associated with the image or make use of the sensitivity in the human visual system (HVS) [2]. Few lossless compressors are run length encoder (RLE), Lempel–Ziv(LZ) compressor and its variants like LZ77 and LZ78, Storer and Szymanski (LZSS), and Lempel, Ziv and Welch (LZW) [1]. Roshal ARchive (RAR) is a popular lossless compressor. Block truncation coding (BTC), vector quantization (VQ), joint photographic expert group (JPEG), etc., are lossy compressors [1]. JPEG is a combination of several algorithms layered one over other to achieve varying degrees of compression.

In images, adjacent pixels have some correlation among them, i.e., it has redundancy. Compression is a process of removing redundancy [1]. Therefore, to achieve compression, this correlation is decorrelated by applying a suitable transform on the input image. The transformed values, called transform coefficients, are decorrelated and contain more energies or information packed in few coefficients and less information in the remaining coefficients. These coefficients can be manipulated independently. This property helps to drop many coefficients with less information and retain only few coefficients containing maximum information to represent the image. This property helps to achieve compression.

Discrete cosine transform (DCT) is a transform technique used for image compression. It is a widely used technique to compress images. DCT is an orthogonal transform and was proposed by Ahmed et al. [3]. For practical applications, a 8×8 transform operator M is formed from an orthogonal polynomial basis function. This transform, when operated on an image block of 8 X 8 pixels, produces an output of the same size 8×8 elements called transform coefficients. These coefficients are decorrelated from each other. It has been shown that few of these coefficients carried the most information and the remaining coefficients carried less information.

By truncating or quantizing the least information carrying coefficients, many coefficients can be removed. This helps to reduce the file size and compress the image. To get back the image, the inverse transform is applied on the coefficients. This transform performed very efficiently and is incorporated in the popular compressor JPEG [4, 5]. Since then, a large number of work related to DCT has been carried out. Four distinct DCTs, such as DCT-I, DCT-II, DCT-III, and DCT-IV, have been proposed [1, 4, and 6]. All the four types differ lightly. Of these, DCT-II is very efficient than the other three and is used for many practical applications. Because of its applications in image and video compression, it has been implemented in hardware for fast computation [7].

The development on DCT then focused on increasing the speed of computing. It is well known that processing with integer numbers is faster than that with floating-point numbers in hardware [8]. Several works have been developed in defining the integer discrete cosine transform (IntDCT) in which the transform basis matrix M is formed with integer coefficients [9–12]. Plonka and Tasche [13] proposed an IntDCT for DCT type II and IV using lifting scheme for 8 X 8 basis transform matrix. Cheung et al. [14] have derived an IntDCT for an 8 X 8 transform matrix from DCT Type II and found that the computational complexity is much less than that of DCT, but with a slightly less accuracy. Similar integer transform has also been derived from Tchebichef orthogonal polynomials and is termed as discrete Tchebichef transform (DTT). Krishnamoorthi [15] and Krishnamoorthi et al. [16] have proposed an orthogonal polynomial-based transform in which the transform matrix consists of integers. Prattipatti et al. [17] have reported a comparative study of integer DCT and DTT.

In this article, we propose an integer DCT by converting the DCT type II basis transform matrix to an integer-based transform and study its performance. The remaining part of this article is organized as follows. In Sect. 2, we give a brief overview of the DCT. In Sect. 3, the proposed extended discrete cosine transform (EDCT) is presented. In Sect. 4, the image metrics are given. In Sect. 5, the performance evaluation metrics are given. Results and discussion are presented in Sect. 6, and in Sect. 7, the conclusions are given.

2 Discrete Cosine Transform

In this section, for continuity, we briefly outline the DCT. Usually, the transform is applied on a small block of image, to reduce the computational complexity. Though image block of any size can be used for the transform, usually an 8×8 size image block is considered and is called blocked transform [1].

The orthogonal basis cosine vectors that are used to define the DCT type-2 are given by [1]:

$$\mathrm{DCT}^2_{k,j} = \sqrt{\frac{2}{n}} C_k \cos\left[\frac{k(j+\frac{1}{2})\pi}{n}\right], \quad k, j = 0, 1, \ldots, n-1 \tag{1}$$

where

$$\begin{aligned} C_k &= 1/\sqrt{2}, \text{ if } k = 0 \\ &= 1 \qquad \text{otherwise} \end{aligned}$$

For the basis function given in Eq. (1), for 8 × 8 DCT, the values of k, j are: k, j = 0,1,…,7. The basis function values form the DCT matrix M of size 8 × 8 and are given by

$$\mathrm{M} = \begin{pmatrix} 0.3536 & 0.3536 & 0.3536 & 0.3536 & 0.3536 & 0.3536 & 0.3536 & 0.3536 \\ 0.4904 & 0.4157 & 0.2778 & 0.0975 & -0.0975 & -0.2778 & -0.4157 & -0.4904 \\ 0.4619 & 0.1913 & -0.1913 & -0.4619 & -0.4619 & -0.1913 & 0.1913 & 0.4619 \\ 0.4157 & -0.0975 & -0.4904 & -0.2778 & 0.2778 & 0.4904 & 0.0975 & -0.4157 \\ 0.3536 & -0.3536 & -0.3536 & 0.3536 & 0.3536 & -0.3536 & -0.3536 & 0.3536 \\ 0.2778 & -0.4904 & 0.0975 & 0.4157 & -0.4157 & -0.0975 & 0.4904 & -0.2778 \\ 0.1913 & -0.4619 & 0.4619 & -0.1913 & -0.1913 & 0.4619 & -0.4619 & 0.1913 \\ 0.0975 & -0.2778 & 0.4157 & -0.4904 & 0.4904 & -0.4157 & 0.2778 & -0.0975 \end{pmatrix} \tag{2}$$

The waveforms of these eight basis functions are shown in Fig. 1

As M is formed out of orthogonal polynomials,

$$M^{\mathrm{T}} = M^{-1} \tag{3}$$

where M^{T} is the transpose of M. Further, the following identity holds good.

$$M^{\mathrm{T}} \times M^{-1} = \mathrm{ID} \tag{4}$$

where ID is a diagonal unitary matrix.

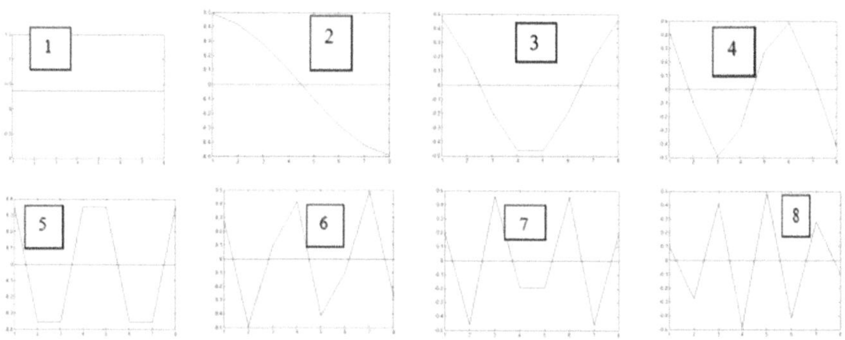

Fig. 1 Waveform of the eight basis functions that form the DCT matrix M

For compression, first, the forward transform of the given image is made. Application of the DCT on a 8 × 8 image block $B(8, 8)$ gives the result $C(8, 8)$, called as transform coefficients, with the same size of the image block.

$$C = M \times B \times M^{\mathrm{T}} \tag{5}$$

The top left coefficient, $C(1,1)$, is called the direct current (DC) coefficient, and the remaining are called the alternating current (AC) coefficients. These coefficients, when arranged in a zigzag manner as shown in Fig. 2, carry information in the decreasing order of weightage. The DC coefficient carries the maximum information about the image. These coefficients in C are then manipulated to get the compressed file C'. The decompressed image block B' is obtained as

$$B' = M^{\mathrm{T}} \times C' \times M \tag{6}$$

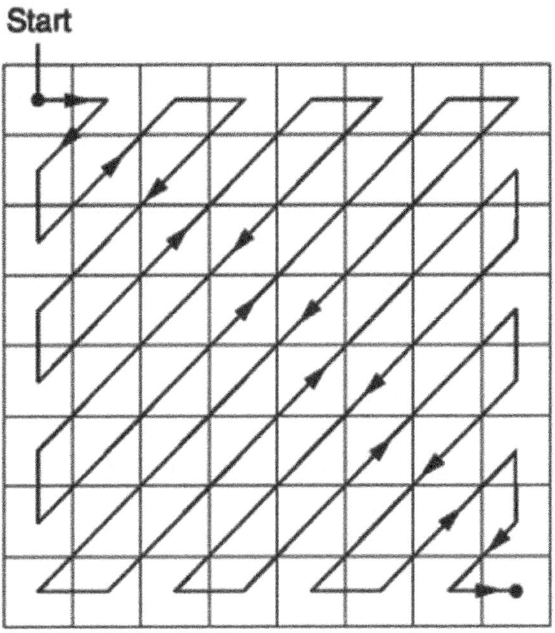

Fig. 2 Zigzag arrangement of 8 × 8 DCT coefficients in decreasing order of weightage

3 The Proposed Method

In this section, we present the new proposed method for DCT matrix. The forward DCT is similar to an integer DCT, and we termed it as extended discrete cosine transform (EDCT). We formulate a process to generate integer element for each of the coefficient in DCT basis matrix M given in Eq. (2). First, find the minimum value of each row of M in Eq. (2) as

$$V(i) = \min(M(i, j)), j = 1, 2 \ldots 8 \tag{7}$$

Then, divide each row of M by the corresponding V and round it to the nearest integer to get the integer transform T as

$$T = \text{round}\ (M(i, j)/V(i)), \qquad i, j = 1, 2 \ldots 8 \tag{8}$$

Using the values of Eq. (2) in Eq. (8), we get the new EDCT transform T as

$$T = \begin{pmatrix} 1 & 1 & 1 & 1 & 1 & 1 & 1 & 1 \\ 5 & 4 & 3 & 1 & -1 & -3 & -4 & -5 \\ 2 & 1 & -1 & -2 & -2 & -1 & 1 & 2 \\ 4 & -1 & -5 & -3 & 3 & 5 & 1 & -4 \\ 1 & -1 & -1 & 1 & 1 & -1 & -1 & 1 \\ 3 & -5 & 1 & 4 & -4 & -1 & 5 & -3 \\ 1 & -2 & 2 & -1 & -1 & 2 & -2 & 1 \\ 1 & -3 & 4 & -5 & 5 & -4 & 3 & -1 \end{pmatrix} \tag{9}$$

It is found in contrast to Eq. (4) that

$$T \times T^{\mathrm{T}} \approx G_D \tag{10}$$

where G_D is a diagonal matrix with four off-diagonal elements having values ± 4. The basis function waveforms for the EDCT in Eq. (9) are given in Fig. 3.

To find the transform of the image, the given image I is divided into a number of non-overlapping image blocks B of size 8 × 8 pixels each, and the forward transform is applied on the image block B to get the transform coefficients C as

$$C(i, j) = T \times B \times T^{\mathrm{T}} \tag{11}$$

where T^{T} is the transpose of T. In order to assess the performance of the transform, the coefficients in C are cut, in the zigzag order starting from the end, as shown in Fig. 2, to give a new coefficient matrix Q which forms the compressed data. The reconstructed image block B_{r} is obtained from the compressed block Q by taking the inverse transform as

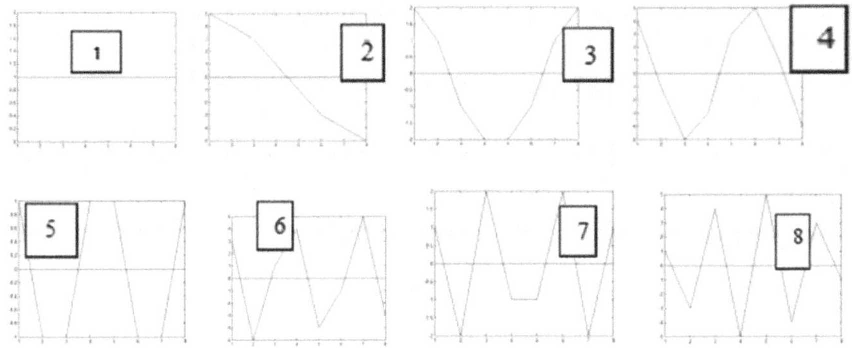

Fig. 3 Waveform of the eight basis functions that form the EDCT

$$B_{\mathrm{r}} = (T)^{-1} \times Q \times \left(T^{\mathrm{T}}\right)^{-1} \tag{12}$$

For comparison, transforms with integer coefficients by few other methods are given below. The transform T generated using the orthogonal polynomials in [15] (ChebyOrtho) is given by

$$T = \begin{pmatrix} 1 & 1 & 1 & 1 & 1 & 1 & 1 & 1 \\ -7 & -5 & -3 & -1 & 1 & 3 & 5 & 7 \\ 7 & 1 & -3 & -5 & -5 & -3 & 1 & 7 \\ -3 & 2 & 3 & 1 & -1 & -3 & -2 & 3 \\ 3 & -5 & -1 & 3 & 3 & -1 & -5 & 3 \\ -1 & 4 & -3 & -3 & 3 & 3 & -4 & 1 \\ 1 & -5 & 9 & -5 & -5 & 9 & -5 & 1 \\ -1 & 7 & -21 & 35 & -35 & 21 & -7 & 1 \end{pmatrix} \tag{13}$$

The integer discrete cosine transform (IntDCT) [14] T is given by

$$T = \begin{pmatrix} 1 & 1 & 1 & 1 & 1 & 1 & 1 & 1 \\ 5 & 3 & 2 & 1 & -1 & -2 & -3 & -5 \\ 3 & 1 & -1 & -3 & -3 & -1 & 1 & 3 \\ 3 & -1 & -5 & -2 & 2 & 5 & 1 & -3 \\ 1 & -1 & -1 & 1 & 1 & -1 & -1 & 1 \\ 2 & -5 & 1 & 3 & -3 & -1 & 5 & -2 \\ 1 & -3 & 3 & -1 & -1 & 3 & -3 & 1 \\ 1 & -2 & 3 & -5 & 5 & -3 & 2 & -1 \end{pmatrix} \tag{14}$$

The integer discrete Tchebichef transform (DTT) [17, 18] is given by

$$T = \begin{pmatrix} 1 & 1 & 1 & 1 & 1 & 1 & 1 & 1 \\ -7 & -5 & -3 & -1 & 1 & 3 & 5 & 7 \\ 7 & 1 & -3 & -5 & -5 & -3 & 1 & 7 \\ -7 & 5 & 7 & 3 & -3 & -7 & -5 & 7 \\ 7 & -13 & -3 & 9 & 9 & -3 & -13 & 7 \\ -7 & 23 & -17 & -15 & 15 & 17 & -23 & 7 \\ 1 & -5 & 9 & -5 & -5 & 9 & -5 & 1 \\ 1 & -7 & 21 & -35 & 35 & -21 & 7 & -1 \end{pmatrix} \tag{15}$$

4 Image Metrics

In the spectral activity measure (SAM) [17], the ratio between arithmetic mean and geometric mean of the transform coefficients of an image is computed as

$$\text{SAM} = \frac{\frac{1}{M \cdot N} \sum_{i=0}^{M-1} \sum_{j=0}^{N-1} |F(i, j)|^2}{\left\{\prod_{i=0}^{M-1} \prod_{j=0}^{N-1} |F(i, j)|^2\right\}^{\frac{1}{M \cdot N}}} \tag{16}$$

where $F(i, j)$ is the original transform coefficients of the whole image, and $M \times N$ is the size of the image. Further, the spatial frequency measure (SFM) of an image $x_{i,j}$ is defined as [17]:

$$\begin{aligned} \text{SFM} &= \sqrt{R_f^2 + C_f^2} \\ R_f &= \sqrt{\frac{1}{M \cdot N} \sum_{i=1}^{M} \sum_{j=2}^{N} \left(x_{ij} - x_{ij-1}\right)^2} \\ C_f &= \sqrt{\frac{1}{M \cdot N} \sum_{j=1}^{M} \sum_{i=2}^{N} \left(x_{ij} - x_{i-1j}\right)^2} \end{aligned} \tag{17}$$

Both SAM and SFM measure the image properties like the variation of the image pixels in the image. These parameters help to infer how much compression can be achieved on them.

5 Performance Evaluation Metrics

In order to assess the performance of the new transform EDCT, we compute the quality of the decompressed image measured using the peak signal-to-noise ratio

(PSNR). Suppose the input image is $I(x, y)$ and the decompressed image is $I'(x, y)$, then the squared error is given by

$$\mathrm{SE}(x, y) = \left(I(x, y) - I'(x, y)\right)^2 \tag{18}$$

The mean squared error (MSE) is computed as

$$\mathrm{MSE} = \Sigma\Sigma\ \mathrm{SE}(x, y)/M \times N, \tag{19}$$

where M and N are the number of rows and columns in the image, respectively.

The PSNR is computed as

$$\mathrm{PSNR} = 10\log_{10}\left(\frac{\mathrm{MAX}_I^2}{\sqrt{\mathrm{MSE}}}\right) \tag{20}$$

where MAX_I is the maximum intensity level in that image. For a gray scale, 8-bit image, it is 255. Higher the value of PSNR, higher will be quality of the decompressed image.

6 Results and Discussion

For our study, we used the standard images, Lena, Mandrill, Peppers, and Boat, each of size 512 × 512 pixel taken from the image database in [19]. But any image can be used for this purpose. The transform coefficients are obtained by applying the transform in Eq. (5) for DCT and Eq. (11) for all other transforms with integer coefficients. We have done experiments on more than 12 images, and only results for four images that represent different image properties in terms of SFM and SAM are given. The computed values of SFM and SAM for the four images, by other methods with integer transform coefficients along with DCT, and the proposed method are given in Table 1.

Table 1 Computed values of SFM and SAM for different methods and images

Image and size (in pixel)	SFM	SAM				
		DCT [1]	EDCT proposed	Cheby Orhto [15]	IntDCT [14]	DTT [18]
Lena—512 × 512	14.0436	47.4550	**4.5041**	1.6694	4.3160	1.3778
Mandrill—512 × 512	36.5194	8.4425	**1.7838**	1.6009	1.6509	1.5267
Peppers—512 × 512	15.9643	34.4740	**3.5546**	2.0687	3.4388	1.5447
Boat—512 × 512	19.8521	28.8138	**3.2973**	1.7711	3.1105	1.4344

The value of SAM lies in the range [0, ∞). Higher values of SAM give higher predictability. Higher value of SAM and smaller values of SFM will give better compression [17]. From Table 1, it can be seen that for Mandrill image, the SFM is the highest among all the images. Therefore, it is harder to compress. The same observation has also been made while compressing it using DCT [20]. The SAM value for the normal DCT is the highest for Lena image and lowest for the Mandrill image. The last four columns in Table 1 are the results for the transform with integer coefficients. Among the four integer type transforms, the proposed transform (EDCT) gives the higher values than the remaining four. The advantage of having transform matrix with integer coefficients is that the computations will be much faster both for hardware and software implementation. Therefore, the proposed transform with integer coefficients can be considered for hardware implementation in the place of other integer coefficient-based transform matrix. The integer-based DTT performed very poorly when compared with all other methods with the lowest values of SAM. Further, it is observed that the SAM values for Mandrill image are lowest by all the methods.

In order to assess the information compaction in various coefficients, n number of coefficients are cut from the 64 transform coefficients arranged in a zigzag manner, from the end of the stream, as shown in Fig. 2. Starting from removing 60 coefficients (only first four coefficients are retained) to removing four coefficients (60 coefficients are retained), the values of PSNR are computed for the four images using the four different methods along with the DCT for comparison. For normal DCT, the coefficients C obtained in Eq. (5) are modified by cutting a number of coefficients as described above and obtain the compressed file D' directly. The reconstruction of the image for DCT is done using Eq. (6), and for other methods, Eq. (12) is used. The computed values of PSNR for different number of coefficients retained for compression are given in Tables 2, 3, 4, and 5, for Lena, Mandrill, Peppers, and Boat by different methods. The best values of PSNR among the four transforms with integer coefficients are highlighted in each table.

6.1 Lena Image

The Lena image has an SFM of 14.0436, the lowest among the four images. The first row, in Table 2, shows the amount of information packed in the first four transform coefficients of the Lena image in different transform methods. All methods are found to give almost the same PSNR values of about 27.5 and therefore contain the same amount of information in these four coefficients. But still there are subtle differences in the value in the following order, DCT > EDCT > ChebyOrtho > DTT > IntDCT.

But when more coefficients are added to the compression file, the DCT and EDCT method fast accumulate more information than the other three methods. With 20 coefficients, the DCT and EDCT methods produce a PSNR value > 36, whereas the other three methods gave values below 36. The PSNR values were in the order DCT > EDCT > ChebyOrtho > IntDCT > DTT. It is to be noted that DCT is the

Table 2 Computed values of PSNR by different methods for Lena image

No of coefficients	PSNR				
	DCT [1]	EDCT proposed	Cheby Orhto [15]	IntDCT [14]	DTT [17, 18]
4	27.5982	**27.5916**	27.5736	27.4688	27.5513
8	31.1738	**31.1071**	30.9604	30.7286	30.8666
12	32.3898	**32.2933**	32.0875	31.8298	31.9934
16	34.6190	**34.4841**	34.1252	34.0114	34.0197
20	36.2878	**36.1191**	35.5679	35.5301	35.5239
24	36.7515	**36.5897**	36.0584	36.1458	36.0105
28	38.3871	**38.2397**	37.6546	37.8025	37.6398
32	39.7405	**39.6063**	38.9734	39.1622	38.9499
36	40.6767	**40.5683**	39.8416	40.1474	39.8405
40	41.5929	**41.5091**	40.7038	41.1064	40.7327
44	43.1501	**43.1116**	42.2657	42.7299	42.2670
48	44.8075	**44.7554**	44.0488	44.4167	44.0834
52	46.1729	**46.1507**	45.4912	45.8874	45.4869
56	48.5975	**48.5707**	48.0022	48.3664	48.0589
60	51.8746	**51.8539**	51.3160	51.6785	51.3831

Table 3 Computed values of PSNR by different methods for Mandrill image

No of coefficients	PSNR				
	DCT [1]	EDCT proposed	Cheby Orhto [15]	IntDCT [14]	DTT [17, 18]
4	21.1520	21.1461	**21.1464**	21.1245	21.1412
8	22.0901	**22.0773**	22.0339	22.0167	22.0180
12	23.4896	**23.4680**	23.3392	23.3526	23.3215
16	24.1562	**24.1346**	23.9658	24.0186	23.9452
20	25.1038	**25.0813**	24.8026	24.9277	24.8043
24	26.4717	**26.4542**	26.0799	26.2300	26.0644
28	27.2059	**27.1956**	26.7675	26.9754	26.7693
32	27.6937	**27.6855**	27.2635	27.4749	27.2682
36	29.2025	**29.1884**	28.7155	28.9592	28.7097
40	30.7772	**30.7546**	30.1747	30.4730	30.1825
44	31.5544	**31.5351**	30.9786	31.2851	30.9710
48	33.0372	**33.0150**	32.4394	32.7986	32.4663
52	35.5356	**35.5061**	34.8010	35.2824	34.8255
56	37.2216	**37.2119**	36.5528	36.9875	36.5992
60	41.9900	**41.9489**	41.1274	41.7367	41.1754

Table 4 Computed values of PSNR by different methods for Peppers image

No. of coefficients	PSNR				
	DCT [1]	EDCT proposed	Cheby Orhto [15]	IntDCT [14]	DTT [17, 18]
4	27.4276	27.4243	**27.4943**	27.3923	27.4903
8	30.3058	30.2223	**30.5323**	30.1233	30.5229
12	32.1093	31.9407	**32.4477**	31.7864	32.4181
16	33.7391	33.5358	**33.9263**	33.4078	33.8665
20	34.6375	34.3948	**34.8624**	34.2672	34.8499
24	35.7987	35.6004	**35.7697**	35.5033	35.7583
28	36.8299	**36.7154**	36.6231	36.5238	36.6217
32	37.5226	**37.4364**	37.2841	37.2800	37.2844
36	38.3394	**38.2904**	38.0707	38.1656	38.0801
40	39.0818	**39.0313**	38.9102	38.9429	38.9292
44	39.7952	**39.7528**	39.6895	39.7285	39.7149
48	40.6388	40.5922	40.6080	40.6140	**40.6480**
52	41.5346	41.4771	41.6660	41.5882	**41.7188**
56	42.9662	42.8963	43.2999	43.0479	**43.3513**
60	45.0097	44.9467	45.6511	45.1269	**45.6826**

Table 5 Computed values of PSNR by different methods for Boat image

No of coefficients	PSNR				
	DCT [1]	EDCT proposed	Cheby Orhto [15]	IntDCT [14]	DTT [17, 18]
4	25.2678	**25.2452**	25.1880	25.0938	25.1511
8	28.1674	**28.1212**	27.8546	27.8091	27.7607
12	29.2979	**29.2460**	28.8995	28.9217	28.8393
16	31.4653	**31.3809**	31.0011	31.0098	30.9627
20	32.7149	**32.6081**	32.0951	32.1383	32.1133
24	33.0178	**32.9230**	32.5128	32.6363	32.5086
28	34.7689	**34.6619**	34.2561	34.3343	34.2482
32	36.5686	**36.4663**	36.0645	36.0997	36.0382
36	36.9249	**36.8508**	36.5143	36.5616	36.4940
40	37.2402	**37.1742**	36.9152	36.9476	36.9054
44	39.6393	**39.5612**	39.3672	39.3314	39.3135
48	40.7545	**40.6755**	40.5804	40.5302	40.5585
52	41.2549	41.1905	**41.2527**	41.1286	41.2107
56	44.1307	44.0567	44.1522	44.0212	**44.1915**
60	45.8285	45.7869	46.0071	45.7625	**46.0762**

traditional one, while EDCT is a transform with integer coefficients. Values for DCT are given for reference only. Therefore, the results of the proposed method EDCT are compared with other three similar transform matrix with integer coefficients. With 40 coefficients, the PSNR values are in the order, DCT > EDCT > IntDCT > DTT > ChebyOrtho. This trend followed up to 60 coefficients. Overall, the proposed method EDCT gave results close that of DCT and performed better than the other three methods for Lena image.

6.2 *Mandrill Image*

The SFM value for Mandrill is 36.5194, the highest among the four images. The computed values of PSNR against the number of coefficients included in the compression file are given in Table 3.

From Table 3, it is observed that with four coefficients, all the methods produced a PSNR value of 21.14, but with minute difference, having values in the order, DCT > ChebyOrtho > EDCT > DTT > IntDCT. ChebyOrtho gave the best result with four coefficients. Further, comparing with Table 3 for Lena, it can be seen that this value is far less in PSNR value. With eight coefficients up to 60, EDCT gave better results than the other three methods. Overall, EDCT performed better than the three integer-based transforms.

6.3 *Peppers Image*

Peppers image has an SFM value of 15.9643, the second lowest among the four images. The computed values of PSNR by different methods for Peppers image are shown in Table 4.

The results (Table 4) for Peppers image presented an interesting aspect. With four coefficients, all methods produced PSNR value of 27.4. The highest value is by ChebyOrhto. With coefficients between 4 and 24 included for compression, ChebyOrtho performed better than the other three. Between 28 and 44 coefficients, EDCT performed better than the other three methods, and between 48 and 60 coefficients, DTT gave the best performance. In fact, it even excelled the values given by DCT.

6.4 *Boat Image*

The estimated values of PSNR for Boat image are given in Table 5. The SFM value of this image is 19.8521 which is higher than Peppers but less than Mandrill. As in the previous image, all methods had almost the same information packed in the

Table 6 Average time taken to compress and decompress the image

Image	Execution time (in seconds)			
	EDCT-proposed	ChebyOrtho [15]	IntDCT [14]	DTT [17, 18]
Lena	0.4683	0.4859	0.4663	0.4670
Mandrill	0.4631	0.4676	0.4662	0.4850
Peppers	0.4672	0.4623	0.4665	0.4669
Boat	0.4632	0.4669	0.4641	0.4703

first four coefficients. The proposed method EDCT has the maximum PSNR value of 25.2452 among the integer coefficient-based transforms.

EDCT produced the highest values of PSNR with 4 to 48 coefficients retained for compression which is close to that of DCT. This is the region where compression is needed with high compression factor. With 52 coefficients, ChebyOrtho method gave the best PSNR values. With 56 and 60 coefficients, DTT method produced the highest values of PSNR even better than the DCT. Thus, Boat image is challenging the different compression methods. DTT seems to perform extremely well with more number of coefficients included in the compression file (low compression factor). Overall, the EDCT performed better than the other three methods in regions which are practically needed.

6.5 Computing Time

To assess the time complexity of the proposed method with that of other integer-based transform methods, we computed the average time taken to compress and decompress the given image. The estimated time is given in Table 6.

From Table 6, it is observed that all the methods take about 0.46 s for compression and decompression. The actual difference can be found when these methods are implemented in hardware. Though the computing time appears to be the same, each method produces images of different quality in terms of PSNR as given in Tables 2, 3, 4, and 5.

7 Conclusions

In this article, we have proposed a novel integer-based transform with integer coefficients, by extending the traditional DCT. It performed better than other transforms with integer coefficients: orthogonal transform (ChebyOrtho), integer DCT, and integer DTT. In general, the integer-based transforms are faster than the floating-point transform, for both hardware implementation and software implementation. Hence, the proposed extended discrete cosine transform (EDCT) will also have similar property. The proposed method gave competitive results as that of normal DCT. The proposed method performed better than the other three integer-based transforms in most regions of compression.

References

1. Salomon, D.: Data Compression, 4th edn. Springer International Edition, New Delhi (2011)
2. Burger, W., Burge, M.J.: Digital Image Processing, 2nd edn. Springer (2016)
3. Ahmed, N., Natarajan, T., Rao, R.K.: Discrete cosine transform. IEEE Trans. Comput. **C-23**, 90–93 (1974)
4. Rao, K.R., Yip, P.: Discrete Cosine Transform—Algorithms, Advantages Applications. Academic Press, San Diego (1990)
5. Wang, R.: Introduction to Orthogonal Transform. Cambridge University Press, Cambridge, UK (2007)
6. Strang, G.: The discrete cosine transform. SIAM Rev. **41**(1), 135–147 (1999)
7. Chang, T.S., Kung, C.S., Jen, C.W.: A simple processor core design for DCT/IDCT. IEEE Trans. Circuits Syst. Video Technol. **10**, 439–447 (2000)
8. Hettiarachchi, D.L.N., Davuluru, V.S.P., Balster, E.J.: Integer vs. floating-point processing on modern FPGA technology. In: 2020 10th Annual Computing and Communication Workshop and Conference (CCWC), Las Vegas, NV, USA, pp. 0606–0612 (2020).https://doi.org/10.1109/CCWC47524.2020.9031118
9. Cham, W.K., Yip, P.C.: Integer sinusoidal transforms for image processing. Int. J. Electron. **70**, 1015–1030 (1991)
10. Chen, Y.J., Oraintara, S., Tran, T.D., Amaratunga, K., Nguyen, T.Q.: Multiplierless approximation of transforms using lifting scheme and coordinate descent with adder constraint. In Proceedings of the IEEE International Conference on Acoustics, Speech, & Signal Processing, vol. 3, pp. 3136–3139 (2002)
11. Cheng, L.Z., Xu, H., Luo, Y.: Integer discrete cosine transform and its fast algorithm. Electronics Letter **37**, 64–65 (2001)
12. Liang, J., Tran, T.D.: Fast multiplierless approximations of the DCT: the lifting scheme. IEEE Trans. Signal Process. **49**, 3032–3044 (2001)
13. Plonka, G., Tasche, M.: Invertible integer DCT algorithms. Appl. Comput. Harmon. Anal. **15**, 70–88 (2003)
14. Cheung, K.M., Pollara, F., Shahshahani, M.: Integer cosine transform for image compression. https://ntrs.nasa.gov/search.jsp?R=19940025116
15. Krishnamoorthi, R.: Transform coding of monochrome images with a statistical design of experiments approach to separate noise. Pattern Recogn. Lett. **28**, 771–777 (2007)
16. Krishnamoorthi, R., Ganesh, M.: Image defect identification with orthogonal polynomial model. Asian J. Inf. Technol. **15**(8), 1389–1395 (2016)
17. Prattipati, S., Swamy, M.N.S., Meher, P.K.: A comparison of integer cosine and Tchebichef transforms for image compression using variable quantization. J. Signal Inf. Process. **6**, 203–216 (2015)
18. Sujatha, I.: Discrete Tchebichef transform and its application to image/video compression. M.S. Thesis, Concordia University, Montreal, Quebec, Canada, June 2008
19. http://sipi.usc.edu/database/database.php?volume=misc
20. Somasundaram, K., Revathy, T.S., Praveenkumar, S., Kalaiselvi, T.: A Study on the effect of truncating the discrete cosine transform (DCT) coefficients for image compression. Int. J. Comput. Sci. Eng. (IJCSE) **7**(4), 23–30 (2018)

Image Reconstruction from Geometric Moments via Cascaded Digital Filters

Mohd Fikree Hassan and Raveendran Paramesran

Abstract Geometric moments belong to the family of non-orthogonal moments that do not have the ability to reconstruct the original image directly from it. Because of this, indirect methods are used to reconstruct images from their geometric moments. In this paper, we introduce a direct reconstruction method that uses cascaded digital filters where each digital filter operates as a subtractor. The proposed method has a less intensive computational process and has a great advantage as it can deal with geometric moments computed from any coordinate format. The proposed method is tested on $N \times N$ binary and gray-scale images, and the performance of the proposed reconstruction process is compared with the existing methods. The results show that the proposed method takes less computational time and reconstruction error when compared with the two current methods.

Keywords Image reconstruction · Geometric moment · Digital filter

1 Introduction

Moment functions are scalar quantities that contain significant features and characteristics of an image. The concept of moment was first introduced and applied in image pattern recognition by Hu [1]. Based on the theory of algebraic invariants, he derived sets of moment invariants with respect to translation, scaling, and rotation [2, 3]. Hu's moment or also known as the geometric moment is one of the non-orthogonal moments. Its invariants are projections of the image intensity function, $f(n, m)$ onto the monomial $n^p m^q$.

Another type of non-orthogonal moment is the complex moment that carries the same amount of information as the geometric moment. Reconstruction of image from

M. F. Hassan · R. Paramesran (✉)
Institute of Computer Science and Digital Innovation, UCSI University, Kuala Lumpur, Malaysia
e-mail: Raveendran@ucsiuniversity.edu.my

M. F. Hassan
e-mail: fikree@ucsiuniversity.edu.my

P. Balasubramaniam et al. (eds.), *Mathematical Modelling and Computational Intelligence Techniques*, Springer Proceedings in Mathematics & Statistics 376,
https://doi.org/10.1007/978-981-16-6018-4_11

non-orthogonal moments is quite difficult [4]. It is due to the non-orthogonality property that produces unnecessary information redundancy in the extracted moments [2, 5].

To solve this shortcoming, orthogonal moments were introduced [2, 5–8]. Although orthogonal moments hold an advantage over non-orthogonal moments due to their orthogonality property, geometric moments are commonly used in image reconstruction and have proven to be the most efficient tool for image analysis [9]. It is because geometric moments can illustrate the characteristics and features of the image. These abilities offer a practical advantage as compared to orthogonal moments. Moreover, the redundancy produced from the non-orthogonality property is useful when dealing with noisy images. In fact, geometric moments have an explicit geometrical and statistical significance defined in statistics which is important in pattern recognition and image processing [10].

Some of the applications that use geometric moments to reconstruct the original image are watermarking [11], fingerprint recognition [12], and medical imaging [13]. However, non-orthogonal moments such as geometric moments do not have direct reconstruction ability [2, 14]. Using the relationship between geometric moments and the family of orthogonal moments such as Zernike, Tchebichef, and Krawchouk moments, an indirect approach to reconstruct the original image can be performed [6].

Ghorbel et al. [14] used the concept of characteristic function proposed in [2] to reconstruct the original image from its geometric moments. In this method, they computed the Discrete Fourier Transform (DFT) of an image from its geometric moments and reconstructed the original image using the Inverse DFT (IDFT). However, when the order of the geometric moments is the same as the size of the image, the computation produces errors [3]. Moreover, IDFT requires intensive computational time, and the approximation in the IDFT produces reconstruction errors [15].

A direct method to reconstruct images from its computed geometric moments was proposed by Flusser et al. in [3]. This method successfully reconstructed the exact image up to 11×11 pixels. However, this method lost its precision for larger images and was unable to reconstruct the exact image. Barmak et al. [15] introduced a novel method of image reconstruction from its geometric moments using Stirling numbers of the first and second kinds. By using the full set of the geometric moments of an image, the original image was reconstructed. However, this method is computationally intensive and has a long computational time.

In this paper, we proposed an image reconstruction method from its geometric moments via cascaded digital filters. The proposed method has been implemented in our recent work on image deblurring [16]. However, we have not included a detailed explanation of the proposed method. Therefore, this paper describes comprehensive details of the proposed method. The proposed method consists of two significant elements; the inverse coefficient matrix and cascaded digital filters as subtractors to restore the original image. It has a less intensive computational process than the other existing methods and can be applied over the intervals of $[-1,1]$ and $[0, N-1]$. The rest of the paper is organized as follows. The proposed method is introduced in Sect. 2. In Sect. 3, experimental results are shown and discussed. Finally, Sect. 4 concludes the paper.

2 Proposed Method

Digital images can be represented by an array of numbers or matrix of the size $M \times N$. The elements in the matrix are called pixel's image intensity function, $f(n, m)$. With these values, geometric moments are computed directly using Eq. (1).

$$M_{pq} = \sum_{n=1}^{N}\sum_{m=1}^{M} f(n, m) n^p m^q \tag{1}$$

These geometric moments can also be computed in a different coordinates ranges. The coordinate ranges used in this paper are defined within interval $[-1, 1]$ and $[0, N - 1]$. Using the same concept as the symmetric transformation in [17], we introduce a normalized image coordinate within the interval of $[-1, 1]$ as shown in Fig. 1b.

The indices of the normalized transformation, n' and m' are given as

$$n' = -1 + \frac{2n - 1}{N} \tag{2}$$

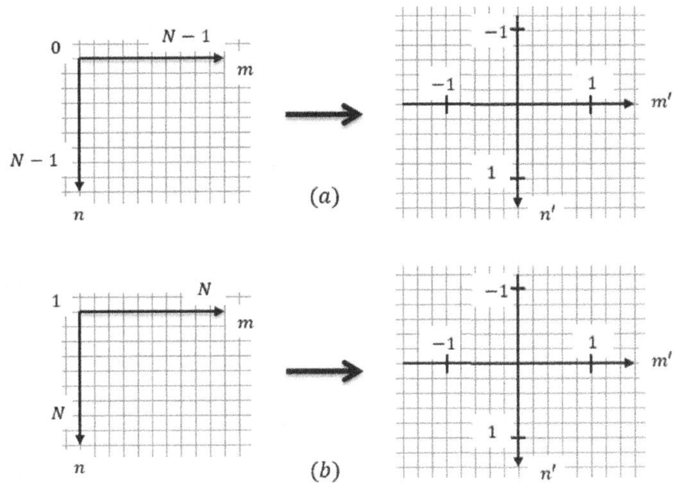

Fig. 1 **a** Symmetric transformation. **b** Normalized transformation

$$m' = -1 + \frac{2m - 1}{N} \quad (3)$$

where $n, m = 1, 2, 3, \ldots, N$. Thus, the relationship between two-dimensional normalized geometric moments and its original geometric moments is represented by the expression

$$M_{pq} = \left(\frac{N}{2}\right)^{p+q} \sum_{r=0}^{p} \sum_{s=0}^{q} \binom{p}{r} \left(\frac{N+1}{N}\right)^{p-r} \binom{q}{s} \left(\frac{N+1}{N}\right)^{q-s} M'_{rs} \quad (4)$$

where M'_{rs} is the two-dimensional normalized geometric moments that can be computed by applying Eqs. (2) and (3) in Eq. (1).

The proposed method consists of two stages; the inverse coefficient matrix and the subtractor circuit. Take note that the proposed inverse method focus on reconstruction of a square image with the size of $N \times N$. Figure 2 shows the complete process flow of the reconstruction of image from geometric moments.

Also, the reconstruction of color image can be performed by converting the color image to a gray-scale image [18]. After the reconstruction process, the grayscale image is converted back to a color image.

2.1 *Inverse Coefficient Matrix*

The first stage uses the inverse matrix of the feedback digital filter's coefficient matrix derived in [19, 20]. The main objective of this process is to generate the outputs of

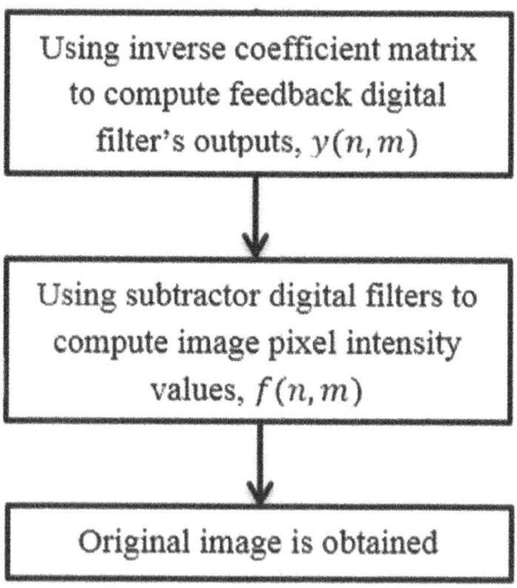

Fig. 2 Process flow of the proposed method

the digital filter from its respective geometric moments. For one-dimensional case, the outputs of the digital filter can be expressed in matrix notation as follows

$$\mathbf{Y} = \mathbf{C}^{-1}\mathbf{M} \tag{5}$$

where **Y** represents the column matrix of the digital filter outputs and **M** represents the column matrix of the geometric moments. The inverse matrix $\mathbf{C}^{-1}$ can be computed from the coefficient matrix of the feed-forward or feedback digital filters.

Due to the improvement provided by the feedback digital filters, the proposed method uses the inverse coefficient matrix of the feedback digital filters. The inverse matrix $\mathbf{C}^{-1}$ of 4-input sequence is given as

$$\mathbf{C}^{-1} = E_{pr} = \begin{bmatrix} 1 & 0 & 0 & 0 \\ 0 & 1 & 0 & 0 \\ 0 & \frac{1}{2} & \frac{1}{2} & 0 \\ 0 & \frac{1}{3} & \frac{1}{2} & \frac{1}{6} \end{bmatrix}$$

Thus, one-dimensional digital filter output, y_p, can be computed using the following equation

$$y_p = \sum_{r=0}^{p} E_{pr} M_r \tag{6}$$

where M_r is the geometric moments and E_{pr} is the inverse coefficient matrix. Extending Eq. (6) to two-dimensional case yields the equation

$$y_{pq} = \sum_{r=0}^{p} \sum_{s=0}^{q} E_{pr} E_{qs} M_{rs} \tag{7}$$

where y_{pq} is the two-dimensional digital filter outputs, M_{rs} is the two-dimensional geometric moments while E_{pr} and E_{qs} are the inverse coefficient matrix. The digital filter outputs obtained are inserted into the subtractor circuit for the second stage.

2.2 *Subtractor Circuit*

In the second stage, the subtractor circuit uses the digital filter outputs to reconstruct the original image. For one-dimensional (1D) case with N-th input sequence, the subtractor circuit is formed by cascading $N \times (N-1)$ subtractor digital filters in two directions; horizontal, S_{fh} and vertical, S_{fv} as shown in Fig. 3. The inputs of the subtractor circuit are the digital filter outputs, y_{N-1}, followed by y_{N-2},y_{N-3} until y_0, and the outputs of the subtractor circuit are the image pixel intensity values, x_0, x_1 until x_{N-1}.

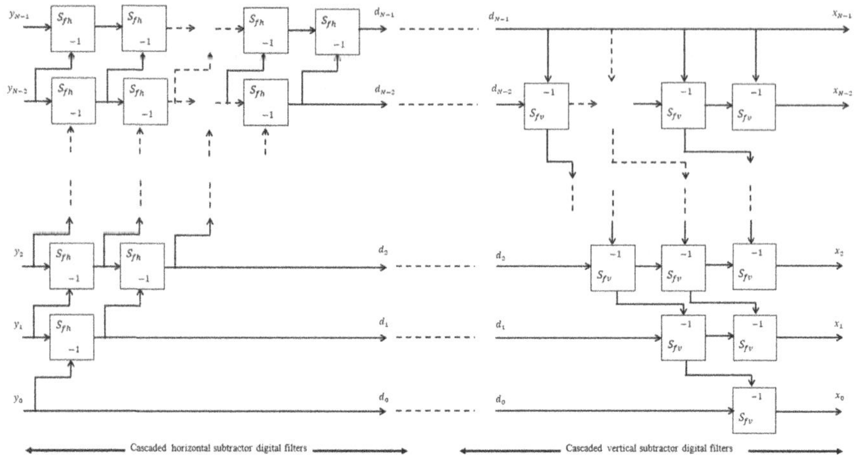

Fig. 3 Block diagram of the one-dimensional (1D) subtractor circuit

From the figure, it can be observed that there are two types of cascaded digital filters; cascaded horizontal subtractor digital filters and cascaded vertical subtractor digital filters. For cascaded horizontal subtractor digital filters, the number of horizontal subtractor digital filters, S_{fh} in each row depends on the input sequence, i.e., for y_0 requires 0 S_{fh}, for y_1 requires 1 S_{fh} and thus, y_{N-1} requires $(N-1)$ S_{fh}.

For cascaded vertical subtractor digital filters, except for the top row which requires 0 vertical subtractor digital filters, S_{fv}, the number of vertical subtractor digital filters, S_{fv} in each row depends on the output sequence, i.e., for x_0 requires 1 S_{fv}, for x_1 requires 2 S_{fv} and, thus, x_{N-2} requires $(N-2+1)$ S_{fv}.

Horizontal Cascaded Subtractor Digital Filters, S_{fh} The basic building block of a horizontal subtractor digital filter is shown in Fig. 4.

The transfer function H_{h0} in the Z-transform domain of the filter structure in Fig. 4 is expressed as

$$H_{h0}(z) = 1 - z^{-1} \tag{8}$$

Fig. 4 Basic building block of horizontal subtractor digital filter, S_{fh}

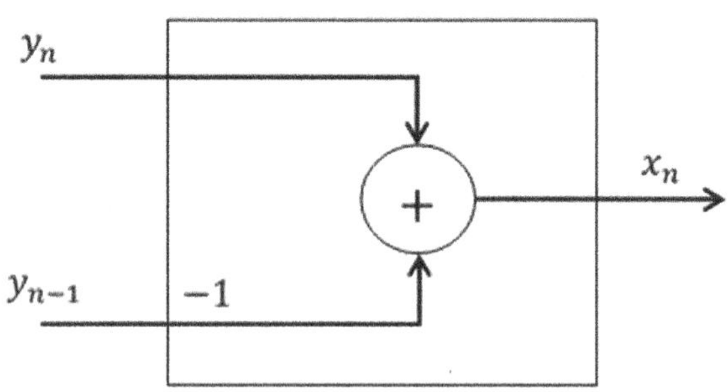

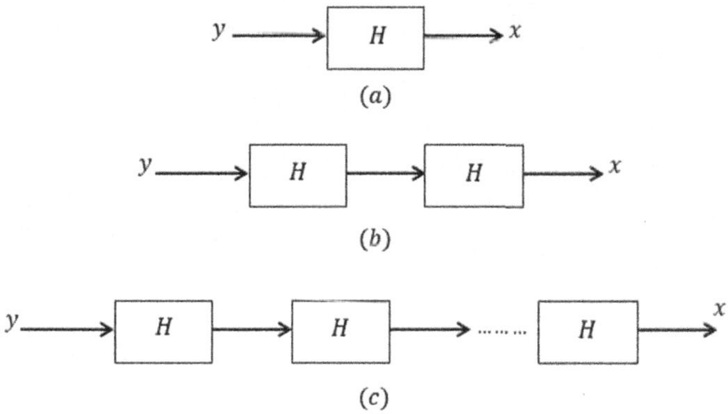

Fig. 5 Discrete-time LTI system with input y and output x

The corresponding impulse response, $h_{h0}(n)$ of Eq. (8) can be expressed as

$$h_{h0}(n) = \delta(n) - \delta(n-1) \quad (9)$$

where $\delta(n)$ is the unit impulse function.

The transfer function of the cascaded horizontal subtractor digital filters of N input sequence can be expressed generally as

$$H_{hN}(z) = (1 - z^{-1})^{N+1} \quad (10)$$

The transfer function H_{hN} is equivalent to the $(M + 1)$ order difference of a discrete-time linear time invariant (LTI) system where $M = 0, 1, 2, 3, \ldots, N - 1$. As shown in Fig. 5a, Eq. (8) can be represented by the first order difference of a discrete time signal $x(n)$ as

$$x(n) = y(n) - y(n-1) \quad (11)$$

The transfer function of two cascaded horizontal subtractor digital filters is as follow

$$H_{h1}(z) = (1 - z^{-1})^2 = 1 - 2z^{-1} + z^{-2} \quad (12)$$

where it is equivalent to the second order difference obtained by taking the first order difference twice as shown in Fig. 5b,

$$x(n) = y(n) - 2y(n-1) + y(n-2) \quad (13)$$

As shown in Fig. 5c, the same process can be applied to obtain higher order difference equations based on their respective transfer function.

Table 1 State table of cascaded horizontal subtractor digital filters

$y_{N-1}(t_0)$	$y_{N-2}(t_0)$ ...	$y_2(t_0)$		$y_1(t_0)$	$y_0(t_0) = d_0$
$y_{N-1}(t_1)$	$y_{N-2}(t_1)$ ...	$y_2(t_1)$	$y_1(t_1) = d_1$		
$y_{N-1}(t_2)$	$y_{N-2}(t_2)$ ... $y_2(t_2) = d_2$				
:	:	:			
$y_{N-1}(t_{N-2})$	d_{N-2}				
d_{N-1}					

To start the second stage, the inputs are inserted simultaneously into the horizontal subtractor digital filters. As represented by the state table in Table 1, the horizontal subtraction process is performed row by row until all the diagonal values are computed. The diagonal values are denoted as d_0 to d_{N-1} in Fig. 3. It can also be observed from Table 1 that y_0 is equal to the first diagonal value, d_0.

Relating the diagonal values with the $(M + 1)$ order difference equations discussed above, the diagonal values for 4-input sequence can be expressed as

$$d_0 = y_0 \tag{14}$$

$$d_1 = y_1 - y_0 \tag{15}$$

$$d_2 = y_2 - 2y_1 + y_0 \tag{16}$$

$$d_3 = y_3 - 3y_2 + 3y_1 - y_0 \tag{17}$$

It can be noted that the coefficients of y_n in the above equations come from Pascal Triangle. Thus, the diagonal values of the 1D horizontal subtractor digital filters can be computed using the general equation

$$d_n = \sum_{r=0}^{n} (-1)^r \binom{n}{r} y_{n-r} \quad \text{for } 0 \leq n \leq N-1 \tag{18}$$

By computing the diagonal values using Eq. (18), the subtractor circuit is simplified, and the number of subtractor digital filters can be reduced to $(N \times (N-1))/2$ which mainly consists of vertical subtractor digital filters.

Vertical Cascaded Subtractor Digital Filters, S_{fv} The basic building block of vertical subtractor digital filter is shown in Fig. 6. The transfer function in the Z-transform domain for the vertical subtractor digital filter is given as

$$H_{v0}(z) = 1 - z \tag{19}$$

The corresponding impulse response, $h_{h0}(n)$ of Eq. (19) can be expressed as

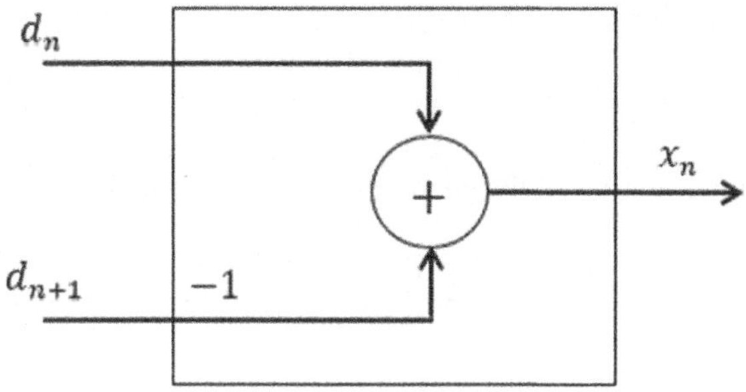

Fig. 6 Basic building block of vertical subtractor digital filter, S_{fv}

Table 2 State table of cascaded vertical subtractor digital filters

				$d_0(t_0)$	x_0
			$d_1(t_0)$	$d_0(t_1)$	x_1
		$d_2(t_0)$	$d_1(t_1)$	$d_0(t_2)$	x_2
	:	:	:	:	:
	$d_{N-2}(t_0)$ ...	$d_2(t_{N-4})$	$d_1(t_{N-3})$	$d_0(t_{N-2})$	x_{N-2}
$d_{N-1}(t_0)$	$d_{N-1}(t_1)$ ...	$d_{N-1}(t_{N-3})$	$d_{N-1}(t_{N-2})$	$d_{N-1}(t_{N-1})$	x_{N-1}

$$h_{h0}(n) = \delta(n) - \delta(n+1) \tag{20}$$

where $\delta(n)$ is the unit impulse function.

The transfer function of the cascaded vertical subtractor digital filters of N input sequence can be expressed generally as

$$H_{vn}(z) = (1-z)^{n+1} \quad \text{for } 0 \leq n < N-1 \tag{21}$$

The vertical subtraction process starts when all the diagonal values obtained from the horizontal cascaded subtractor digital filters are computed. All the diagonal values are inserted into the vertical subtractor digital filters as shown in Fig. 3. Here, the subtraction process is performed until all the image pixel intensity values are computed. The image pixel intensity values are denoted as x_0 to x_{N-1}. The last image pixel intensity value, x_{N-1} is the same as the last diagonal value, d_{N-1}. Moreover, for N input sequence, to compute x_n of the image pixel intensity value where $0 \leq n < N-1$, we need $n+1$ cascaded vertical subtractor digital filters as shown in Fig. 3.

As represented by the state table in Table 2, the vertical subtraction process is perform column by column. Thus, to obtained the correct results, the values in most left column must be computed first in order to start the vertical subtraction process for the next column. Moreover, the values in each column are generated at different time intervals.

Using all the information thus far, and the fact that each diagonal value is related to their corresponding image pixel intensity, i.e., d_2 related to x_2; by expanding Eq. (21) according to the number of cascaded vertical subtractor digital filters and relate it with difference equation between the diagonal value, d_{N-1} and the image pixel intensity, x_{N-1}, the general equation to compute each image pixel intensity is given as follow

$$x_n = \sum_{r=0}^{n} (-1)^r \binom{n}{r} d_n(t_r) \quad \text{for } 0 \leq n < N - 1 \tag{22}$$

where $d_n(t_r)$ is the diagonal value at the time interval t_r.

2-dimensional Subtractor Circuit For image reconstruction, the two-dimensional (2D) subtractor circuit is formed by arranging multiple 1D subtractor circuits horizontally and vertically. The number of 1D subtractor circuit required are depending on the size $N \times N$ of the square image used, i.e., a 4×4 image reconstruction requires 4 vertically arranged 1D subtractor circuits, S_v and 4 horizontally arranged 1D subtractor circuits, S_h. Both S_v and S_h in this case are similar to the 1D subtractor circuit shown in Fig. 3.

In general, the number of subtractor digital filters for 2D subtractor circuit is represented by $2N^2(N-1)$, and the block diagram is shown in Fig. 7. The number of subtractor digital filters can be reduced to $N^2(N-1)$ by applying Eq. (18) to calculate the diagonal values for each 1D subtractor circuit.

3 Experimental Results and Discussion

In this section, our proposed method was validated and evaluated through a series of experiments that were carried out on a system running Windows 10 64-bit with Matlab version R2103a with an AMD Ryzen 3 3200U CPU operating at 2.60GHz with a physical memory of 8.00GB. The reconstruction process was conducted in the first experiment where three types of images were reconstructed; artificial, binary [14, 15] and gray-scale images [21]. The reconstruction error and the CPU elapsed time of the reconstruction of images were evaluated in the second experiment. In addition, the performance of the proposed method was compared with the DFT method [14] and the Stirling method [15].

3.1 Image Reconstruction from Geometric Moments

In this experiment, first, we reconstructed an artificial test image. Then, binary and gray-scale images were used in the reconstructing process using the proposed method.

Artificial test image An artificial test image with the following image pixel intensity function $f(m, n)$ was considered.

$$f(m,n) = \begin{bmatrix} 27 & 26 & 25 & 24 \\ 18 & 17 & 16 & 15 \\ 20 & 21 & 22 & 23 \\ 14 & 13 & 12 & 11 \end{bmatrix}$$

The geometric moments of the artificial test images are given as followed. Using Eq. (7), the digital filter outputs $y(m,n)$ were computed.

$$y(m,n) = \begin{bmatrix} 304 & 750 & 1490 & 2597 \\ 692 & 1710 & 3400 & 5929 \\ 1316 & 3250 & 6460 & 11263 \\ 2226 & 5490 & 10905 & 19005 \end{bmatrix}$$

Using the 2D subtractor circuit as shown in Fig. 7, the first row values of the digital filter outputs matrix were inserted into the first 1D vertical subtractor digital circuit $S_{v,0}$, the second row values of the digital filter outputs matrix were inserted into the second 1D vertical subtractor digital circuit $S_{v,1}$ and so on. All these values underwent the subtraction process and the resulting values were inserted into the horizontal subtractor circuit.

The first output values from each 1D vertical subtractor digital circuit were inserted into the first 1D horizontal subtractor digital filter $S_{h,0}$, the second output values into the second 1D horizontal subtractor digital filter and so on. The final values obtained from the subtraction process were the recovered artificial test image pixel intensity, $\hat{f}(m,n)$ as below.

$$\hat{f}(m,n) = \begin{bmatrix} 27 & 26 & 25 & 24 \\ 18 & 17 & 16 & 15 \\ 20 & 21 & 22 & 23 \\ 14 & 13 & 12 & 11 \end{bmatrix}$$

It can be observed that the recovered artificial test image pixel intensity values, $\hat{f}(m,n)$ were the same with the original pixel intensity values of the artificial image, $f(m,n)$.

Real test image In this experiment, we performed the reconstruction process to binary [14, 15] and gray-scale images [21] as shown in Figs. 8 and 9. The reconstruction process was performed to the complete set of the respective geometric moments. The geometric moments of the images were computed and used as the input to the reconstruction process. Figures 8b and 9b are the reconstructed images of Figs. 8a and 9a, respectively.

3.2 Reconstruction Error and CPU Elapsed Time

This experiment was performed on binary [14, 15] and gray-scale images [21] shown in Figs. 8 and 9 with the image size of 32 × 32 pixels, 128 × 128 pixels and 256

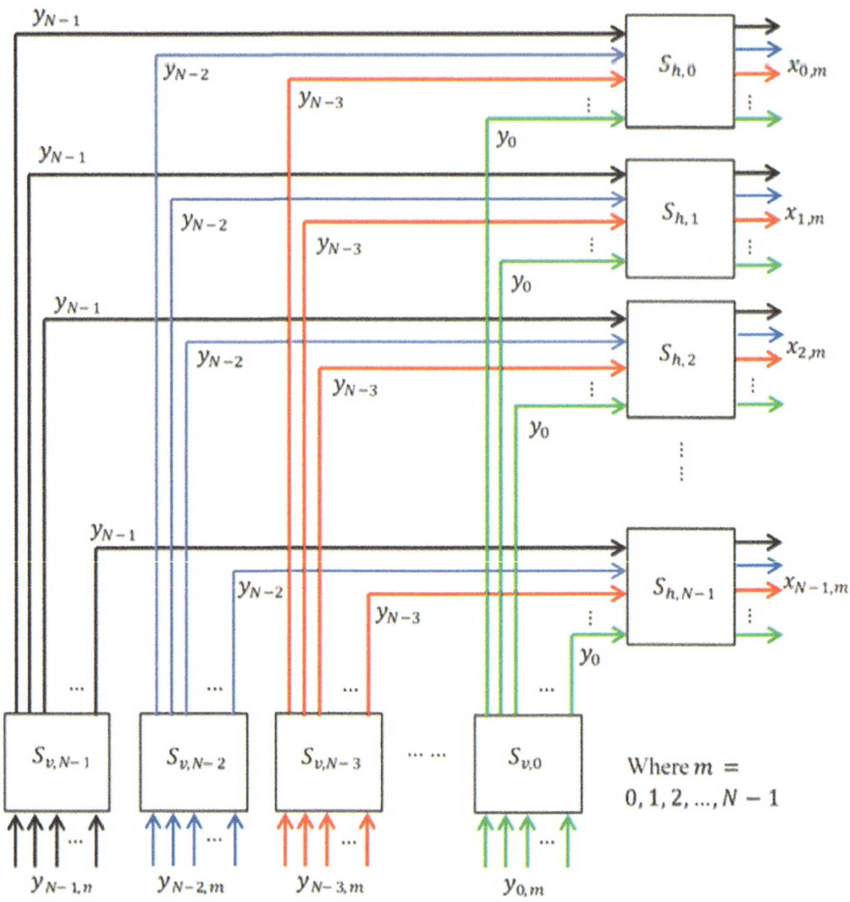

Fig. 7 Block diagram of the 2-dimensional (2D) subtractor circuit for $N \times N$ image

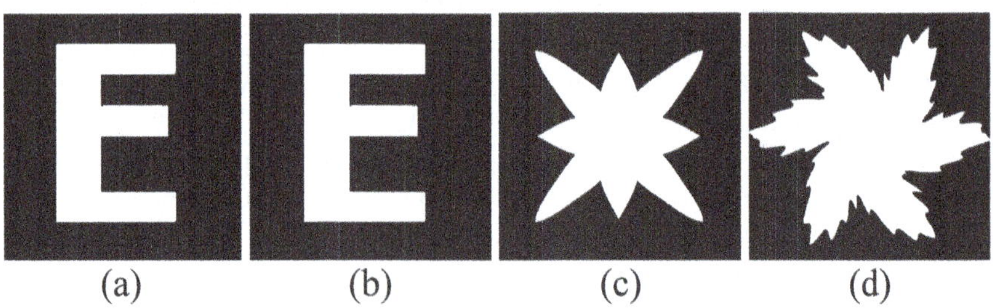

Fig. 8 Binary images **a** E; **b** reconstructed E image; **c** star; **d** spiral

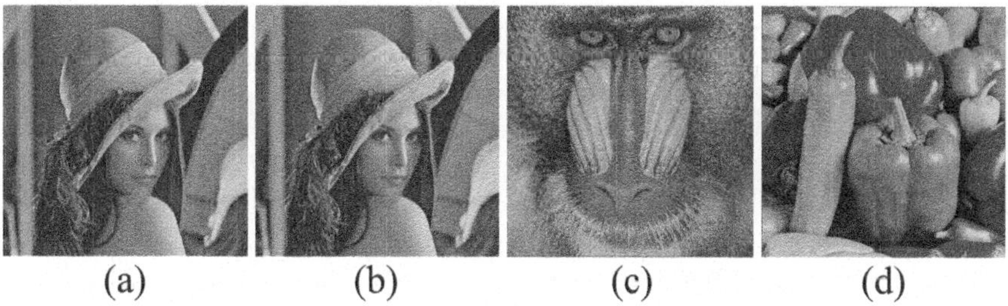

Fig. 9 Gray-scale images **a** Lena; **b** reconstructed Lena image; **c** baboon; **d** pepper

Table 3 Reconstruction error and CPU elapsed time of the reconstructed binary and gray-scale images

Image	Size: 32 × 32		Size: 128 × 128		Size: 256 × 256	
	Reconstruction error, ϵ	CPU elapsed time	Reconstruction error, ϵ	CPU elapsed time	Reconstruction error, ϵ	CPU elapsed time
E	0.0	0.0784	0.0	1.2546	0.0	5.0180
Star	0.0	0.0780	0.0	1.2483	0.0	4.9970
Spiral	0.0	0.0722	0.0	1.1558	0.0	4.6213
Lena	0.0	0.0848	0.0	1.3562	0.0	5.4275
Baboon	0.0	0.0902	0.0	1.4332	0.0	5.7730
Pepper	0.0	0.0910	0.0	1.4558	0.0	5.8310

× 256 pixels. To measure the reconstruction error of our proposed method, we used statistical normalization image reconstruction error (SNIRE) [22]. It is given as

$$\epsilon = \sum_{m=0}^{N-1} \sum_{n=0}^{N-1} \left\{ \frac{\left[f(m,n) - \hat{f}(m,n) \right]^2}{[f(m,n)]^2} \right\} \tag{23}$$

where $f(m,n)$ and $\hat{f}(m,n)$ are the pixel intensity of the original and reconstructed images, respectively. In addition, the CPU elapsed time was also computed. Each image underwent the reconstruction process repeatedly for five times, and the average time was computed.

The experiment results are shown in Table 3. The CPU elapsed times in Table 3 includes the time taken to compute the geometric moments from the original images and all the processes of the proposed method. It can be observed that the CPU elapsed times for reconstruction of the binary, and gray-scale images are almost similar for every group of image size. It is because the proposed method has less intensive computational process, and it depends on the size of the image matrix and not the element values of the image matrix.

We also compared the performance of the proposed method to two existing methods; the Stirling method [15] and the DFT method [14]. The results in Table 4 show

Table 4 Comparison between the proposed method and two existing methods

Image and image size	Proposed method		Stirling method [15]		DFT method [14]	
	Reconstruction error, ϵ	CPU elapsed time	Reconstruction error, ϵ	CPU elapsed time	Reconstruction error, ϵ	CPU elapsed time
E	0.0	0.0778	0.0	4.2386	0.8743	10.2094
32×32						
Lena	0.0	21.7092	0.0	23.4051	0.7429	48.4988
512×512						

that the reconstruction error of the proposed method is better than the DFT method. Both the proposed method and the Stirling methods achieved zero reconstruction error. However, in terms of the CPU elapsed time for reconstruction, the proposed method performed better.

The DFT method [14] requires higher order moments to compute the image pixel intensity values. DFT method only provides an approximation instead of an exact values of the image pixel intensity. Meanwhile, the Stirling numbers of the first and second kinds used in [15] increase as the size of image increases. Therefore, in this experiment, the Stirling method requires six computational loops to compute the image pixel intensity values which increased the computational complexity and time.

Using the proposed method, the digital filter outputs were computed directly using the geometric moments and the inverse coefficient matrix. Moreover, subtraction operation used in the second stage reduced larger values to smaller values. Thus, the computational complexity is reduced since the proposed method uses direct and simple computation as compared to the two existing methods.

Although the proposed method provides simpler and faster computational process, precision of the computed digital filter output values plays an essential role for the second process of the proposed method. The digital filter output values computed from Eq. (7) must produce the exact digital filter outputs. If the values obtained are not the exact digital filter outputs, the reconstruction process produces reconstruction error.

4 Conclusion

In this paper, a direct computational method has been proposed to reconstruct images from geometric moments via cascaded digital filters. The proposed method consists of two stages. The first stage uses the inverse coefficient matrix of the feedback digital filter to compute the digital filter outputs from its respective geometric moments. In the second stage, the digital filter outputs are inserted into the cascaded digital filters that act as subtractor circuit, and the image pixel intensity values are computed.

These processes are direct and has less computational complexity as compared to the existing methods. To evaluate the proposed method, binary and gray-scale images were reconstructed and compared with two existing methods. The results showed that the proposed method performed better in terms of the computational time and reconstruction error.

References

1. Hu, M.K.: Visual pattern recognition by moment invariants. IRE Trans. Inf. Theory **8**, 179–187 (1962)
2. Teague, M.R.: Image analysis via the general theory of moments. J. Opt. Soc. Am. **70**, 920–930 (1980)
3. Flusser, J., Zitova, B., Suk, T.: Moments and Moment Invariants in Pattern Recognition. Wiley Publishing (2009)
4. Teh, C.-H., Chin, R.T.: On image analysis by the methods of moment. IEEE Trans. Pattern Anal. Mach. Intell. **10**, 496–513 (1988)
5. Hmimid, A., Sayyouri, M., Hassan.: Image classification using novel set of Charlier moment invariants. WSEAS Trans. Sig. Process. **10**, 156–167 (2014)
6. Mukundan, R., Ong, S.H., Lee, P.A.: Dicrete vs. continuous orthogonal moments for image analysis. In: International Conference on Imaging Science, Systems, and Technology, pp. 23–29 (2001)
7. Mukundan, R., Ong, S.H., Lee, P.A.: Image analysis by Tchebichef moments. IEEE Trans. Image Process. **10**, 1357–1364 (2001)
8. Yap, P.T., Paramesran, R., Ong, S.-H.: Image analysis by Krawtchouk moments. IEEE Trans. Image Process. **12**, 1367–1377 (2003)
9. Bimbo, A.D.: Visual Information Retrieval. Morgan Kaufmann (1999)
10. Shen, J., Wang, P. S.-P., Zhang, T.: Multispectral Image Processing and Pattern Recognition. World Scientific (2001)
11. Li, Z., Kwong, S., Wei, G.: Geometric moment in image watermarking. Proc. Int. Symp. Circ. Syst. **2**, 932–935 (2003)
12. Yang, J., Xiong, N., Vasilakos, A.V., Fang, Z., Park, D., Xu, X., Yoon, S., Xie, S., Yang, Y.: A fingerprint recognition scheme based on assembling invariant moments for cloud computing communications. IEEE Syst. J. **5**(4), 574–583 (2011)
13. Bhagat, A., Atique, M.: Web based medical image retrieval system using fuzzy connectedness image segmentation and geometric moments. Int. Conf. Comput. Sci. Comput. Intell. **1**, 208–214 (2014)
14. Ghorbel, F., Derrode, S.; Dhahbi, S., Mezhoud, R.: Reconstructing with geometric moments. In: International Conference on Machine Intelligence (ACIDCA-ICMI'05) (2005)
15. Honarvar, B., Paramesran, R., Lim, C.-L.: Image reconstruction from a complete set of geometric and complex moments. Sig. Process. **98**, 224–232 (2014)
16. Kumar, A., Hassan, M.F., Paramesran, R.: Learning based restoration of Gaussian blurred images using weighted geometric moments and cascaded digital filters. Appl. Soft Comput. **62**, 124–138 (2018)
17. Wee, C.-Y., Paramesran, R., Mukundan, R.: Fast computation of geometric moments using a symmetric kernel. Pattern Recogn. **41**, 2369–2380 (2008)
18. Sowmya, V., Govind, D., Soman, K.P.: Significance of incorporating chrominance information for effective color-to-grayscale image conversion. Sig. Video Image Process. **11**, 129–136 (2017)
19. Al-Rawi, M.: Fast Zernike moments. J. Real-Time Image Proc. **3**, 89–96 (2008)

20. Wong, W.H., Siu, W.C.: Improved digital filter structure for fast moments computation. IEEE Proc. Vis. Image Sig. Process. **146**, 73–79 (1999)
21. USC-SIPI Image Database. http://sipi.usc.edu/database/ (2016)
22. Sheng, Y., Shen, L.: Orthogonal Fourier-Mellin moments for invariant pattern recognition. J. Opt. Soc. Am. **11**, 1748–1757 (1994)

Background Preserved and Feature-Oriented Contrast Improvement Using Weighted Cumulative Distribution Function for Digital Mammograms

Senguttuvan Dhamodharan and Shanmugavadivu Pichai

Abstract Digital mammography is an inevitable source for the early detection of breast cancer. The limitations of this imaging modality tend to impede the contrast and brightness of digital mammograms, which may adversely affect the accuracy of breast cancer diagnosis. Many of the existing contrast enhancement methods pose the challenges of over-transformation of background, noise amplification, and lack of details preservation. The proposed technique named Background Preserved and Feature-Oriented Contrast Improvement (BPFO-CI) using Weighted Cumulative Distribution Function (WCF) for digital mammograms aims to solve these limitations of contrast enhancement techniques. The BPFO-CI uses a threshold to control the background-over-transformation, and further, it divides the remaining dynamic gray levels of the foreground into two symmetric regions. The first region is partitioned by its mean gray level, and the second region is partitioned by its median, generating four distinct ranges of gray levels. The algorithm further computes the weighted cumulative distribution function using a local intensity adjustment factor which is the weight of each range that controls over-transformation of each partitioned range of intensities. Finally, the original intensities of the input mammogram are mapped with the computed gray levels. The performance measures of BPFO-CI were recorded as Mean Brightness Error (MBE), Structural SIMilarity index (SSIM), and Peak Signal Noise Ratio (PSNR) by experimenting on 322 mini-MIAS mammograms. The quantitative measures of this proposed technique are confirmed to be better than the existing methods. The qualitative measure, namely human visual perception also endorses its merit.

Keywords Digital mammogram · Breast cancer diagnosis · Contrast enhancement · Adaptive histogram equalization · Controlled contrast improvement · Histogram partitioning

---

S. Dhamodharan · S. Pichai (✉)
Department of Computer Science and Applications, Gandhigram Rural Institute (Deemed to be University), Gandhigram, Dindigul, Tamil Nadu 624302, India

P. Balasubramaniam et al. (eds.), *Mathematical Modelling and Computational Intelligence Techniques*, Springer Proceedings in Mathematics & Statistics 376,
https://doi.org/10.1007/978-981-16-6018-4_12

1 Introduction

Across the world, 71% of death is due to the Non-Communicable Diseases (NCD). Cancer is one of the NCD that causes 9% of deaths among 63% of NCD victims [1]. Breast cancer claims the lives of 1 in 29 females in India [2]. Early detection of breast cancer is confirmed to increase the chances of survival of the victims [3, 4]. Mammograms that visualize the breast abnormalities play a vital role in breast cancer screening either by human visual perception, or by computer vision is in vogue[5, 6].

Automatic screening of mammograms for breast cancer diagnosis assumes paramount importance in assisting the physicians/radiologists to predict the cancer abnormalities with higher accuracy [3–5]. The low contrast/brightness of mammograms puzzles the automatic detection of abnormalities. It is a well-known fact that such low-contrast images affect the performance accuracy of the image processing elements such as edge/boundary detection, segmentation, and feature extraction. Any manual or automatic diagnosis of medical images is also misled by the poor quality medical images. Computer-Aided Diagnosis (CAD) is confirmed to be efficient in cancer detection and prediction [1, 5, 6]. The contrast enhancement algorithms play an important role in adjusting the contrast of a given image that pays-off in the subsequent steps of image processing such as image segmentation and classification. Hence, the development of efficient contrast enhancement algorithms is a valuable research contribution that directly finds application during acquisition or there-after [7–11].

The review of the literature highlights that the classical Histogram Equalization (HE) and its variants are largely used for the contrast enhancement of the images. The HE technique primarily computes the frequency of the intensities of an image, within the dynamic range of an input image. The Probability Density Function (PDF) is computed for each intensity, based on which the Cumulative Distribution Function (CDF) is computed. Further, the histogram of the entire image is redistributed to obtain better contrast, by means of intensity mapping. Despite the computational advantages of HE, it poses major disadvantages such as magnification of noise, imbalance enhancement among the foreground and background of the image [12]. Moreover, HE is unable to preserve the mean brightness of the input image in the enhanced image, which indicates its poor edge and details preservation ability.

1.1 HE Partition-Based Methods

To overcome the demerits of HE, researchers have been developing a spectrum of novel approaches since three decades. Histogram partition-based methods (Brightness Preserving Bi-Histogram Equalization (BBHE) and Dualistic Sub-Image Histogram Equalization (DSIHE)) tend to improve the contrast and preserving brightness [13]. Some of its variants, Minimum Mean Brightness Error Bi-HE method

(MMBEBHE), Recursive Mean-Separate HE method (RMSHE), and Mean Brightness Preserving Histogram Equalization (MBPHE) attempted to address the ill-effects of its predecessors, with a tradeoff on contrast improvement [7–9, 12, 14–17]. These methods typically use the statistical measures viz., mean and median to divide the input histogram either recursively or non-recursively for contrast adjustment. Though these methods have exhibited better contrast/brightness enhancement than HE and its counterparts, their ability on mean brightness preservation and suppression of noise was minimum. A variant of DSIHE is modified-DSIHE which improves the contrast of fatty type mammogram images with better brightness preservation. However, it fails to improve the contrast of the other types of features in mammogram images [18].

1.2 Adaptive Histogram Equalization (AHE) and Its Variants

Another variant of the HE is Adaptive Histogram Equalization (AHE). It stretches the contrast within the range of the image mask to improve the contrast, but it fails to preserve the mean brightness of an image than the histogram partition-based methods. Hence, contrast limited AHE and its variants attempt to clip the excess higher frequency gray levels and then compute intensity transformation/equalization [19–22]. These methods control the over-enhancement of the background with better brightness preservation. But, their performance on mammograms was not appreciable due to the principles of these techniques reduce the tissue gray levels or cloud the tissues, as confirmed by their assessment with SSIM [23, 24].

1.3 Unsharp Masking (UM)-Based Methods

Some of the researchers have used Unsharp Masking (UM) method and its variants for contrast adjustment [25, 26]. These algorithms have shown promising results in controlling the noise amplification during intensity transformation. This mechanism may alter the gray levels of regions depicting suspicious tumors, which may reduce the accuracy of the subsequent operations such as segmentation, feature identification, and extraction.

Review of literature on mammogram enhancement confirms that researchers have used CLAHE and the variants of HE in traditional computing approaches, machine learning, and deep learning models [27–40]. These techniques are applied either adaptively or non-adaptively as required. The suitability of the CLAHE and HE-based methods for mammogram enhancement in pectoral muscle removal/segmentation [27–30], mass detection/segmentation/classification [31–39], and micro-calcification detection [40] reiterate the merits of these techniques. Despite their computational advantages, they suffer from the ill-effects of over-transformation of background and internal tissues gray levels, noise amplification, and failed to retain the internal breast structure. This observation has motivated the researchers to develop a contrast enhancement method termed as Background Preserved and Feature-Oriented Contrast Improvement (BPFO-CI) using Weighted Cumulative distribution Function (WCF) is designed and developed with a twin objective of offering a viable solution for contrast enhancement and to address the drawbacks of CLAHE, HE, and its variants.

In this paper, the implementation mechanism and performance assessment of the proposed method BPFO-CI are presented in Sects. 2 and 3, respectively. The conclusion on this research work is presented in Sect. 4.

2 The Proposed Background Preserved and Feature-Oriented Contrast Improvement (BPFO-CI) Method

The proposed Background Preserved and Feature-Oriented Contrast Improvement (BPFO-CI) is objectively developed to enhance the image characteristics such as background and foreground features using its statistics. First, the BPFO-CI computes the background preserving threshold gray level. Then, it finds a symmetric dynamic gray level division point which is a difference between background preserving threshold gray level and maximum gray level of the input mammogram. Now, the dynamic gray level range of input mammogram image is divided into three ranges, namely background preserving and two symmetric ranges. The first symmetric range is further divided using mean and the second one by median. The entire mammogram dynamic gray level range of the input mammogram is segmented into five ranges. The BPFO-CI assigns numeric value 0 to all the gray levels that are less than on or equal to the background preserving threshold value to achieve background preservation. Then, it computes the PDF and CDF to all other remaining gray level ranges. While computing the CDF, the proposed method uses a weighted intensity adjustment factor, which is drawn from the respective partitioned intensity range. This weighted Intensity Adjustment Factor (IAF) is observed to control the computation of CDF. The IAF of each intensity range is a difference between the initial and final gray levels of the respective ranges, except the last gray level range. The last range's IAF is computed as a difference between L-1 and its starting gray level value. The gray level ranges of WCFs are merged into a single WCF. Finally, the intensity mapping of the WCF gray level values with respective input mammogram gray levels produces the resultant enhanced mammogram.

Figure 1a and b depict the partitioning mechanism of BPFO-CI on the histogram of input mammogram image. The first division on the histogram is performed using Background Preserving threshold (BP_{th}) gray level value. Now the histogram is partitioned into two parts. The first one is background preserving gray level range, and the second is the remaining dynamic gray level range. The second part of this histogram is divided into two symmetric ranges by the Symmetric Division point

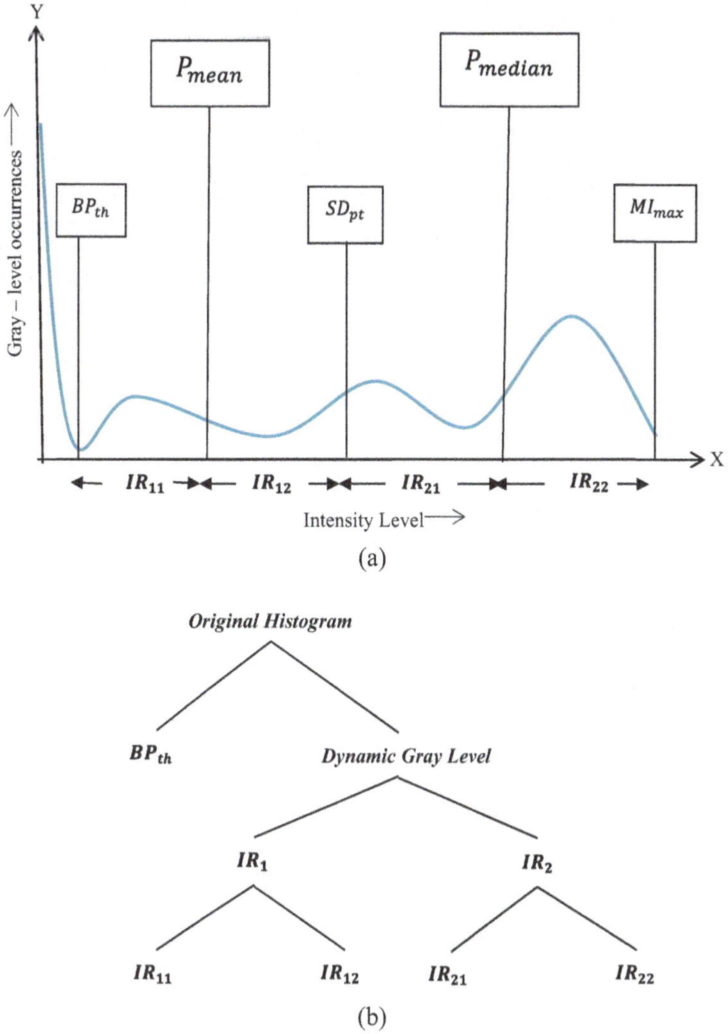

Fig. 1 **(a)** Illustration of the proposed BPFO-CI method's histogram partition, **(b)** Hierarchical representation of the proposed BPFO-CI method's histogram partition

(SD_{pt}) of the histogram, namely Intensity Range1 (IR_1) and Intensity Range2 (IR_2). The mean intensity of IR_1 representing [P_{mean}] and median intensity of IR_2 representing [P_{median}] are used to divide these respective ranges. Finally, the histogram is partitioned into five ranges, namely background preserving, IR_{11}, IR_{12}, IR_{21}, and IR_{22}. IR_1 ranges of histogram gray levels are in a similar gray level occurrence patterns. Hence, mean (P_{mean}) is taken as threshold to divide the range of gray levels in a balanced manner. The median (P_{median}) as a threshold is used for IR range to divide the gray levels due to its irregular gray level occurrence pattern in the histogram.

2.1 Implementation Mechanism of BPFO-CI

Let MI be an input mammogram image, $\mathrm{MI}_k(x, y)$ and k represent pixel at Cartesian coordinate system and its gray level in MI, respectively. The intensity gray level of the image lies between 0 to $(L - 1)$ and 'b'' bits are used store the intensity. In $\mathrm{MI}_k(x, y)$, k denotes an intensity that lies between the gray level range.

$$L = 2^b \tag{1}$$

The implementation of BPFO-CI is described by the following steps:

Step 1: Define a background preserving threshold $\mathrm{BP}_{\mathrm{th}}$ and a symmetric division point $\mathrm{SD}_{\mathrm{pt}}$ of the dynamic gray level range of the MI histogram by the following equations.

$$\mathrm{BP}_{\mathrm{th}} = \mathrm{MI}_{\mathrm{min}} + 5 \tag{2}$$

$$\mathrm{SD}_{\mathrm{pt}} = \left\lceil \frac{(\mathrm{BP}_{\mathrm{th}} + 1) + \mathrm{MI}_{\mathrm{max}}}{2} \right\rceil \tag{3}$$

where $\mathrm{MI}_{\mathrm{min}}$ and $\mathrm{MI}_{\mathrm{max}}$ are minimum and maximum gray level of the input MI. The experimental results confirm that intensity value 5 can be chosen as the background range adjustment factor which is taken based on our experimental Mini-MIAS dataset.

The entire MI dynamic gray level range is divided by $\mathrm{BP}_{\mathrm{th}}$. Now, the dynamic gray level range is partitioned into two parts:

(i) The background preservation range (lies from $\mathrm{MI}_{\mathrm{min}}$ to $\mathrm{BP}_{\mathrm{th}}$) and
(ii) Remaining dynamic range (lies from $\mathrm{BP}_{\mathrm{th}} + 1$ to $\mathrm{MI}_{\mathrm{max}}$).
The remaining dynamic range is symmetrically divided into two symmetric ranges (IR_1 and IR_2) using $\mathrm{SD}_{\mathrm{pt}}$ and these two ranges further segmented by P_{mean} and P_{median}, respectively. Finally, the remaining dynamic range is converted into four ranges (IR_{11}, IR_{12}, IR_{21}, and IR_{22}).

Step 2: The computation of MI gray levels' probability is computed as:

$$P(\mathrm{MI}_k) = \frac{n_k}{R \times C}, \quad k = 0, 1, \ldots, \mathrm{MI}_{\mathrm{max}} \tag{4}$$

where n_k is the number of k gray level pixels in MI, R, and C are row and column limit of MI for computing $R \times C$ total number of pixels.

Step 3: Calculation of the Weighted Cumulative distribution Function ($WCF(MI_k)$) of the gray levels of MI.

$$WCF_{BP}(MI_k) = 0, k = MI_{min}, .., BP_{th} \quad (5)$$

where $WCF_{BP}(MI_k)$ represents the computation of $WCF(MI_k)$ from MI_{min} to BP_{th} gray level. Computing the $WCF(MI_k)$ for the remaining four ranges of gray level using equations from (6) to (9), respectively, as follows:

$$WCF_{IR_{11}}(MI_k) = BP_{th} + IAF_{11} \times \sum_{k=BP_{th}+1}^{P_{mean}} P(MI_k); \quad IAF_{11} = P_{mean} - BP_{th} \quad (6)$$

The above equation calculates $WCF_{IR_{11}}(MI_k)$ gray levels range between $BP_{th} + 1$ and P_{mean} using product of IAF_{11} with cumulative distribution function of the respective gray level range. Similarly, the WCF of remaining gray level ranges computed by the following equations.

$$WCF_{IR_{12}}(MI_k) = P_{mean} + IAF_{12} \times \sum_{k=P_{mean}+1}^{SD_{pt}} P(MI_k); \quad IAF_{12} = SD_{pt} - P_{mean} \quad (7)$$

Equation (7) calculates $WCF_{IR_{12}}(MI_k)$ gray levels range between $P_{mean} + 1$ and SD_{pt}.

$$WCF_{IR_{21}}(MI_k) = SD_{pt} + IAF_{21} \times \sum_{k=SD_{pt}+1}^{P_{median}} P(MI_k); \quad IAF_{21} = P_{median} - SD_{pt} \quad (8)$$

Equation (8) calculates $WCF_{IR_{21}}(MI_k)$ gray levels range between $SD_{pt} + 1$ and P_{median}.

$$WCF_{IR_{22}}(MI_k) = P_{median} + IAF_{22} \times \sum_{k=P_{median}+1}^{MI_{max}} P(MI_k); \quad (9)$$

where $IAF_{22} = (L - 1) - P_{median}$

Equation (9) calculates $WCF_{IR_{22}}(MI_k)$ gray levels range between $P_{median} + 1$ and MI_{max}.

Step 4: Merge WCF of all the gray level ranges into a single WCF. Finally, the whole $WCF(MI_k)$ map with the input mammogram image pixels $MI_k(x, y)$ with respect to their gray level k. The resultant Enhanced Mammogram Image (EMI) pixel is denoted as $EMI_k(x, y)$.

$$EMI_k(x, y) = WCF(MI_k), k = 0, 1, 2, \ldots, MI_{max} \quad (10)$$

2.2 Algorithm for Background Preserved and Feature-Oriented Contrast Improvement Using Weighted Cumulative Distribution Function

Input: Digital Mammogram Image (MI)
Output: Enhanced Digital Mammogram Image

Step 1: Find minimum (MI_{min}) and the maximum (MI_{max}) gray level of the input mammogram. Compute the background preserving threshold (BP_{th}) and symmetric division point (SD_{pt}) to the mammogram image MI using Eqs. (2) and (3), respectively.
Step 2: Partitioning the histogram of input mammogram into five regions is Background Preserving range using BP_{th}, divide the remaining dynamic range into two symmetric ranges, namely IR_1 and IR_2 using SD_{pt}. Segment the IR_1 range into IR_{11} and IR_{12} ranges by mean value of the gray level range between BP_{th} and SD_{pt}, similarly segment the IR_2 range into IR_{21} and IR_{22} by median value between the SD_{pt} and MI_{max} gray level range.
Step 3: Compute the probabilities of all MI gray level ranges using Eq. (4).
Step 4: Calculate Weighted Cumulative distribution Function (WCF) for the background preserving, IR_{11}, IR_{12}, IR_{21}, and IR_{22} ranges gray levels using Eqs. (5)–(9).
Step 5: Merge all the gray level range of WCF into a single WCF.
Step 6: Map the single WCF of each gray level value with respective input mammogram gray level value of pixels.

The proposed BPFO-CI method ideally proposes solutions to the limitations of reported methods, such as over-transformation and noise amplification of the mammogram background. It improves the contrast between tumor suspicious regions and their surrounding tissues and also improves the contrast of breast skin contour.

3 The Experimental of BPFO-CI

In this experiment, 322 number of 8-bit mini-MIAS mammogram images were taken [41]. The mini-MIAS database has different types of breast images such as fatty, fatty glandular, and dense-glandular. The dynamic gray level range of these images is [0,255]. The proposed BPFO-CI method was applied on all these types of images using MATLAB R2018b programming tool with system configuration of 8 GB RAM, Intel(R) core (IM) i5 processor. Metrics chosen for the evaluation were Mean Brightness Error (MBE), Structural SIMilarity index (SSIM), and Peak Signal Noise Ratio (PSNR) [42–44]. The results of BPFO-CI were compared with Unsharp Masking (UM), Histogram Equalization (HE), Brightness Preserving Bi-Histogram Equalization (BBHE), Dualistic Sub-Image Histogram Equalization (DSIHE), Contrast Limited Adaptive Histogram Equalization (CLAHE), and Modified-Dualistic Sub-Image Histogram Equalization (M-DSIHE) for performance evaluation and analysis.

3.1 Results and Discussion

The merits of BPFO-CI was evaluated in terms of quantitative and qualitative analysis. The observations reveal that the proposed method exhibits better performance that endorses the suitability of BPFO-CI in mammogram image enhancement.

3.1.1 Quantitative Analysis

The quantitative assessment of proposed BPFO-CI with existing methods was done using MBE, SSIM, and PSNR metrics. The MBE is the mean difference from enhanced image mean to the original image, where the negative (−) sign depicts that the resultant enhanced image reduces the original image mean brightness and the positive sign represents enhanced image increases the mean brightness. The lower quantity value of MBE depicts that the enhanced image preserved the original image mean brightness. The SSIM analyzes the structural differences between original and enhanced images with a range from 0 to 1 where the high value of SSIM conforms to the enhancement method retains the input image structure. PSNR measures noise variations between the original and enhanced images. The high value of PSNR reveals low noise impact in the enhanced image.

Table 1 contains MBE values of all seven methods including the proposed method. From this table, the proposed BPFO-CI method preserves the original mean brightness than the other methods. In some cases, the CLAHE performed better than the proposed method, which evident from the recorded results. Table 4 shows the average performance of the proposed methods and the other chosen enhancement methods.

Though the quantitative assessment results of UM is better than all other analyzed methods, it does not make any contrast changes on input mammogram that is proven from its resultant images.

From Tables 2 and 4, the proposed method's SSIM values are more than 0.9 which assures that the enhanced original mammogram of BPFO-CI retains the original

Table 1 Comparison of Mean Brightness Error (MBE)

S. No.	Image ID	UM	HE	BBHE	DSIHE	CLAHE	M-DSIHE	BPFO-CI (Proposed)
1	mdb025	**0.0577**	90.5009	33.5922	18.2243	12.1050	4.1358	**0.9668**
2	mdb053	**0.0694**	152.8094	15.2805	13.4749	3.1212	5.7121	**0.4292**
3	mdb060	**0.0351**	148.3069	24.0310	25.7236	7.9156	6.3221	**−0.1963**
4	mdb105	**0.0618**	87.3560	26.5843	−4.1183	2.2844	3.1141	**0.7769**
5	mdb171	**0.0647**	102.6394	28.4123	−7.7707	1.9883	2.1253	**0.4061**
6	mdb179	**0.0425**	170.2312	18.5001	−4.0237	2.7535	2.5962	**−0.3277**
7	mdb189	**0.0675**	103.4411	41.6272	23.6259	13.2196	1.1585	**−0.4572**

Bold values indicate the best outcome among the methods and the proposed method results

Table 2 Structural SIMilarity index (SSIM)

S. No.	Image ID	UM	HE	BBHE	DSIHE	CLAHE	M-DSIHE	BPFO-CI (Proposed)
1	mdb025	**0.9907**	0.4113	0.4662	0.5127	0.5503	0.7959	**0.9370**
2	mdb053	**0.9966**	0.1522	0.1701	0.6898	0.4238	0.6305	**0.9238**
3	mdb060	**0.9945**	0.1811	0.2917	0.7298	0.4958	0.6593	**0.9332**
4	mdb105	**0.9945**	0.3916	0.4404	0.7979	0.5628	0.7961	**0.9743**
5	mdb171	**0.9949**	0.3410	0.3843	0.8971	0.5334	0.7723	**0.9778**
6	mdb179	**0.9980**	0.1518	0.1652	0.8341	0.4170	0.6756	**0.9771**
7	mdb189	**0.9913**	0.3710	0.4195	0.4828	0.5529	0.7695	**0.9121**

Bold values indicate the best outcome among the methods and the proposed method results

Table 3 Peak Signal Noise Ratio (PSNR)

S. No.	Image ID	UM	HE	BBHE	DSIHE	CLAHE	M-DSIHE	BPFO-CI (Proposed)
1	mdb025	**44.6055**	8.8179	16.5856	20.1089	19.4893	26.5530	**26.5813**
2	mdb053	**44.5125**	4.0439	18.0519	18.2128	**29.0256**	23.3131	**28.83171**
3	mdb060	**48.1903**	4.6495	18.0609	15.3759	**26.0745**	24.9592	**25.5967**
4	mdb105	**45.6765**	8.3392	15.9128	26.6621	21.8327	31.5626	**35.2695**
5	mdb171	**43.9876**	6.8577	14.9601	23.2815	23.1392	33.6205	**34.6027**
6	mdb179	**46.7696**	2.9580	16.4926	22.0768	26.8133	27.7216	**30.5165**
7	mdb189	**44.0072**	7.7989	14.4950	16.3620	20.7436	22.3246	**22.5148**

Bold values indicate the best outcome among the methods and the proposed method results

mammogram structure. Other enhancement methods have lower SSIM values than proposed except UM. However, it failed to improve the contrast of mammograms. From Table 3, the proposed method's higher values of the PSNR conforms that noise amplification of BPFO-CI is lower than other methods except UM. In some cases, CLAHE is also performed well on reducing noise amplification.

Time consumption of all the methods on whole 322 mammogram images and average time taken for a mammogram is provided in 5th and 6th column of Table 4. The proposed BPFO-CI method was taken 43.6010 s for whole 322 images and 0.1354 s for a mammogram. The time consumption of BPFO-CI is better than BBHE, DSIHE, and M-DSIHE. However, other three methods were taken shortened time than the BPFO-CI.

3.1.2 Qualitative Analysis

Figure 2 contains 8 rows and each row contains 5 columns of images. The 1st row consists of input mammograms (mdb025, mdb060, mdb189, mdb053, mdb179),

Table 4 Average performance on MIAS database

Method	MBE	SSIM	PSNR	Time taken for whole 322 images (s)	Average time taken for an image (s)
UM	**0.0537**	**0.9936**	**45.4075**	**3.8147**	**0.0118**
HE	106.1443	0.3230	7.5454	10.3771	0.0322
BBHE	26.3676	0.3750	17.5522	90.7552	0.2818
DSIHE	12.0253	0.7009	20.9631	44.9446	0.1395
CLAHE	**5.1242**	0.5439	23.4314	9.2878	0.0288
M-DSIHE	10.1240	0.7051	23.4084	44.2174	0.1395
BPFO-CI (Proposed)	**5.3288**	**0.9139**	**25.9639**	**43.6010**	**0.1354**

Bold values indicate the best outcome among the methods and the proposed method results

denoted as A1–A5; 2nd row labeled as B1–B5 are the UM-applied mammograms; 3rd row are the images enhanced by HE and are denoted by C1–C5. The 4th row shows the BBHE applied images, indicated as D1–D5, 5th row depicts the enhanced images by DSIHE, marked as E1–E5, 6th row is CLAHE applied mammograms namely F1–F5, 7th row labeled as G1–G5, are the M-DSIHE-enhanced images. The last row images given as H1–H5 are the proposed BPFO-CI mammograms.

In Fig. 2, 1st and 2nd columns are fatty types, 3rd column is a fatty glandular type, and the 4th and 5th columns are of dense-glandular type mammograms. UM method has not exhibited any contrast improvement on all the types of mammograms, as shown in 2nd row (B1–B5). The HE produced poor contrast images and washed out mammograms that are evident from 3rd row C1–C5 images. The 4th row is BBHE which addresses the demerits of HE. However, it could not control the over-transformation of mammogram background and has failed to improve the contrast among normal breast tissues with suspicious abnormalities and breast skin contour. From the 5th row, the DSIHE reduces the over-transformation of mammogram background. However, it fails to address the demerits of BBHE's contrast enhancement and noise amplification. The CLAHE shown in the 6th row solves the issues of over-transformation and noise amplification. However, it fails to improve the contour of the breast skin and reduces the normal tissue brightness. In some cases, it does not differentiate the surrounding tissues of suspicious breast abnormalities region. The 7th row is M-DSIHE's enhancement and its performance is similar to that of CLAHE. However, it fails to address the noise amplification which is evident from G2 and G4. The last row of this figure depicts the results of the proposed BPFO-CI method. It is apparent that these resultant images depict substantial improvement on the contrast between breast tissues and tumor suspicious regions, and suppression of noise amplification. This method also controls the over-transformation of mammogram background. In the dense-glandular mammograms, all other methods failed to improve the contrast of breast skin except the proposed BPFO-CI method which is evident from H4 to H5. The quantitative and qualitative analysis confirms that the BPFO-CI outperforms other selected methods. Moreover, it is reliable on fatty and

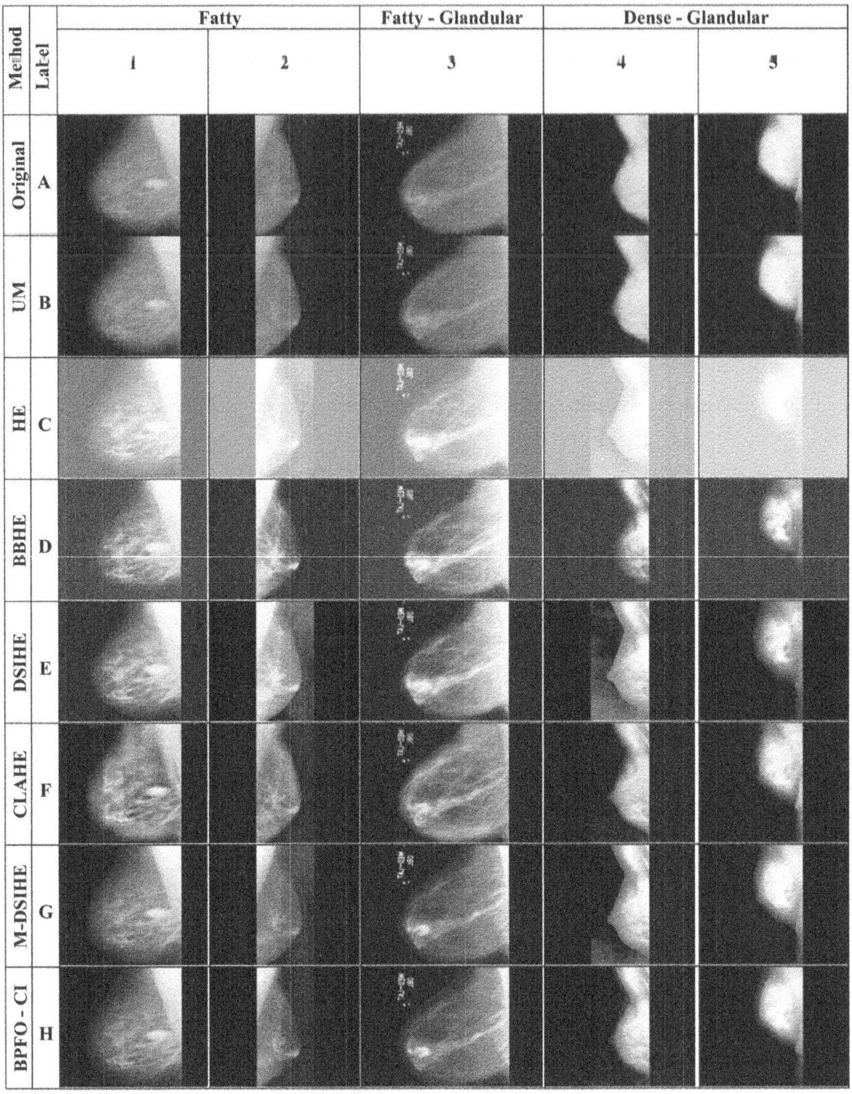

Fig. 2 Result of enhanced mammograms. Row 1: Original image; Row 2: UM-enhanced; Row 3: HE-enhanced; Row 4: BBHE-enhanced, Row 5: DSIHE-enhanced; Row 6: CLAHE-enhanced; Row 7: M-DSIHE–Enhanced and Row 8: Proposed BPFO-CI

fatty-glandular type mammograms. However, the BPFO-CI partially performed well on dense-glandular type mammograms, where it improved the contrast of breast skin contour, and it failed to improve the contrast of internal breast components. But other methods are completely failed on dense-glandular mammograms. These observations endorse the merits of the BPFO-CI over the competitive enhancement methods.

4 Conclusion

The proposed image enhancement method BPFO-CI is computationally simple, and it aims to improve the perception of mammogram images, especially on fatty and fatty-glandular type breast mammograms. The merits of BPFO-CI are duly vouched by the quantitative metrics, namely MBE, PSNR, and SSIM as well as by the qualitative assessment using human visual perception. The BPFO-CI method finds application in the mammogram segmentation and classification of breast tumors. The preprocessing of mammograms with BPFO-CI shall attribute to the accuracy in breast cancer detection and/or classification. It is confirmed to perform well on the fatty and fatty-glandular type mammograms and has exhibited better contrast distinction between the tumor abnormalities and the surrounding tissues as well as the background. However, BPFO-CI method fails to improve the contrast between tumor and normal tissues in dense-glandular type mammograms on which other contemporary methods have too had failed. These benefits of BPFO-CI shall pave the way to the furtherance of this research work in terms of augmenting its performance and applications.

Acknowledgements The authors thank Department of Health Research—Indian Council of Medical Research (DHR-ICMR), India for the financial support under the funded research project (No. GIA/80/DHR/2014 and V.25011/50/2015-GIA/HR). They also thank the reviewers for valuable comments and suggestions.

References

1. World Health Organisation: https://www.who.int/news-room/fact-sheets/detail/noncommunicable-diseases. Last accessed 2018
2. Mathur, P., Sathishkumar, K., Chaturvedi, M., Das, P., Sudarshan, K.L., Santhappan, S., Nallasamy, V., John, A., et al.: Cancer statistics, 2020: Report from National cancer registry programme, India. JCO Glob. Oncol. **6**, 1063–1075 (2020)
3. Gupta, B., Tiwari, M., Singh, L.S.: Visibility improvement and mass segmentation of mammogram images using quantile separated histogram equalisation with local contrast enhancement. CAAI Trans. Intell. Technol. **4**(2), 73–79 (2019)
4. Rangaraj, M.R., Liang, S., Yiping, S., Leo, J.E.D., Heather, B., Timothy, J.T., Natalka, H., Sarah, M.R.: Improvement of sensitivity of breast cancer diagnosis with adaptive neighborhood contrast enhancement of mammograms. IEEE Trans. Inf. Technol. Biomed. **1**(3), 161–170 (1997)
5. Hamouda, S.K.M., Abo El-Ezz, R.H., Wahed, M.E.: Enhancement accuracy of breast tumor diagnosis in digital mammograms. J. Biomed. Sci. **6**(4), 1–8 (2017)
6. Shanmugavadivu, P., Sivakumar, V.: Comparative analysis of microcalcifications detected in mammogram images by edge detection using fractal Hurst co-efficient and fudge factor. In

IEEE Proceedings of the International Conference on Emerging Trends in Science Engineering and Technology: Recent Advancements on Science and Engineering Innovation, pp. 174–179. IEEE, INCOSET (2012)
7. Omer, A.M., Elfadil, M.: Preprocessing of digital mammogram image based on Otsu's threshold. Am. Sci. Res. J. Eng. Technol. Sci. **37**(1), 220–229 (2017)
8. Charate, A.P., Jamge, S.B.: The preprocessing methods of mammogram images for breast cancer detection. Int. J. Recent Innov. Trends Comput. Commun. **5**(1), 261–264 (2017)
9. Singh, S., Bovis, K.: An evaluation of contrast enhancement techniques for mammographic breast masses. IEEE Trans. Inf. Technol. Biomed. **9**(1), 109–119 (2005)
10. Patel, B.C., Sinha, G.R.: Gray level clustering and contrast enhancement (GLC–CE) of mammographic breast cancer images. CSI Trans. ICT **2**(4), 279–286 (2015)
11. Shanmugavadivu, P., Sivakumar, V., Sudhir, R.: Fractal dimension-bound spatio-temporal analysis of digital mammograms. Eur. Phys. J.: Special Topics **225**(1), 137–146 (2016)
12. Sundaram, M., Ramar, K., Arumugam, N., Prabin, G.: Histogram based contrast enhancement for mammogram images. In: 2011 International Conference on Signal Processing, Communication, Computing and Networking Technologies, pp. 842–846. IEEE (2011)
13. Wang, Y., Chen, Q., Zhang, B.: Image enhancement based on equal area dualistic sub-image histogram equalization method. IEEE Trans. Consum. Electron. **45**(1), 68–75 (1999)
14. Kaur, M., Kaur, J., Kaur, J.: Survey of contrast enhancement techniques based on histogram equalization. Int. J. Adv. Comput. Sci. Appl. **2**(7), 137–141 (2011)
15. Raju, A., Dwarakish, S., Reddy, G., Venkat, D.: A comparative analysis of histogram equalization based techniques for contrast enhancement and brightness preserving. Int. J. Signal Process. Image Process. Pattern Recogn. **6**(5), 353–366 (2013)
16. Surya Gowri, D., Amudha, T.: A review on mammogram image enhancement techniques for breast cancer detection. In: Proceedings 2014 International Conference on Intelligent Computing Applications, pp. 47–51. IEEE (2014)
17. Gupta, B., Tiwari, M.: A tool supported approach for brightness preserving contrast enhancement and mass segmentation of mammogram images using histogram modified grey relational analysis. Multidimension. Syst. Signal Process. **28**(4), 1549–1567 (2017)
18. Dhamodharan, S., Shanmugavadivu, P.: Brightness preserving contrast enhancement of digital mammogram using modified-dualistic sub-image histogram equalization. Int. J. Comput. Sci. Eng. **6**(4), 257–261 (2018)
19. Pisano, E.D., Zong, S., Hemminger, B.M., Deluca, M., Johnston, R.E., Muller, K., Braeuning, M.P., Pizer, S.M.: Contrast limited adaptive histogram equalization image processing to improve the detection of simulated spiculations in dense mammograms. J. Digit. Imaging **11**(4), 193–200 (1998)
20. Anand, S., Gayathri, S.: Mammogram image enhancement by two-stage adaptive histogram equalization. Optik (Stuttg) **126**(21), 3150–3152 (2015)
21. Cunha Carneiro, P., Lemos Debs, C., Oliveira Andrade, A., Patrocinio, A.C.: CLAHE parameters effects on the quantitative and visual assessment of dense breast mammograms. IEEE Latin Am. Trans. **17**(5), 851–857 (2019)
22. Sheeba, J., Parasuraman, S., Kadirvelu, A.: Contrast enhancement and brightness preserving of digital mammograms using fuzzy clipped contrast-limited adaptive histogram equalization algorithm. Appl. Soft Comput. **42**, 167–177 (2016)
23. Akila, K., Jayashree, L.S., Vasuki, A.: Mammographic image enhancement using indirect contrast enhancement techniques—a comparative study. Procedia Comput. Sci. **47**, 255–261 (2015)
24. Gandhamal, A., Talbar, S., Gajre, S., Hani, A.F.M., Kumar, D.: Local gray level S-curve transformation—a generalized contrast enhancement technique for medical images. Comput. Biol. Med. **83**, 120–133 (2017)
25. Bhateja, V., Misra, M., Urooj, S.: Human visual system based unsharp masking for enhancement of mammographic images. J. Comput. Sci. **21**, 387–393 (2017)
26. Patrick, N., Chan, H.P., Sahiner, B., Wei, D.: An adaptive density-weighted contrast enhancement filter for mammographie breast mass detection. IEEE Trans. Med. Imaging **15**(1), 59–67 (1996)

27. Slavković-Ilić, M., Gavrovska, A., Milivojević, M., Reljin, I., Reljin, B.: Breast region segmentation and pectoral muscle removal in mammograms. Telfor J. **8**(1), 50–55 (2016)
28. Alam, N., Islam, M.J.: Pectoral muscle elimination on mammogram using K-means clustering approach. Int. J. Comput. Vision Signal Process. **4**(1), 11–21 (2014)
29. Maitra, I.K., Nag, S., Bandyopadhyay, S.K.: Technique for preprocessing of digital mammogram. Comput. Methods Programs Biomed. **107**(2), 175–188 (2012)
30. Angayarkanni, S.P., Kamal, N.B., Thangaiya, R.J.: Dynamic graph cut based segmentation of mammogram. Springerplus **4**(1), 1–9 (2015)
31. Anitha, J., Peter, J.D., Pandian, S.I.A.: A dual stage adaptive thresholding (DuSAT) for automatic mass detection in mammograms. Comput. Methods Programs Biomed. **138**, 93–104 (2017)
32. Neto, O.P.S., Silva, A.C., Paiva, A.C., Gattass, M.: Automatic mass detection in mammography images using particle swarm optimization and functional diversity indexes. Multimedia Tools Appl. **76**(18), 19263–19289 (2017)
33. Mughal, B., Sharif, M., Muhammad, N.: Bi-model processing for early detection of breast tumor in CAD system. Eur. Phys. J. Plus **132**(6), 1–14 (2017)
34. Salazar-Licea, L.A., Pedraza-Ortega, J.C., Pastrana-Palma, A., Aceves-Fernandez, M.A.: Location of mammograms ROI's and reduction of false-positive. Comput. Methods Programs Biomed. **143**, 97–111 (2017)
35. Makandar, A., Halalli, B.: Mammography image analysis using wavelet and statistical features with SVM classifier. In: Proceedings of International Conference on Cognition and Recognition, pp. 371–382. Springer, Singapore (2018)
36. Sampaio, W.B., Diniz, E.M., Silva, A.C., De Paiva, A.C., Gattass, M.: Detection of masses in mammogram images using CNN, geostatistic functions and SVM. Comput. Biol. Med. **41**(8), 653–664 (2011)
37. Moh'd Rasoul, A., Al-Gawagzeh, M.Y., Alsaaidah, B.A.: Solving mammography problems of breast cancer detection using artificial neural networks and image processing techniques. Indian J. Sci. Technol. **5**(4), 2520–2528 (2012)
38. Taghanaki, S.A., Liu, Y., Miles, B., Hamarneh, G.: Geometry-based pectoral muscle segmentation from mlo mammogram views. IEEE Trans. Biomed. Eng. **64**(11), 2662–2671 (2017)
39. Rahmati, P., Adler, A., Hamarneh, G.: Mammography segmentation with maximum likelihood active contours. Med. Image Anal. **16**(6), 1167–1186 (2012)
40. Bougioukos, P., Glotsos, D., Kostopoulos, S., Daskalakis, A., Kalatzis, I., Dimitropoulos, N., Cavouras, D.: Fuzzy c-means-driven fhce contextual segmentation method for mammographic microcalcification detection. Imaging Sci. J. **58**(3), 146–154 (2010)
41. Suckling, J., Parker, J., Dance, D., Astley, S., Hutt, I., C. Boggis, Ricketts, I., et al.: Mammographic Image Analysis Society (MIAS) Database, 21. [Online]. Available: https://www.repository.cam.ac.uk/handle/1810/250394 (2015). Last accessed 28 Aug 2015
42. Shanmugavadivu, P.: Modified histogram equalization for image contrast enhancement using particle swarm optimization. Int. J. Comput. Sci. Eng. Inf. Technol. **1**(5), 13–27 (2011)
43. Tiwari, M., Gupta, B., Shrivastava, M.: High-speed quantile-based histogram equalisation for brightness preservation and contrast enhancement. IET Image Process. **9**(1), 80–89 (2015)
44. Shanmugavadivu, P., Balasubramanian, K., Somasundaram, K.: Median adjusted constrained PDF based histogram equalization for image contrast enhancement. In: D. Nagamalai, E. Renault, M. Dhanushkodi (Eds.): CCSEIT 2011, CCIS 204, pp. 244–253. Springer-Verlag Berlin Heidelberg (2011)

Control Theory and Its Applications

Finite-Time Passification of Fractional-Order Recurrent Neural Networks with Proportional Delay and Impulses: An LMI Approach

N. Padmaja and P. Balasubramaniam

Abstract This manuscript deals with the problem of passification of fractional-order neural networks (FONNs) with proportional delays and impulses in finite time. The primary contribution of this work lies in the fact that the concept of finite-time passivity (FTP) is extended for FONNs with and without impulses for the first time. At the fore set, the notion of FTP in terms of Lyapunov function is extended to fractional-order systems. Based on the proposed definition of FTP, sufficient conditions in terms of LMI are derived to ensure the FTP of the considered FONNs. Further to this, a new lemma using comparison principle is derived using which the definition of FTP for FONNs with impulses is put forth. From these results, some new set of LMI conditions for the considered system to be finite-time stable (FTS) is derived. Finally, theoretical results are verified via a numerical example.

Keywords Finite-time passivity · Finite-time stability · Fractional-order neural networks · Delay-dependent LMI · Lyapunov functional

1 Introduction

Fractional calculus is the branch of pure mathematics which generalizes the ordinary integer-order differentiation/integration to any arbitrary real numbers. Its significance over the traditional calculus lies in the fact that it can model many real-life applications/phenomena, especially biological systems, with more precision compared to the latter [1]. This together with their hereditary and memory property have paved way for FONNs which are found to be more efficient tools in system identification [2]. It is a well-known fact that any recurrent NNs are usable for an application only if it has stable dynamics. Hence, a vast number of works could be traced on stability to multi-stability, bifurcation, etc. of FONNs (e.g. [3–5]).

N. Padmaja · P. Balasubramaniam (✉)
Department of Mathematics, The Gandhigram Rural Institute (Deemed to be University), Gandhigram, Tamil Nadu 624302, India

P. Balasubramaniam et al. (eds.), *Mathematical Modelling and Computational Intelligence Techniques*, Springer Proceedings in Mathematics & Statistics 376,
https://doi.org/10.1007/978-981-16-6018-4_13

Due to the limited amplifier speed, delays are quite common in the hardware implementation of NNs. Depending on the nature of the application, there are many kinds of delays like bounded time-varying delay, constant delay, distributed delay, leakage delay, etc. Of these, proportional delay is a special kind of variable delay in which the delay function takes the form $\tau(t) = (1 - p)t$, where p is called as proportional delay factor whose value lies between 0 and 1. This kind of delay usually occurs in modelling of biological/physical systems, routing problems, etc. Due to the spatial nature of neural networks, it was suggested by Zhou in [6] that it is desirable to model a cellular NNs with proportional delay. This kicked off the research on recurrent NNs with proportional delays [7]. Also, from the definition of the delay function, it is clear that, this kind of delay is unbounded and monotonically increasing for $t > 0$. This characteristic allows us to adjust the running time of the network depending on the maximum allowable delay bound. Hence, the reader could find a wide range of results on stability analysis through different methods in [8–10], for both integer- and fractional-order NNs with proportional delays.

Just like delays, sudden external disturbance in the communication between two neurons leads to change in their behaviour. This abrupt change is termed as impulse. The presence of impulse can modify the stability properties of a system sometimes leading even to instability or delayed convergence of states to its equilibrium. Thus, it is more practical to analyse the system properties for an impulsive system as is done in [11, 12].

Passivity, a concept that arose from circuit theory, is a very significant and effective tool in the dynamic analysis of control systems. A passive system is the one in which the variation of internal energy does not exceed the external energy supplied to it thereby keeping the system stable internally. It is an efficient method to analyse stability of a system with external disturbances. Further, the property that feedback connection of two passive systems is passive has particular importance in analysing stability of large-scale systems. Its popularity also stems from the fact that it tells about the behaviour of a system based on its input–output dynamics in terms of storage function, thereby giving a clear interpretation of energy changes in a system. This fact could be particularly exploited in neuro-control systems when NNs are used to identify an unknown nonlinear system with only its input–output characteristics. Thus, it is necessary to study passivity and passification of NNs. For more details on notable results in this field, the readers can refer to [13–15].

However, all the above-mentioned studies deal with passivity analysis over an infinite time range. Recently, the concept of finite-time control is gaining attention because of the applications such as finite-time attitude tracking problem [16], robotics [17], fixed-time synchronization [18], etc. requiring the corresponding system to become stable within a finite time. As seen in [19], passivity is also used in synchronization problems. Thus, it becomes essential to study the concept of finite-time passivity (FTP) (for definition of FTP, the readers may refer to [20]). According to this proposed concept of FTP, the considered Lyapunov function decays to zero within finite-time interval (see [21, 22] for details). However, regretfully, the notion of FTP in terms of energy function still needs to be explored for FONNs. Based on the above facts, in this work we discuss the finite time passivity of FONNs with proportional delay in terms of energy function.

The following are the key contributions of this work:

1. The concept of FTP in terms of energy function is extended to FONNs for the first time.
2. A new lemma extending Razumikhin's method for studying FTS of impulsive FONNs is proved. Based on this lemma, the concept of FTP is proposed for FONNs with impulses.
3. The results are provided in terms of LMIs which involve the delay factors, fractional-order of the system and impulse gain matrix values.
4. Numerical simulations are given to validate the proposed theory as well as to showcase the advantage of using fractional order in NNs.

Notations: For any matrix $B > 0$ or $B < 0$ indicates that B is a symmetric positive definite or negative definite matrix. $\mathbb{R}$ and $\mathbb{R}_+$ denote the set of all real number and real positive numbers, respectively, while $\mathbb{Z}$, $\mathbb{Z}^+$ denote the set of integers. $\mathbb{R}^n$ depicts the n-dimensional vector space over real numbers with Euclidean norm. For any matrix D, $\lambda_{\min}(D)$ and $\lambda_{\max}(D)$ are, respectively, the minimum and maximum eigenvalues of D.

2 Model Description

In this paper, the FTP of the following FONNs model with proportional delay is analysed.

$$\begin{aligned} {}_{t_0}^{C}D_t^{\beta}\mathcal{Z}(t) &= -A\mathcal{Z}(t) + B\mathcal{G}(\mathcal{Z}(t)) + B_d\mathcal{G}(\mathcal{Z}(\xi t)) + i(t) + u(t) \\ \mathcal{Z}(t) &= \phi(t),\ t \in [\bar{\xi}t_0, t_0];\ \ y(t) = O_f\mathcal{G}(Z(t)) + O_i i(t) \end{aligned} \tag{1}$$

where ${}_{t_0}^{C}D_t^{\beta}\mathcal{Z}(t)$ denotes the Caputo fractional differential operator of order β whose definition is subsequently defined in Definition 1, $0 < \beta < 1$ is the commensurate fractional order of the system, $\mathcal{Z}(t) = (\mathcal{Z}_1(t), \mathcal{Z}_2(t), \ldots, \mathcal{Z}_N(t))^T$ depicts the neuronal state vector at instant t; $\mathcal{G}(\mathcal{Z}(t)) = (\mathcal{G}_1(\mathcal{Z}_1(t)), \mathcal{G}_2(\mathcal{Z}_2(t)), \ldots, \mathcal{G}_N(\mathcal{Z}_N(t)))^T$ is the neuron activation function; $\xi = (\xi_1, \ldots, \xi_N)$ is the proportional delay factors; $\xi t = t - (1-\xi)t$ is the proportional delay value; $\mathcal{G}(\mathcal{Z}(\xi t)) = (\mathcal{G}_1(\mathcal{Z}_1(\xi_1 t)), \mathcal{G}_2(\mathcal{Z}_2(\xi_2 t)), \ldots, \mathcal{G}_N(\mathcal{Z}_N(\xi_N t)))^T$ is the delayed activation function; $i(t)$ is the external input; $u(t)$ is the control input; $y(t)$ is the output. Besides in (1), $A > 0$ is a positive diagonal matrix which gives the self-decay rate of each neuron, B and B_d are interconnection and delayed interconnection weight matrices, respectively. O_f, and O_i are also weight matrices of appropriate dimensions. Finally, $\phi(t)$ is a differentiable function which gives the initial values of the states and $\bar{\xi} = \min_{1\le i\le N} \xi_i$.

Further, the controller considered in this work is given below,

$$u(t) = -K\mathcal{Z}(t) - k_1 \mathrm{sign}(\mathcal{Z}(t))|\mathcal{Z}(t)|^{\eta} - \mathrm{sign}(\mathcal{Z}(t))B_d M \tag{2}$$

where $\text{sign}(\mathcal{Z}(t)) = \text{diag}\{\text{sign}(\mathcal{Z}_1(t)), \ldots, \text{sign}(\mathcal{Z}_N(t))\}$, $|\mathcal{Z}(t)| = (|\mathcal{Z}_1(t)|, \ldots, |\mathcal{Z}_N(t)|)^T$; K is a real matrix of appropriate dimension to be estimated. It is called as controller gain matrix. The real constants $k_1 > 0$ and $0 < \eta < 2\beta - 1$ are known as tunable control parameters which determine the settling time and at last, $M = \text{diag}\{M_1, \ldots, M_N\}$.

Remark 1 This is a type of delay-independent controller considered in [21]. The first term $-K\mathcal{Z}(t)$ is the state-feedback control term. The second and third terms $-k_1\text{sign}(\mathcal{Z}(t))|\mathcal{Z}(t)|^{\eta}$, $-\text{sign}(\mathcal{Z}(t))B_d M$ are included for faster convergence of the system. The parameters k_1 and η determine the time of convergence of the system. The advantage of this control over the simple state-feedback controller is that the convergence is faster and adjustable. Also, as it is delay-independent, one need not know the information of delay in the system to use this controller.

The following are the assumptions made on the above NNs system.

Assumption 1 For all $i \in \{1, 2, \ldots, N\}$, the neuron activation function $\mathcal{G}_i(.)$ satisfies the conditions below:

(i) $\mathcal{G}_i(0) = 0$
(ii) $\mathcal{G}_i(.)$ is a bounded function, i.e. $|\mathcal{G}_i(x)| \le M_i, \forall\, x \in \mathbb{R}$ where $M_i > 0$ is a real constant
(iii) $c_i^- \le \frac{\mathcal{G}_i(x) - \mathcal{G}_i(y)}{x - y} \le c_i^+, \forall x \ne y \in \mathbb{R}$ with c_i^- and c_i^+ being any real constants.

3 Basic Results and Definitions

In this section, we recall all the definitions and lemmas which are needed to derive the main results.

Definition 1 (*Caputo Fractional Derivative* [1]) The Caputo's fractional derivative of $f(t) \in C^m([0, \infty), \mathbb{R})$ is defined by

$$ {}^{C}_{t_0}D_t^{\beta} f(t) = \frac{1}{\Gamma(m - \beta)} \int_{t_0}^{t} \frac{f^{(m)}(s)}{(t - s)^{\beta + 1 - m}} \, \mathrm{d}s $$

where $m - 1 < \beta < m$ is the order of derivative, $m \in \mathbb{Z}^+$.

Definition 2 (*Fractional Integral Operator* [1]) The fractional integral operator of order $\beta \in \mathbb{R}$ for an integrable function $\mathcal{G}(t) \in C([0, \infty), \mathbb{R}])$ is defined as,

$$ {}_aI_t^{\beta}\mathcal{G}(t) = \frac{1}{\Gamma(\beta)} \int_{a}^{t} (t - \gamma)^{\beta - 1}\mathcal{G}(\gamma)\, \mathrm{d}\gamma $$

where $\Gamma(\cdot)$ is Euler gamma function and $a \in [0, \infty)$.

Definition 3 (*Finite-Time Stability* [20]) The trivial solution of system (1) with zero external disturbance input (i.e. $i(t) \equiv 0$) is said to be FTS if

(i) it is stable in Lyapunov sense and
(ii) any solution $\mathcal{Z}(t, \mathcal{Z}_0)$ of (1) converges to the point of equilibrium within some finite time. That is, for each initial value $\mathcal{Z}_0 \in \mathbb{R}^n$, there exists $T(\mathcal{Z}_0) \in \mathbb{R}_+ \cup \{0\} \ni: \mathcal{Z}(t, \mathcal{Z}_0) = 0 \forall\, t \geq T(\mathcal{Z}_0)$.

This function $T : \mathbb{R}^n \to \mathbb{R}_+ \cup \{0\}$ is called as settling time function.

Note: In the above definition, $\mathcal{Z}(t, \mathcal{Z}_0)$ denotes the solution of a system with initial value $\mathcal{Z}_0$.

Lemma 1 ([23]) *Assume that a continuous positive definite functional $\mathcal{V}(t)$ satisfies the following fractional-order differential inequality*

$${}_{t_0}^{C}D_t^{\beta}\mathcal{V}(t) \leq -\mathrm{d}\,\mathcal{V}^{\gamma}(t)$$

where $d > 0, 0 < \gamma < \beta$ are constants. Then,

$$\mathcal{V}^{\beta-\gamma}(t) \leq \mathcal{V}^{\beta-\gamma}(t_0) - \frac{\mathrm{d}\,\Gamma(1+\beta-\gamma)}{\Gamma(1+\beta)\Gamma(1-\gamma)}(t-t_0)^{\beta}, \quad t_0 < t < t^*$$

and $\mathcal{V}(t) = 0, \forall t \geq t^*$. Here, t^* is called as the settling time and its estimation is

$$t^* = t_0 + \left(\frac{\Gamma(1+\beta)\Gamma(1-\gamma)\mathcal{V}^{\beta-\gamma}(t_0)}{\mathrm{d}\,\Gamma(1+\beta-\gamma)}\right)^{1/\beta}.$$

Based on this lemma, the following definition for FTP of fractional order system is extracted here in

Definition 4 (*Finite-Time Passivity*) A system of fractional-order $0 < \beta < 1$ with output $y(t) \in \mathbb{R}^n$ and input $i(t) \in \mathbb{R}^n$ is said to be FTP if there exists a non-negative function $\mathcal{V}(t)$ satisfying the following inequality

$${}_{t_0}^{C}D_t^{\beta}\mathcal{V}(t) + a\,\mathcal{V}^{\eta}(t) \leq y^T(t)i(t), \quad \forall t \in [t_0, \infty)$$

where $a > 0$ and $0 < \eta < \beta$ are constants.

Remark 2 FTP means that a system is passive in addition to being finite-time stable. In many practical applications like attitude control of spacecraft, missile guidance system, unmanned underwater vehicle systems, etc., the corresponding system has to converge to its steady state within a fixed-time interval. That is, the system is FTS. However, disturbances in each of these systems are not avoidable. Thus, to analyse the stability of these systems with external disturbances, it is a known fact that passivity is an effective tool. Therefore, FTP is an important concept needed for many such practical applications.

Lemma 2 ([24, 25]) *The fractional integral operator and Caputo fractional derivative satisfy the following properties.*

(i) *For any* $\mathcal{G}(t) \in C^m([0,\infty),\mathbb{R})$ *and* $0 < \beta < 1$,

$$ {}_aI_t^{\beta}({}_a^CD_t^{\beta}\mathcal{G}(t)) = \mathcal{G}(t) - \mathcal{G}(a), \quad a \in [0,\infty) $$

(ii) *For any* $\mathcal{G}(t) \subset C^m([0,\infty),\mathbb{R})$ *and* $0 < \beta < 1$,

$$ {}_a^CD_t^{\beta}({}_aI_t^{\beta}\mathcal{G}(t)) = \mathcal{G}(t) $$

(iii) ${}^CD_t^{\beta}(X^T(t)RX(t)) \leq 2X^T(t)R\,{}^CD_t^{\beta}X(t)$, *for any* $X(t) \in \mathbb{R}^n$, $\beta \in (0,1)$, $R > 0$ *is a symmetric matrix of order n.*

(iv) *For* $\beta \in (0,1)$, $\mu \in \mathbb{R}$, ${}^CD_t^{\beta}X^{\mu}(t) = \frac{\Gamma(1+\mu)}{\Gamma(1+\mu-\beta)}X^{\mu-\beta}(t){}^CD_t^{\beta}X(t)$.

Lemma 3 ([26]) *If* $a_1, a_2, \ldots, a_m \geq 0$, $0 < n_2 < n_1$ *are real numbers, then*

$$ \left(\sum_{i=1}^{m} a_i^{n_1}\right)^{1/n_1} \leq \left(\sum_{i=1}^{m} a_i^{n_2}\right)^{1/n_2} $$

Lemma 4 (Comparison Lemma [27]) *Suppose that*

(i) $g : \mathbb{R}_+ \times \mathbb{R} \to \mathbb{R}$ *is continuous in each set* $(t_{k-1}, t_k] \times \mathbb{R}$, $\psi_k : \mathbb{R} \to \mathbb{R}$, *is a non-decreasing function, for each* $k = 1, 2, \ldots$ *There is a function* $m : [t_0, T) \to \mathbb{R} \ni$: *for* $0 < q < 1$,

$$ \begin{aligned} D^q m(t) &\leq g(t, m(t)), \quad t \neq t_k, \\ m(t_k^+) &\leq \psi_k(m(t_k)), \quad t = t_k, \\ m(t_0) &\leq u_0. \end{aligned} $$

(ii) $u(t)$ *is the solution of the system,*

$$ \begin{aligned} D^q u(t) &= g(t, m(t)), \quad t \neq t_k, \\ u(t_k^+) &= \psi_k(u(t_k)), \quad t = t_k, \\ u(t_0) &= u_0. \end{aligned} $$

existing on $[t_0, T)$. *Then* $m(t) \leq u(t)$, $t_0 \leq t < T$.

4 Theoretical Results

4.1 Analysis for FONNs Without Impulses

This part of the manuscript compiles the conditions obtained for the system (1) to be FTP under the controller (2).

Theorem 1 *The system* (1) *is FTP if there exists real matrix $C > 0$, diagonal matrices $R > 0$, $G > 0$ and $\tilde{K}$ of appropriate dimensions such that,*

$$\begin{bmatrix} \Xi & RB + \frac{(C^+ + C^-)G}{2} & R \\ * & -G + \hat{\xi}C & -O_f^T/2 \\ * & * & -O_i \end{bmatrix} < 0 \tag{3}$$

where $\Xi = -RA - A^T R - \tilde{K} - \tilde{K}^T - C^+GC^- + \hat{\xi}C$; $\hat{\xi} = \sum_{i=1}^{n} \frac{(1-\xi_i)^\beta}{\beta}$; $C^+ = \text{diag}\,\{c_1^+, c_2^+, \ldots, c_n^+\}$ *and* $C^- = \text{diag}\{c_1^-, c_2^-, \ldots, c_n^-\}$. *Then, the controller gain is given by* $K = R^{-1}\tilde{K}$.

Proof Define positive definite storage function as

$$\mathcal{V}(t) = \mathcal{Z}^T(t) R \mathcal{Z}(t)$$

where $R > 0$. Then taking Caputo fractional derivative, by using Lemma 2(iii) and Eqs. (1)–(2) we obtain,

$$\begin{aligned}
{}_{t_0}^{C}D_t^\beta \mathcal{V}(t) &\le 2\mathcal{Z}^T(t) R\, {}^{C}{}_{t_0}D_t^\beta \mathcal{Z}(t) \\
&= 2\mathcal{Z}^T(t) R[-A\mathcal{Z}(t) + B\mathcal{G}(\mathcal{Z}(t)) + B_d\mathcal{G}(\mathcal{Z}(\xi t)) + i(t) + u(t)] \\
&\le 2\mathcal{Z}^T(t)[-RA - RK]\mathcal{Z}(t) + 2\mathcal{Z}^T(t) RB\mathcal{G}(\mathcal{Z}(t)) + 2|\mathcal{Z}^T(t)| RB_d M \\
&\quad + 2\mathcal{Z}^T(t) Ri(t) - 2k_1\lambda_{\min}(R)|\mathcal{Z}^T(t)||\mathcal{Z}(t)|^\eta - 2|\mathcal{Z}^T(t)| RB_d M \\
&= 2\mathcal{Z}^T(t)[-RA - RK]\mathcal{Z}(t) + 2\mathcal{Z}^T(t) RB\mathcal{G}(\mathcal{Z}(t)) + 2\mathcal{Z}^T(t) Ri(t) \\
&\quad - 2k_1\lambda_{\min}(R) \sum_{i=1}^{N} |\mathcal{Z}_i(t)|^{\eta+1}.
\end{aligned} \tag{4}$$

Now from Assumption 1, there exists diagonal matrix $G \ni$:

$$(C^+\mathcal{Z}(t) - \mathcal{G}(\mathcal{Z}(t)))^T G(\mathcal{G}(\mathcal{Z}(t)) - C^-\mathcal{Z}(t)) \ge 0. \tag{5}$$

Further from Lemma 3 and the fact that $0 < \eta < 2\beta - 1$, it follows that,

$$-\sum_{i=1}^{N} |\mathcal{Z}_i(t)|^{\eta+1} \le -\left(\sum_{i=1}^{N} |\mathcal{Z}_i(t)|^{2\beta}\right)^{\frac{\eta+1}{2\beta}} \le -\left(\sum_{i=1}^{N} |\mathcal{Z}_i(t)|^{2}\right)^{\frac{\eta+1}{2}}. \tag{6}$$

Now taking $RK = \tilde{K}$ Eqs. (4)–(6) boils down to the following inequality.

$$
\begin{aligned}
{}_{t_k}^{C}D_t^{\beta}\mathcal{V}(t) &\leq \chi^T(t)\Omega'\chi(t) - 2k_1\lambda_{\min}(R)\left((\lambda_{\max}(R))^{-1}\mathcal{V}(t)\right)^{\frac{\eta+1}{2}} \\
&= \chi^T(t)\Omega'\chi(t) - 2k_1\lambda_{\min}(R)\lambda_{\max}(R)^{-(\frac{\eta+1}{2})}(\mathcal{V}(t))^{\frac{\eta+1}{2}},
\end{aligned} \tag{7}
$$

where $\chi(t) = \left[\mathcal{Z}^T(t)\ \mathcal{G}^T(\mathcal{Z}(t))\ i^T(t)\right]^T$ and $\Omega' = \begin{bmatrix} \Phi & RB + \frac{(C^+ + C^-)G}{2} & R \\ * & -G & 0 \\ * & * & 0 \end{bmatrix}$ with

$\Phi = -RA - A^T R - \tilde{K} - \tilde{K}^T - C^+GC^-$.

Finally for any $C > 0$, taking $e(t) = \left[\mathcal{Z}^T(t)\ \mathcal{G}^T(\mathcal{Z}(t))\right]^T$ we have,

$$
\sum_{i=1}^{N}\left[\frac{(1-\xi_i)^{\beta}}{\beta} - \int_{\xi_i}^{1}(1-s)^{\beta-1}\,\mathrm{d}s\right] e^T(t)\begin{bmatrix} C & 0 \\ 0 & C \end{bmatrix} e(t) = 0. \tag{8}
$$

On combining (7) and (8) we can obtain,

$$
\begin{aligned}
{}_{t_0}^{C}D_t^{\beta}\mathcal{V}(t) - y^T(t)i(t) \leq\ & \chi^T(t)\Omega'\chi(t) + \sum_{i=1}^{N}\frac{(1-\xi_i)^{\beta}}{\beta}[\mathcal{Z}^T(t)C\mathcal{Z}(t) \\
& + \mathcal{G}^T(\mathcal{Z}(t))C\mathcal{G}(\mathcal{Z}(t))] - \sum_{i=1}^{N}\int_{\xi_i}^{1}(1-s)^{\beta-1}e^T(t)\hat{C}e(t)\,\mathrm{d}s \\
& - 2k_1\lambda_{\min}(R)\,(\lambda_{\max}(R))^{-\frac{\eta+1}{2}}\,(\mathcal{V}(t))^{\frac{\eta+1}{2}} \\
& - \mathcal{G}^T(\mathcal{Z}(t))O_f^T\,i(t) - i^T(t)O_i^T\,i(t) \\
=\ & \chi^T(t)\Omega\chi(t) - \sum_{i=1}^{N}\int_{\xi_i}^{1}(1-s)^{\beta-1}e^T(t)\hat{C}e(t)\,\mathrm{d}s \\
& - 2k_1\lambda_{\min}(R)(\lambda_{\max}(R))^{-\frac{\eta+1}{2}}(\mathcal{V}(t))^{\frac{\eta+1}{2}} \\
\leq\ & \chi^T(t)\Omega\chi(t) - 2k_1\lambda_{\min}(R)(\lambda_{\max}(R))^{-\frac{\eta+1}{2}}(\mathcal{V}(t))^{\frac{\eta+1}{2}},
\end{aligned}
$$

where $\Omega = \begin{bmatrix} -RA - A^T R - \tilde{K} - \tilde{K}^T - C^+GC^- + \xi C & RB + \frac{(C^+ + C^-)G}{2} & R \\ * & -G + \xi C & -O_f^T/2 \\ * & * & -O_i \end{bmatrix}$,

$\hat{C} = \mathrm{diag}\{C, C\}$.

From (3), it is clear that $\Omega < 0$. Hence

$$
{}_{t_0}^{C}D_t^{\beta}\mathcal{V}(t) - y^T(t)i(t) \leq -2k_1\lambda_{\min}(R)(\lambda_{\max}(R))^{-\frac{\eta+1}{2}}(\mathcal{V}(t))^{\frac{\eta+1}{2}}.
$$

Thus,

$$ {}_{t_0}^{C}D_t^{\beta}\mathcal{V}(t) + a(\mathcal{V}(t))^q \leq y^T(t)i(t), $$

where $a = 2k_1\lambda_{\min}(R)(\lambda_{\max}(R))^{-\frac{\eta+1}{2}} > 0$ and $0 < q = \frac{\eta+1}{2} < \beta$. Therefore, under condition (3), FONNs (1) is FTP according to Definition 4.

Corollary 1 *The system* (1) *with* $y(t) \equiv 0$ *and* $i(t) \equiv 0$ *is FTS under the control scheme* $u(t)$ *given in* (2)*, if there exist symmetric matrices* $R > 0, C > 0$*, diagonal matrices* $G > 0$ *and* $\tilde{K}$ *of appropriate dimensions such that,*

$$ \begin{bmatrix} \Xi & RB + \frac{(C^+ + C^-)G}{2} \\ * & -G + \hat{\xi}C \end{bmatrix} < 0, \tag{9} $$

where $\Xi, \hat{\xi}, C^+$ *and* C^- *are described in Theorem 1. Then, the controller gain matrix can be found as* $K = R^{-1}\tilde{K}$ *and the upper bound of the settling time could be calculated as*

$$ T^* = t_0 + \left[\frac{\Gamma(1+\beta)\Gamma\left(\frac{1-\eta}{2}\right)\mathcal{V}^{\beta-\frac{\eta+1}{2}}(1)}{2k_1\lambda_{\min}(R)(\lambda_{\max}(R))^{-\frac{\eta+1}{2}}\Gamma\left(\beta+\frac{1-\eta}{2}\right)} \right]^{1/\beta}. $$

Proof Considering the same LKF and following the same steps as presented in the Theorem 1, one could easily prove this result. The upper bound of settling time is also calculated with the aid of Lemma 1.

4.2 Analysis for FONNs with Impulses

If the system (1) include impulsive perturbation, then it gets transformed to the following system (10).

$$ \begin{aligned} {}_{t_k}^{C}D_t^{\beta}\mathcal{Z}(t) &= -A\mathcal{Z}(t) + B\mathcal{G}(\mathcal{Z}(t)) + B_d\mathcal{G}(\mathcal{Z}(\xi t)) + i(t) + u(t), \\ & \quad t \in (t_k, t_{k+1}],\ t \neq t_k, \\ \mathcal{Z}(t_k^+) &= \mathcal{Z}(t_k) = M_k\mathcal{Z}(t_k^-),\ k \in \mathbb{N}, \quad \mathcal{Z}(t) = \phi(t),\ t \in [\bar{\xi}t_0, t_0]; \\ y(t) &= O_f\mathcal{G}(\mathcal{Z}(t)) + O_i i(t) \end{aligned} \tag{10} $$

where $\{t_k\}_{k=1}^{\infty}$ denotes the sequence of impulse points such that $0 < t_1 < t_2 < \cdots < t_k < t_{k+1} < \cdots$ and $\lim_{t\to\infty} t_k = \infty$. Further, the neuronal states are assumed to be right continuous at each impulse point t_k. The rest of the parameters take the same meaning as in (1).

The following Lemma 5 is an extension of Lemma 7 in [23] for FONNs with impulses. This lemma is needed to propose the definition of FTP for impulsive fractional-order differential systems.

Lemma 5 *Assume that a positive definite functional $\mathcal{V}(t)$ satisfies the following fractional-order differential inequality,*

$$ {}_{t_k}^{C}D_t^{\beta}\mathcal{V}(t) \leq -\mathrm{d}\mathcal{V}^{\gamma}(t), \quad t \in (t_k, t_{k+1}] \tag{11} $$

$$ \mathcal{V}(t_k) \leq \mathcal{V}(t_k^-), \quad k \in \mathbb{N}. \tag{12} $$

where $d > 0$, $0 < \gamma < \beta$ are all some constants. Then, $\exists\, t^* \in \mathbb{R}_+$ such that $\mathcal{V}(t) = 0$ for all $t \geq t^*$. Here, t^* is called as settling time, and its estimate is given by,

$$ t^* = t_0 + \left(\frac{\Gamma(1+\beta)\Gamma(1-\gamma)\mathcal{V}^{\beta-\gamma}(t_0)}{d\Gamma(1+\beta-\gamma)}\right)^{1/\beta}. $$

Proof We prove this result using comparison principle for impulsive fractional differential equations. Now, consider the following comparison system.

$$ {}_{t_k}^{C}D_t^{\beta}r(t) = -\mathrm{d}\, r^{\gamma}(t), \quad t \in (t_k, t_{k+1}],\ t \neq t_k, \tag{13} $$

$$ r(t_0) = \mathcal{V}(t_0), \quad r(t) \geq 0, \quad r(t_k) = r(t_k^-),\ k \in \mathbb{N}. \tag{14} $$

Multiplying both sides of (13) by $\frac{\Gamma(1+\beta-\gamma)}{\Gamma(1-\gamma)}r^{-\gamma}(t)$ yields,

$$ \frac{\Gamma(1+\beta-\gamma)}{\Gamma(1-\gamma)}r^{-\gamma}(t)\,{}_{t_k}^{C}D_t^{\beta}r(t) = -\mathrm{d}\frac{\Gamma(1+\beta-\gamma)}{\Gamma(1-\gamma)}. \tag{15} $$

From Lemma 2(iv), (15) can be rewritten as follows,

$$ {}_{t_k}^{C}D_t^{\beta}r^{\beta-\gamma}(t) = -\mathrm{d}\frac{\Gamma(1+\beta-\gamma)}{\Gamma(1-\gamma)}. $$

Applying fractional integration on either sides and simplifying, the following Eq. (16) could be obtained.

$$ \begin{aligned} r^{\beta-\gamma}(t) &= r^{\beta-\gamma}(t_k) - \mathrm{d}\frac{\Gamma(1+\beta-\gamma)}{\Gamma(1-\gamma)\Gamma(\beta)}\int_{t_k}^{t}\frac{1}{(t-s)^{(1-\beta)}}\,\mathrm{d}s \\ &= r^{\beta-\gamma}(t_k) - \mathrm{d}\frac{\Gamma(1+\beta-\gamma)}{\Gamma(1-\gamma)\Gamma(\beta+1)}(t-t_k)^{\beta}. \end{aligned} \tag{16} $$

Again repeating this procedure for $t = t_k$ and using Eqs. (13)–(14) we obtain,

$$ r^{\beta-\gamma}(t_k) = r^{\beta-\gamma}(t_{k-1}) - \mathrm{d}\frac{\Gamma(1+\beta-\gamma)}{\Gamma(1-\gamma)\Gamma(1+\beta)}(t_k - t_{k-1})^{\beta}, \quad \forall k \in \mathbb{N}. \tag{17} $$

Equations (16) and (17) together lead to,

$$r^{\beta-\gamma}(t) = r^{\beta-\gamma}(t_{k-1}) - \mathrm{d}\frac{\Gamma(1+\beta-\gamma)}{\Gamma(1-\gamma)\Gamma(1+\beta)}[(t-t_k)^{\beta} + (t_k - t_{k-1})^{\beta}]$$

$$\vdots$$

$$= r^{\beta-\gamma}(t_0) - \mathrm{d}\frac{\Gamma(1+\beta-\gamma)}{\Gamma(1-\gamma)\Gamma(1+\beta)}[(t-t_k)^{\beta} + (t_k - t_{k-1})^{\beta} + \cdots + (t_1 - t_0)^{\beta}].$$

Employing Lemma 3 with $n_1 = 1$ and $n_2 = \beta$ to the last equation we get,

$$r^{\beta-\gamma}(t) \le r^{\beta-\gamma}(t_0) - \mathrm{d}\frac{\Gamma(1+\beta-\gamma)}{\Gamma(1-\gamma)\Gamma(1+\beta)}[(t-t_0)^{\beta}].$$

Let $h(t) = r^{\beta-\gamma}(t_0) - \mathrm{d}\frac{\Gamma(1+\beta-\gamma)}{\Gamma(1-\gamma)\Gamma(1+\beta)}[(t-t_0)^{\beta}]$. Then, clearly, $h(t)$ is a strictly decreasing function of t and $r^{\beta-\gamma} \le h(t)$. Now, $h(t) = 0$ iff

$$t = \left[\frac{\Gamma(1-\gamma)\Gamma(1+\beta)}{\mathrm{d}\Gamma(1+\beta-\gamma)} r^{\beta-\gamma}(t_0)\right]^{1/\beta}_{+t_0} = \left[\frac{\Gamma(1-\gamma)\Gamma(1+\beta)}{\mathrm{d}\Gamma(1+\beta-\gamma)} \mathcal{V}^{\beta-\gamma}(t_0)\right]^{1/\beta}_{+t_0}$$
$$= t^*(\text{say}).$$

As $h(t)$ is a decreasing function, clearly, $r^{\beta-\gamma}(t) \le h(t) \le 0$, $\forall t \ge t^*$. Further from (14), we know that, $r(t) \ge 0$. Thus, $r(t) = 0 \forall t \ge t^*$. The comparison lemma (Lemma 4) together with (11) and (12) imply that,

$$\mathcal{V}(t) \le r(t).$$

As $r(t) = 0$ for every $t \ge t^*$, and $\mathcal{V}(t)$ is a positive definite functional, we get that $\mathcal{V}(t) = 0$ for all $t \ge t^*$. This proves Lemma 5.

With Lemma 5 the definition of FTP of impulsive fractional order systems is proposed herein Definition 5.

Definition 5 The system (10) with input $\omega(t)$ and output $y(t)$ is said to be FTP if there exists a positive definite functional $\mathcal{V}(t)$ such that,

$$\begin{aligned} {}^{C}_{t_k}D^{\beta}_{t}\mathcal{V}(t) + a(\mathcal{V}(t))^{q} &\le y^{T}(t)\omega(t), \quad \forall t \in (t_k, t_{k+1}] \\ \mathcal{V}(t_k) &\le \mathcal{V}(t_k^{-}), \quad \forall k \in \mathbb{N}. \end{aligned}$$

The following theorem gives the sufficient conditions for system (10) to be FTP under the same control input $u(t)$ defined in (2).

Theorem 2 *The system* (10) *is FTP if there exists real matrix $C > 0$, diagonal matrices $R > 0$, $G > 0$ and $\tilde{K}$ of appropriate dimensions such that,*

$$\begin{bmatrix} \Xi & RB + \frac{(C^+ + C^-)G}{2} & R \\ * & -G + \hat{\xi}C & -O_f^T/2 \\ * & * & -O_i \end{bmatrix} < 0 \tag{18}$$

and

$$M_k^T P M_k - P < 0, \quad k \in \mathbb{N} \tag{19}$$

where $\Xi = -RA - A^T R - \tilde{K} - \tilde{K}^T - C^+ G C^- + \hat{\xi}C$; $\hat{\xi} = \sum_{i=1}^{n} \frac{(1-\xi_i)^\beta}{\beta}$; $C^+ = \text{diag}\{c_1^+, c_2^+, \ldots, c_n^+\}$ *and* $C^- = \text{diag}\{c_1^-, c_2^-, \ldots, c_n^-\}$. *Then, the controller gain is given by* $K = R^{-1}\tilde{K}$.

Proof Taking the positive definite LKF to be $\mathcal{V}(t) = \mathcal{Z}^T(t) R \mathcal{Z}(t)$ and proceeding as in Theorem 1, we obtain,

$$ {}_{t_k}^{C}D_t^\beta \mathcal{V}(t) \leq \chi^T(t)\Omega'\chi(t) - 2k_1\lambda_{\min}(R)\left(\sum_{i=1}^{N} |\mathcal{Z}_i(t)|^2\right)^{\frac{\eta+1}{2}} \tag{20}$$

where $\chi(t) = \left[\mathcal{Z}^T(t)\ \mathcal{G}^T(\mathcal{Z}(t))\ i^T(t)\right]^T$ and $\Omega' = \begin{bmatrix} \Xi & RB + \frac{(C^+ + C^-)G}{2} & R \\ * & -G + \hat{\xi}C & 0 \\ * & * & 0 \end{bmatrix}$ with $\Xi = -RA - A^T R - \tilde{\mathrm{K}} - \tilde{\mathrm{K}}^T - C^+ G C^- + \hat{\xi}C$ and $\tilde{\mathrm{K}} = RK$. Therefore,

$$\begin{aligned} {}_{t_k}^{C}D_t^\beta \mathcal{V}(t) - y^T(t)i(t) &\leq \chi^T(t)\Omega\chi(t) - 2k_1\lambda_{\min}(R)\left(\lambda_{\max}(R)\right)^{-\frac{\eta+1}{2}}\left(\mathcal{V}(t)\right)^{\frac{\eta+1}{2}} \\ {}_{t_k}^{C}D_t^\beta \mathcal{V}(t) + a(\mathcal{V}(t))^q &\leq y^T(t)i(t) \end{aligned} \tag{21}$$

where Ω is the matrix given in (18). Further by (19),

$$\mathcal{V}(t_k) = \mathcal{Z}^T(t_k)R\mathcal{Z}(t_k) = \mathcal{Z}^T(t_k^-)M_k^T R M_k \mathcal{Z}(t_k^-) \leq \mathcal{Z}^T(t_k^-)R\mathcal{Z}(t_k^-) = \mathcal{V}(t_k^-) \tag{22}$$

Thus by (21) and (22), the impulsive FONNs (10) is FTP under the controller $u(t)$ defined in (2) when the LMIs (18) and (19) hold.

Similar to Corollary 1, one can derive the following Corollary 2 which gives the sufficient conditions for FONNs (10) under zero input to be FTS with controller (2).

Corollary 2 *The system* (10) *with* $y(t) \equiv 0$ *and* $i(t) \equiv 0$ *is FTS under the control scheme* $u(t)$ *given in* (2), *if there exist symmetric matrices* $R > 0, C > 0$, *diagonal matrices* $G > 0$ *and* $\tilde{K}$ *of appropriate dimensions such that,*

$$\begin{bmatrix} \Xi & RB + \frac{(C^+ + C^-)G}{2} \\ * & -G + \hat{\xi}C \end{bmatrix} < 0,$$

and

$$M_k^T R M_k - R < 0, \quad \forall t \in \mathbb{N}.$$

where $\Xi, \hat{\xi}, C^+$ and C^- are described in Theorem 1. Then, the controller gain matrix can be found as $K = R^{-1}\tilde{K}$ and the upper bound of the settling time could be calculated as

$$T^* = t_0 + \left[\frac{\Gamma(1+\beta)\Gamma\left(\frac{1-\eta}{2}\right) \mathcal{V}^{\beta-\frac{\eta+1}{2}}(1)}{2k_1\lambda_{\min}(R)(\lambda_{\max}(R))^{-\frac{\eta+1}{2}}\Gamma(\beta+\frac{1-\eta}{2})} \right]^{1/\beta}.$$

Proof Using Lemma 4 and using the same LKF as in Theorem 2, one could prove this result. As the proof is very similar to that of Theorem 2, it is omitted here.

5 Numerical Simulations

The aim of this section is to validate the propounded results. The results are simulated in MATLAB, and the state trajectories are plotted by implementing the algorithm based on Grunwald–Letnikov (G-L) definition for numerical solution of fractional-order delayed systems [28]. A reason for choosing this algorithm is that it is, efficient and simple for implementation.

Example 1 Consider the FONNs of order $\beta = 0.8$ with the following parameters

$$A = \begin{bmatrix} 5.7 & 0 \\ 0 & 5.6 \end{bmatrix}; \quad B = \begin{bmatrix} 2.3 & -0.5 \\ 2.3 & 2.1 \end{bmatrix}; \quad B_d = \begin{bmatrix} 3.9 & 1 \\ -1 & 3 \end{bmatrix};$$
$$O_f = \begin{bmatrix} 0.1 & 0.2 \\ 0.2 & 0.1 \end{bmatrix}; \quad O_i = \begin{bmatrix} 0.5 & 0 \\ 0 & 2 \end{bmatrix}; \quad \xi = (0.5, 0.5).$$

Select $\mathcal{G}(\mathcal{Z}(t)) = \tanh(\mathcal{Z}(t))$. Then $C^+ = \begin{bmatrix} 1 & 0 \\ 0 & 1 \end{bmatrix}$, $C^- = \begin{bmatrix} -1 & 0 \\ 0 & -1 \end{bmatrix}$ and $M = \begin{bmatrix} 1 & 0 \\ 0 & 1 \end{bmatrix}$.

Without Impulse: With these values, it is found from MATLAB LMI toolbox that the inequality (3) is feasible and the control gain matrix is obtained as $K = \begin{bmatrix} 15.8462 & 0 \\ 0 & 28.6780 \end{bmatrix}$.

Now according to the Theorem 1, the considered FONNs are FTP under the control input

$$u(t) = -K\mathcal{Z}(t) - 0.6\text{sign}(\mathcal{Z}(t))|\mathcal{Z}(t)|^{0.3} - \text{sign}(\mathcal{Z}(t))B_d M.$$

Figure 1 shows the trajectories of the neuron states with disturbance input $i(t) = [10\sin(t); 8\cos(t)]$.

According to the Corollary 1, this system with the same initial values will converge to its equilibrium point 0 within $T^* = 8.8957$ s. This conclusion is validated

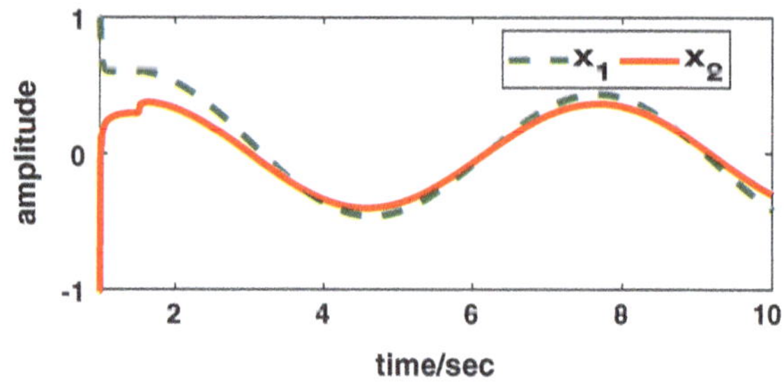

Fig. 1 State trajectories of the FONNs in Example 1 with $i(t) = [10\sin(t); 8\cos(t)]$

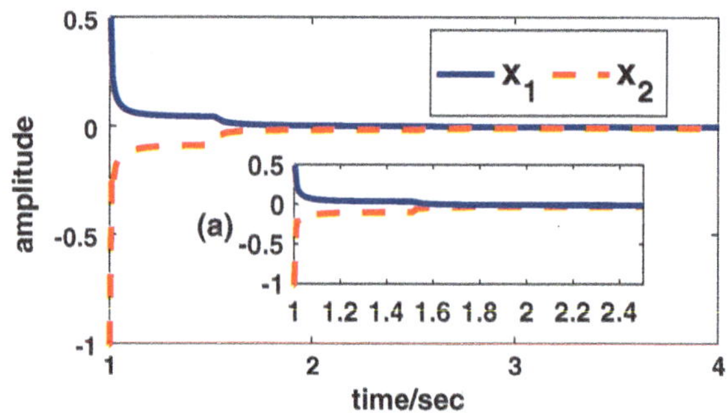

Fig. 2 State trajectories of FONNs in Example 1 with initial value $\mathcal{Z}_0 = [0.5; -1]$; **a** shows the portion of the graph between 1 and 2.5

by Fig. 2 which shows the state trajectory of the above said FONNs under zero disturbance input. It could be observed that the state reaches zero around 1.8 s. This clearly proves the validity of the obtained theoretical results.

With Impulses: For the same parameters and controller chosen for FONNs (1), the LMIs in Theorem 2 are found to be feasible with impulse gain matrix $M_k = 0.45\ I_{2\times 2}$ and the control gain matrix is found as $K = \begin{bmatrix} 6.8282 & 0 \\ 0 & 14.0113 \end{bmatrix}$. Hence, according to Theorem 2, the FONNs (10) is FTP. Further, as LMIs in Corollary 2 are also feasible, this FONN is also FTS with settling time 7.5964s. This conclusion is validated by plotting the trajectories as shown in Fig. 3. It could be observed that the states converge to its equilibrium point zero a bit later than that without impulse.

Remark 3 The model presented in this work is that a fractional-order Hopfield neural networks. The FONNs (1) without delays could be realized using the circuit given in Fig. 4 [29]. This circuit consists of resistor, capacitor and operational amplifiers. The resistors R_{ij} connecting the output of jth fractional neuron to the input of ith fractional neuron simulate the interconnection synaptic weights (given by the matrices A, B in Example 1) between the corresponding neurons. Operational amplifiers are used to perform the action of nonlinear activation functions. F_i is the fractance of ith fractor. F_i in parallel connection with R_i represents ith fractional neuron of the FONNs considered here. The voltage $Z_i(t)$ across this connection varies with time and is the neuron state value. $y_i(t)$ is the output voltage. Finally, N is the number of neurons in this model.

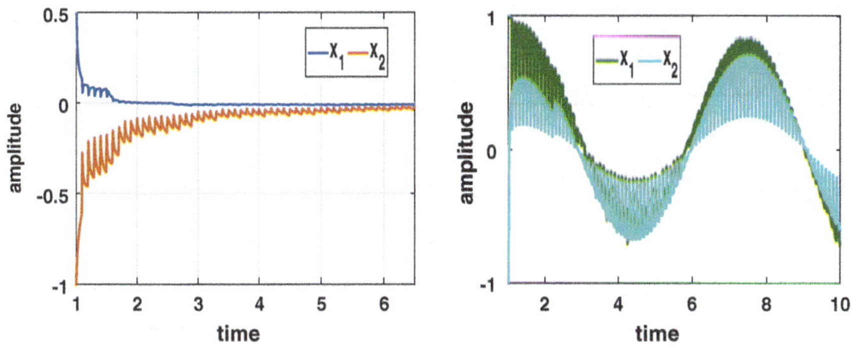

(a) Without input (b) With input

Fig. 3 State trajectories for the impulsive FONNs (10)

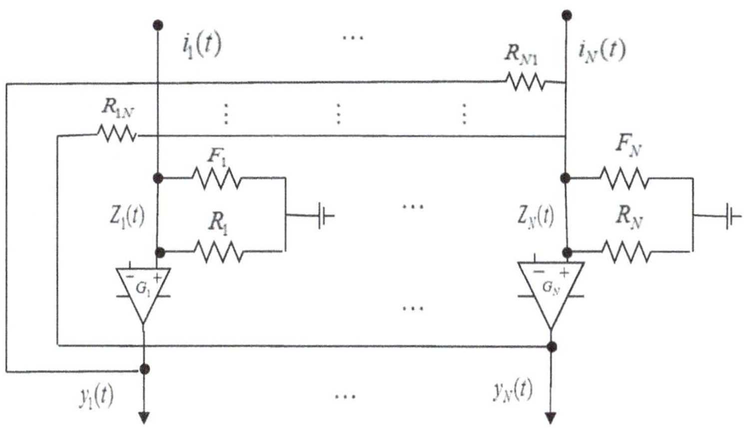

Fig. 4 Circuit model of FONNs (1) without delays

Remark 4 In order to expedite the advantage of modelling NNs with fractional derivative, we herein plot the trajectories for the corresponding integer-order NNs and FONNs (1) and (10) with $\beta = 0.9$ and $\beta = 1$. It could be observed that convergence time varies with fractional-order and for integer-order system the convergence is slightly longer than that of the fractional-order system. Thus, fractional order of the system provides additional advantage of adjusting the convergence time of the system (Fig. 5).

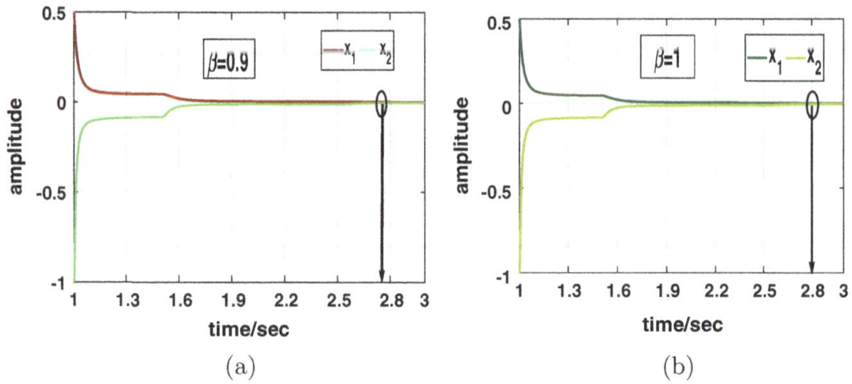

Fig. 5 State trajectory of FONNs (1) with **a** $\beta = 0.9$ **b** $\beta = 1$

6 Conclusion and Future Directions

In this work, FTP analysis of FONNs with proportional delays under a suitable control scheme has been done for the first time in the literature. New set of sufficient conditions for the considered FONNs to be FTS along with an estimate for the settling time have been derived. Finally, an example has been demonstrated to show the validity of the propounded results. However, in this work, an estimate of settling time depends on initial conditions and does not depend on the delay values. This drawback will be overcome in our future research works.

Acknowledgements *Funding*: This work was supported by UGC-SAP (DSA-I), New Delhi, India, File No.: F.510/7/DSA-1/2015(SAP-I).

References

1. Kilbas, A., Srivastava, H., Trujillo, J.: Theory and Applications of Fractional Differential Equations. Elsevier, New York (2006)
2. Jahanbakhti, H.: A novel fractional-order neural network for model reduction of large-scale systems with fractional-order nonlinear structure. Soft Comput. **24**, 13489–13499 (2020)
3. Wang, H., Yu, Y., Wen, G.: Stability analysis of fractional order Hopfield neural networks with time delays. Neural Netw. **55**, 98–109 (2014)
4. Wan, L., Liu, Z.: Multiple $O(t^{-\alpha})$ stability for fractional-order neural networks with time-varying delays. J. Franklin Inst. **357**(17), 12742–12766 (2020)
5. Celik, V.: Bifurcation analysis of fractional order single cell with delay. Int. J. Bifurcat. Chaos **25**(02), 1550020 (2015)
6. Zhou, L.: On the global dissipativity of a class of cellular neural networks with multi-pantograph delays. Adv. Artif. Neural Syst. **2011**, 941426 (2011)
7. Zhou, L.: Delay-dependent and delay-independent passivity of a class of recurrent neural networks with impulse and multi-proportional delays. Neurocomputing **308**, 235–244 (2018)

8. Zhou, L.: Dissipativity of a class of cellular neural networks with proportional delays. Nonlinear Dyn. **73**, 1895–1903 (2013)
9. Zhou, L.: Delay-dependent exponential stability of cellular neural networks with multi-proportional delays. Neural Process. Lett. **38**, 321–346 (2013)
10. Xiong, X., Tang, R., Yang, X.: Finite-time synchronization of memristive neural networks with proportional delay. Neural Process. Lett. **50**, 1139–1152 (2019)
11. Hu, J., Sui, G.: Fixed-time control of static impulsive neural networks with infinite distributed delay and uncertainty. Commun. Nonlinear Sci. Numer. Simul. **78**, 104848 (2019)
12. Li, H., Li, C., Huang, T., Zhang, W.: Fixed-time stabilization of impulsive Cohen-Grossberg BAM neural networks. Neural Netw. **98**, 203–211 (2018)
13. Wu, A., Zeng, Z.: Passivity analysis of memristive neural networks with different memductance functions. Commun. Nonlinear Sci. Numer. Simul. **19**, 274–285 (2014)
14. Wang, S., Cao, Y., Huang, T., Wen, S.: Passivity and passification of memristive neural networks with leakage term and time-varying delays. Appl. Math. Comput. **361**, 294–310 (2019)
15. Velmurugan, G., Rakkiyappan, R., Lakshmanan, S.: Passivity analysis of memristor based complex valued neural networks with time-varying delays. Neural Process. Lett. **42**, 517–540 (2015)
16. Jiang, B., Hu, Q., Friswell, M.I.: Fixed-time attitude control for rigid spacecraft with actuator saturation and faults. IEEE Trans. Control Syst. Technol. **24**, 1892–1898 (2016)
17. Polyakov, A., Efimov, D., Perruquetti, W.: Robust stabilization of MIMO systems in finite/fixed-time. Int. J. Robust Nonlinear Control **26**, 69–90 (2016)
18. Peng, X., Wu, H., Cao, J.: Global nonfragile synchronization in finite time for fractional order discontinuous neural networks with nonlinear growth activations. IEEE Trans. Neural Netw. Learn. Syst. **30**(7), 2123–2137 (2018)
19. Wang, J., Wu, H., Huang, T.: Passivity based synchronization of a class of complex dynamic networks with time-varying delays. Automatica **56**(1), 105–112 (2010)
20. Hou, M., Tan, F., Duan, G.: Finite-time passivity of dynamic systems. J. Franklin Inst. **353**(18), 4870–4884 (2016)
21. Xiao, J., Zeng, Z.: Finite-time passivity of neural networks with time-varying delay. J. Franklin Inst. **357**(4), 2437–2456 (2020)
22. Wang, Z., Cao, J., Lu, G., Abdel-Aty, M.: Fixed-time passification analysis of interconnected memristive reaction-diffusion neural networks. IEEE Trans. Netw. Sci. Eng. **7**(3), 1814–1824 (2020)
23. Li, H.-L., Cao, J., Jiang, H., Alsaedi, A.: Graph theory based finite-time synchronization of fractional-order complex dynamical networks. J. Franklin Inst. **355**(13), 5771–5789 (2018)
24. Li, C., Deng, W.: Remarks on fractional derivatives. Appl. Math. Comput. **187**, 777–784 (2007)
25. Duarte-Mermoud, M.A., Aguila-Camacho, N., Gallegos, J.A., Castro-Linares, R.: Using general quadratic Lyapunov functions to prove Lyapunov uniform stability for fractional order systems. Commun. Nonlinear Sci. Numer. Simul. **22**, 650–659 (2015)
26. Wang, L., Shen, Y., Ding, Z.: Finite-time stabilization of delayed neural networks. Neural Netw. **70**, 74–80 (2015)
27. Wu, R., Fečkan, M.: Stabilty analysis of impulsive fractional-order systems by vector comparison principle. Nonlinear Dyn. **82**, 2007–2019 (2015)
28. Wang, Z., Huang, X., Zhou, J.: A numerical method for delayed fractional-order differential equations: based on G-L definition. Appl. Math. Inf. Sci. **7**(21), 525–529 (2013)
29. Pu, Y.-F., Yi, Z., Zhou, J.-L.: Fractional Hopfield neural networks: fractional dynamic associative recurrent neural networks. IEEE Trans. Neural Netw. Learn. Syst. **28**(10), 2319–2333 (2017)

Synchronization of Delayed Fractional-Order Memristive BAM Neural Networks

M. Shafiya and G. Nagamani

Abstract In this paper, the global Mittag-Leffler synchronization problem for a class of fractional-order memristive bi-directional associative memory (FOMBAM) neural networks (NNs) with time-varying delays has been investigated. The motive of this study is to establish an efficient slave system for the considered master system by proving that the neuron states of the synchronization error signal is globally Mittag-Leffler stable. By utilizing the master–slave synchronization technique and Lyapunov approach, the desired global Mittag-Leffler synchronization criteria for the addressed FOMBAM NNs are derived in the form of linear matrix inequalities (LMIs). Finally, one numerical example is also presented to illustrate the feasibility of the proposed theoretical results.

Keywords Caputo's fractional derivative · Memristor · Neural networks · Lyapunov function · Linear matrix inequalities

1 Introduction

Artificial NNs are the computational models that are designed to perform the information process in the same way as human brain and these are inspired from biological NNs. They have self-learning capabilities that enable them to produce better result due to the availability of more data. Each neuron can make simple decisions, and feed those decisions to other neurons, systematized in interconnected layers. There are many types of NNs, namely Cohen–Grossberg NNs, recurrent NNs, bi-directional associative memory (BAM) NNs, and so on [1–3]. Every types of NNs has its own advantages over the other NNs. Among them, BAM NNs are a type of recurrent NNs which was first introduced by Kosko in 1987 [4, 5]. The remarkable feature of BAM NNs is its close relationship of the neurons between the two layers. Because of this specific characteristics of BAM NNs, their dynamical behavior have been

M. Shafiya · G. Nagamani (✉)
Department of Mathematics, The Gandhigram Rural Institute (Deemed to be University), Gandhigram, Tamil Nadu, India

P. Balasubramaniam et al. (eds.), *Mathematical Modelling and Computational Intelligence Techniques*, Springer Proceedings in Mathematics & Statistics 376,
https://doi.org/10.1007/978-981-16-6018-4_14

extensively studied and successfully applied in the fields of pattern recognition, signal and image processing, solving optimization problems and automatic control engineering. On the other hand, in 1971, Leon O. Chua reasoned from physical symmetry arguments that, apart from the resistor, capacitor, and inductor, there should exist a fourth circuit component, called a memristor (as a contraction of memory and resistor) [6, 7]. Memristor is a passive two-terminal electronic device described by the nonlinear relationship between the electric charge and flux. Owing to these reasons, the resistors can be replaced by memristors to build some new models of NNs, namely the memristor-based NNs (MNNs), where the connection weights changes with state. Due to this reason, in recent years, many researchers have studied the dynamical behavior of MNNs because of its effective computing in various fields, namely image processing and secure communication [8]. By virtue of the characteristics of memristor and BAM NNs, the dynamical behavior of memristor-based or memristive BAM NNs has been extensively applied in the field of pattern recognition and neural learning.

In 1695, the theory of fractional calculus is introduced by Leibniz [9, 10]. The field of fractional calculus is the generalization of the traditional calculus. The fractional-order derivative has non-local property and infinite memory. As derivative operators in artificial NNs are usually implemented by capacitors, researchers tried to replace the ideal integer-order capacitors with fractional-order capacitors, and hence constructed fractional-order NNs (FONNs). Compared with integer-order NNs, FONNs are more suitable to emulate biological NNs. Considering the fact that memristor-based synapses can work in a similar way to synapses in human brains, it is necessary to incorporate memristor-based synapses into FONNs. Due to these unique features of fractional derivatives and memristor, up to now, many authors have investigated various sorts of dynamical behaviors for memristor-based FONNs (FOMNNs) [7, 11].

After the pioneering work made by Pecora and Carroll [12], the study of chaos synchronization has received rapid attention because of its wide applications in the fields of information processing and secure communication. In [13], it has been stated that FOMNNs can exhibit chaos, which brought remarkable research interest in the field of synchronization of FOMNNs and many efficient controllers for synchronization of FOMNNs, namely linear feedback control, switching control and adaptive control have been designed [14–18]. Owing to these facts, in our paper, for the first time, we have introduced the global synchronization of delayed FOMBAM NNs in the Mittag-Leffler sense through LMI technique.

The merits of this paper are listed as follows:

(i) In this paper, on the light of the bilayer structure of BAM NNs and the memristor characteristics, a class of FOMBAM NNs has been proposed by taking the time-varying delays into account.
(ii) The linear feedback controller is designed for the considered FOMBAM NNs and with reference to the error analysis technique, the desired Mittag-Leffler synchronization criteria for the FOMBAM NNs has been derived through the framework of Lyapunov stability theory in terms of LMIs.
(iii) To facilitate the effectiveness and validity of the proposed theoretical results, the numerical example along with simulation results is presented.

The rest of this paper is structured as follows: Some of the basic and fundamental definitions related to fractional calculus have been discussed in Sect. 2. For the addressed delayed FOMBAM NNs, a sufficient LMI criterion has been developed to analyze the synchronization behavior in Sect. 3. To testify the validity of the established results, one numerical example with simulations have been demonstrated in Sect. 4. Finally, some general concluding remarks are given in Sect. 5.

Notations: Let the set $\mathcal{I}_n$ ($n \in \mathbb{N}$) be defined as $\mathcal{I}_n = \{1, 2, \dots, n\}$. 0 denotes the zero matrix, I denotes the identity matrix, diag$\{\cdots\}$ represents a block diagonal matrix. Given a matrix $Z = (z_{ij})_{n\times n}$, $\lambda_{\max}(Z)$ and $\lambda_{\min}(Z)$ stand for the maximal and minimal eigenvalues of Z, respectively; $Z > 0\,(Z < 0)$ suggest that Z is positive definite (negative definite). Let $C^q([\eta, \infty), \mathbb{R})$, $(\eta \geq 0)$ stands for the space of all q-order continuous and differentiable functions from $[\eta, \infty)$ to $\mathbb{R}$.

2 System Formulation and Preliminaries

In this section, we recall some basics of fractional calculus, definitions and lemmas that are necessary in deriving the main results.

Definition 1 ([10]) For two parameters $a > 0$ and $b > 0$, the Mittag-Leffler function is defined as

$$E_{a,b}(z) = \sum_{k=0}^{\infty} \frac{z^k}{\Gamma(ak + b)},$$

where $z \in \mathbb{C}$. When $b = 1$, its one-parameter form can be rewritten as

$$E_a(z) = E_{a,1}(z) = \sum_{k=0}^{\infty} \frac{z^k}{\Gamma(ak + 1)}.$$

Especially, $E_{1,1}(z) = e^z$.

Definition 2 ([10]) For the function $x(t) \in C^m([t_0, \infty), \mathbb{R})$, the αth-order Caputo's fractional-order derivative ${}_{t_0}^{C}D_t^{\alpha}$ is given by

$${}_{t_0}^{C}D_t^{\alpha} x(t) = \frac{1}{\Gamma(m - \alpha)} \int_{t_0}^{t} (t - s)^{m-\alpha-1} x^{(m)}(s)\mathrm{d}s$$

in which $t \geq t_0$, α is a positive scalar and integer m satisfies $0 < m - 1 < \alpha < m$.

For notational convenience, the operator D^{α} will be considered instead of ${}_{t_0}^{C}D_t^{\alpha}$ throughout this sequel.

Remark 1 Caputo's derivative demands higher conditions of regularity for differentiability: to compute the fractional derivative of a function in the Caputo sense, we must first calculate its derivative. Caputo derivatives are defined only for differentiable functions while functions that have no first-order derivative might have fractional derivatives of all orders less than one in the Riemann–Liouville sense. Even though, the fractional derivatives have disadvantages, the definition of Riemann–Liouville plays an important role mainly in the development of the theory of fractional derivatives and integrals and for its application in pure mathematics (solution of integer-order differential equations, definitions of some new function classes and so on) [10]. However, applied problems require proper definitions of fractional derivatives which can provide initial conditions with clear physical interpretation for the differential equations of fractional order. This makes the Caputo fractional derivative more suitable to be applied. Owing to these facts, throughout this paper, the fractional-order derivative has been considered as Caputo's fractional-order derivative.

In this paper, the following class of FOMBAM NNs with time-varying delays has been considered:

$$\begin{cases} D^{\alpha}x(t) = -A_x x(t) + B_x f_x(y(t)) + C_x f_x(y(t-h_y(t))) + J_x(t), \\ D^{\alpha}y(t) = -A_y y(t) + B_y f_y(x(t)) + C_y f_y(x(t-h_x(t))) + J_y(t), \end{cases} \tag{1}$$

where $x(t)$ and $y(t)$ represent the neuron state vectors comprised of n neurons; $f_x(y(t))$ and $f_y(x(t))$ represent the neuron activation functions; $f_x(y(t-h_y(t)))$ and $f_y(x(t-h_x(t)))$ represent the neuron activation functions in correspondence with time-varying delay; $J_x(t)$ and $J_y(t)$ represent the external inputs. $A_x = \text{diag}\{a_{x_1}, a_{x_2}, \ldots, a_{x_n}\}$ and $A_y = \text{diag}\{a_{y_1}, a_{y_2}, \ldots, a_{y_n}\}$ are the positive definite state feedback matrices; $B_x = (b_{x_{ij}})_{n\times n}$ and $B_y = (b_{y_{ij}})_{n\times n}$ are the connection weight matrices, and $C_x = (c_{x_{ij}})_{n\times n}$ and $C_y = (c_{y_{ij}})_{n\times n}$ are the connection weight matrices in correspondence with time-varying delay; $h_x(t)$ and $h_y(t)$ denote the delay functions satisfying $0 \le h_x(t) \le h_x$ and $0 \le h_y(t) \le h_y$, respectively. The initial conditions of the delayed FOMBAM NNs (1) are assumed by the condition $x(s) = \phi(s)$ and $y(s) = \psi(s)$, $s \in [-h, 0]$, where $h = \max\{h_x, h_y\}$ and the initial value functions $\phi(\cdot)$ and $\psi(\cdot)$ from $[-h, 0]$ to $\mathbb{R}$ are continuous functions.

Motivated by the characteristics and features of memristor and current-voltage and based on Kirchoff's current law, the matrix entries $a_{x_i}(t)$, $a_{y_i}(t)$, $b_{x_{ij}}(t)$, $b_{y_{ij}}(t)$, $c_{x_{ij}}(t)$ and $c_{y_{ij}}(t)$ $(i, j \in \mathcal{I}_n)$, are defined as follows, since they represent the memristive synaptic weights:

$$a_{x_i}(t) = \frac{1}{C_{x_i}}\left[\sum_{j=1}^{n}\left(\frac{1}{R_{b_{x_{ij}}}} + \frac{1}{R_{c_{x_{ij}}}}\right)\text{sign}_{ij} + \frac{1}{R_{x_i}}\right] = \begin{cases} \widehat{a}_{x_i} & |x_i| > \kappa_{x_i}, \\ \widetilde{a}_{x_i} & |x_i| \le \kappa_{x_i}, \end{cases}$$

$$a_{y_i}(t) = \frac{1}{C_{y_i}}\left[\sum_{j=1}^{n}\left(\frac{1}{R_{b_{y_{ij}}}} + \frac{1}{R_{c_{y_{ij}}}}\right)\text{sign}_{ij} + \frac{1}{R_{y_i}}\right] = \begin{cases} \widehat{a}_{y_i} & |y_i| > \kappa_{y_i}, \\ \widetilde{a}_{y_i} & |y_i| \le \kappa_{y_i}, \end{cases}$$

$$h_{x_{ij}}(t) = \frac{\text{sign}_{ij}}{C_{x_i} R_{b_{x_{ij}}}} = \begin{cases} \widehat{b}_{x_{ij}} & |x_i| > \kappa_{x_i}, \\ \widetilde{b}_{x_{ij}} & |x_i| \le \kappa_{x_i}, \end{cases} \quad b_{y_{ij}}(t) = \frac{\text{sign}_{ij}}{C_{y_i} R_{b_{y_{ij}}}} = \begin{cases} \widehat{b}_{y_{ij}} & |y_i| > \kappa_{y_i}, \\ \widetilde{b}_{y_{ij}} & |y_i| \le \kappa_{y_i}, \end{cases}$$

$$c_{x_{ij}}(t) = \frac{\text{sign}_{ij}}{C_{x_i} R_{c_{x_{ij}}}} = \begin{cases} \widehat{c}_{x_{ij}} & |x_i| > \kappa_{x_i}, \\ \widetilde{c}_{x_{ij}} & |x_i| \le \kappa_{x_i}, \end{cases} \quad c_{y_{ij}}(t) = \frac{\text{sign}_{ij}}{C_{y_i} R_{c_{y_{ij}}}} = \begin{cases} \widehat{c}_{y_{ij}} & |y_i| > \kappa_{y_i}, \\ \widetilde{c}_{y_{ij}} & |y_i| \le \kappa_{y_i}, \end{cases}$$

where $\text{sign}_{kl} = \begin{cases} 1 & k \ne l, \\ -1 & k = l, \end{cases}$; R_{x_i} and R_{y_i} $(i \in \mathcal{I}_n)$ denote the parallel memristor in correspondence with the capacitors C_{x_i} and C_{y_i}, respectively; $R_{b_{x_{ij}}}$, $R_{b_{y_{ij}}}$, $R_{c_{x_{ij}}}$, and $R_{c_{y_{ij}}}$ denote the memristor that establishes the connection between the state vectors and the activation functions with the switching jumps $\kappa_{x_i} > 0$ and $\kappa_{y_i} > 0$; $\widehat{a}_{x_i}$, $\widetilde{a}_{x_i}$, $\widehat{a}_{y_i}$, $\widetilde{a}_{y_i}$, $\widehat{b}_{x_{ij}}$, $\widetilde{b}_{x_{ij}}$, $\widehat{b}_{y_{ij}}$, $\widetilde{b}_{y_{ij}}$, $\widehat{c}_{x_{ij}}$, $\widetilde{c}_{x_{ij}}$, $\widehat{c}_{y_{ij}}$, and $\widetilde{c}_{y_{ij}}$ are known constants.

Quite often, while the implementation of NNs in the field of electronics, the uncertainties may occur. Due to this occurrence, there may be changes in the memristive synaptic weights. Extending the assumptions for MNNs in [19] to the memristor-based BAM NNs, the uncertainties, namely $\Delta A_x(t)$, $\Delta B_x(t)$, $\Delta C_x(t)$, $\Delta A_y(t)$, $\Delta B_y(t)$, and $\Delta C_y(t)$ are given below:

$$\Delta A_x(t) = GM_{x_1}(t)N_{x_1}, \quad \Delta B_x(t) = GM_{x_2}(t)N_{x_2}, \quad \Delta C_x(t) = GM_{x_3}(t)N_{x_3},$$
$$\Delta A_y(t) = GM_{y_1}(t)N_{y_1}, \quad \Delta B_y(t) = GM_{y_2}(t)N_{y_2}, \quad \Delta C_y(t) = GM_{y_3}(t)N_{y_3},$$

where the matrices G, N_{x_1}, N_{x_2}, N_{x_3}, N_{y_1}, N_{y_2}, and N_{y_3} are known.

Consequently, $M_{x_1}(t)$, $M_{x_2}(t)$, $M_{x_3}(t)$, $M_{y_1}(t)$, $M_{y_2}(t)$, and $M_{y_3}(t)$ are the unknown time-varying matrices satisfying $M_{x_i}^T(t)M_{x_i}(t) \le I$ and $M_{y_i}^T(t)M_{y_i}(t) \le I$ $(i \in \mathcal{I}_3)$. With the above discussion, the system (1) is transformed into

$$\begin{cases} D^\alpha x(t) = -(\widetilde{A}_x + \Delta A_x(t))x(t) + (\widetilde{B}_x + \Delta B_x(t))f_x(y(t)) \\ \qquad\qquad +(\widetilde{C}_x + \Delta C_x(t))f_x(y(t - h_y(t))) + J_x(t), \\ D^\alpha y(t) = -(\widetilde{A}_y + \Delta A_y(t))y(t) + (\widetilde{B}_y + \Delta B_y(t))f_y(x(t)) \\ \qquad\qquad +(\widetilde{C}_y + \Delta C_y(t))f_y(x(t - h_x(t))) + J_y(t). \end{cases} \tag{2}$$

In order to establish the main results, we make the following assumption on neuron activation function throughout this paper.

Assumption 1 For any ν_1 and $\nu_2 \in \mathbb{R}$ with $\nu_1 \ne \nu_2$, we have,

$$l_{x_i}^- \le \frac{f_{x_i}(\nu_1) - f_{x_i}(\nu_2)}{\nu_1 - \nu_2} \le l_{x_i}^+ \quad \text{and} \quad l_{y_i}^- \le \frac{f_{y_i}(\nu_1) - f_{y_i}(\nu_2)}{\nu_1 - \nu_2} \le l_{y_i}^+ \ (i \in \mathcal{I}_n),$$

where $l_{x_i}^-$, $l_{x_i}^+$, $l_{y_i}^-$, and $l_{y_i}^+$ $(i \in \mathcal{I}_n)$ are scalars to be determined. Moreover, the activation functions $f_{x_i}(\cdot)$ and $f_{y_i}(\cdot)$ $(i \in \mathcal{I}_n)$ of $x(t)$ and $y(t)$ are bounded and continuous.

Remark 2 In Assumption 1, the constants $l_{x_i}^-$ and $l_{x_i}^+$ can be positive, negative, or zero according to [20]. Specifically, when $l_{x_i}^- = 0$ and $l_{x_i}^+ > 0$, Assumption 1 implies that

$f_{x_i}(\cdot)$ $(i \in \mathcal{I}_n)$ are a class of monotonically non-decreasing activation functions satisfying global Lipschitz condition. When $l_{x_i}^+ > l_{x_i}^- > 0$, Assumption 1 gives a description of the class of monotonically increasing functions that their derivative has upper and lower bounds [21]. It is obvious that many monotonic or non-monotonic activation functions satisfy Assumption 1, and the commonly used sigmoid-type activation functions and piecewise linear functions are the special cases of this condition. Compared with the previous assumptions such as monotonically non-decreasing functions, common Lipschitz continuous functions, Assumption 1 is more general and applicable. Similarly, the same holds for the constants $l_{y_i}^-$ and $l_{y_i}^+$ corresponding to the function f_{y_i} $(i \in \mathcal{I}_n)$.

The master–slave synchronization technique has been taken into consideration throughout this paper. Globally, Mittag-Leffler synchronization will be achieved if all of the states of the synchronization error dynamics approach zero as the time t tends to infinity. For analyzing the synchronization characteristics of the master system (2), the corresponding slave system has been constructed as follows:

$$\begin{cases} D^\alpha \bar{x}(t) = -(\widetilde{A}_x + \Delta A_x(t))\bar{x}(t) + (\widetilde{B}_x + \Delta B_x(t)) f_x(\bar{y}(t)) \\ \qquad\qquad +(\widetilde{C}_x + \Delta C_x(t)) f_x(\bar{y}(t - h_y(t))) + J_x(t) + U_x(t), \\ D^\alpha \bar{y}(t) = -(\widetilde{A}_y + \Delta A_y(t))\bar{y}(t) + (\widetilde{B}_y + \Delta B_y(t)) f_y(\bar{x}(t)) \\ \qquad\qquad +(\widetilde{C}_y + \Delta C_y(t)) f_y(\bar{x}(t - h_x(t))) + J_y(t) + U_y(t). \end{cases} \tag{3}$$

Here, the linear feedback controllers are chosen as $U_x(t) = F_x(x(t) - \bar{x}(t))$ and $U_y(t) = F_y(y(t) - \bar{y}(t))$, where F_x and F_y are the control gain matrices.

With reference to the error analysis technique and by defining the synchronization error signal as $e_x(t) = x(t) - \bar{x}(t)$ and $e_y(t) = y(t) - \bar{y}(t)$, the following error system has been obtained from master system (2) and slave system (3):

$$\begin{cases} D^\alpha e_x(t) = -(\widetilde{A}_x + \Delta A_x(t) + F_x) e_x(t) + (\widetilde{B}_x + \Delta B_x(t)) \hat{f}_x(e_y(t)) \\ \qquad\qquad +(\widetilde{C}_x + \Delta C_x(t)) \hat{f}_x(e_y(t - h_y(t))), \\ D^\alpha e_y(t) = -(\widetilde{A}_y + \Delta A_y(t) + F_y) e_y(t) + (\widetilde{B}_y + \Delta B_y(t)) \hat{f}_y(e_x(t)) \\ \qquad\qquad +(\widetilde{C}_y + \Delta C_y(t)) \hat{f}_y(e_x(t - h_x(t))), \end{cases} \tag{4}$$

where $\hat{f}_x(e_y(\cdot)) = f_x(y(\cdot)) - f_x(\bar{y}(\cdot))$ and $\hat{f}_y(e_x(\cdot)) = f_y(x(\cdot)) - f_y(\bar{x}(\cdot))$.

In order to obtain the main results of this paper, the following definition and lemma are necessary.

Remark 3 In [22], the definition of the Mittag-Leffler stability has been introduced for the Caputo's fractional-order system involving one variable. In addition, the authors of [23] introduced the stabilization criterion for FOBAM NNs in the Mittag-Leffler sense. Following these definitions, in our paper, we have introduced the definition for global Mittag-Leffler synchronization for the FOMBAM NNs (2).

Definition 3 The fractional master system (2) is said to be synchronized with the fractional slave system (3), if the error system (4) is globally Mittag-Leffler stable.

That is, for any $\zeta > 0$ and $\varsigma > 0$, if there exists a function $\mathcal{H}((\zeta, \varsigma)) > 0$ (satisfying the locally Lipschitz condition with Lipschitz constant $\mathcal{H}_0$) and constants $\beta > 0$ and $\rho > 0$ such that the following condition holds:

$$\|e_x(t)\| + \|e_y(t)\| \leq [\mathcal{H}(x_0, y_0)E_\alpha(\rho(t-t_0)^\alpha)]^\beta.$$

Lemma 1 ([23]) *If there exists positive constants $\alpha_1,\ \alpha_2,\ \alpha_3,\ \upsilon$ and ω, along with a function $V(t, x(t), y(t)) : [0, \infty) \times D_1 \times D_2 \to \mathbb{R}$, locally Lipschitz in the variables x and y and having continuous derivatives (D_1 and $D_2 \subset \mathbb{R}^n$ are the domains around the origin) such that the following condition holds:*

$$\alpha_1\Big(\|x(t)\|^\upsilon + \|y(t)\|^\upsilon\Big) \leq V(t, x(t), y(t)) \leq \alpha_2\Big(\|x(t)\|^{\upsilon\omega} + \|y(t)\|^{\upsilon\omega}\Big), \quad (5)$$

$$D^\alpha V(t, x(t), y(t)) \leq -\alpha_3\Big(\|x(t)\|^{\upsilon\omega} + \|y(t)\|^{\upsilon\omega}\Big), \quad (6)$$

where $t \geq 0,\ \alpha \in (0, 1)$, then the equilibrium point of the fractional-order system is Mittag-Leffler stable. If the assumptions (5) and (6) hold globally on $\mathbb{R}^n$, then the equilibrium point $x^* = 0$ is globally Mittag-Leffler stable.

Lemma 2 ([24]) *Let $\Lambda_1, \Lambda_2, \Lambda_3$, and $F(t)$ be real matrices of proper dimensions with Λ_1 satisfying $\Lambda_1 = \Lambda_1^T$ and $F^T(t)F(t) \leq I$. Then, the inequality $\Lambda_1 + \Lambda_2F(t)\Lambda_3 + (\Lambda_2F(t)\Lambda_3)^T < 0$ holds if and only if there exists a positive scalar ϵ such that $\Lambda_1 + \epsilon\Lambda_2\Lambda_2^T + \epsilon^{-1}\Lambda_3^T\Lambda_3 < 0$ holds or equivalently,*

$$\begin{bmatrix} \Lambda_1 & \epsilon\Lambda_2 & \Lambda_3^T \\ * & -\epsilon I & 0 \\ * & * & -\epsilon I \end{bmatrix} < 0.$$

3 Main Results

In this section, a new synchronization criteria has been developed in the aspect of Mittag-Leffler stability of the considered error system (4) under the linear feedback controller.

Theorem 1 *Suppose the Assumption 1 holds, then the fractional slave system (3) is globally Mittag-Leffler synchronized with the fractional master system (2) under the linear feedback controller, if for some symmetric matrices $P_x > 0$, $P_y > 0$, diagonal matrices R_i $(i \in \mathcal{I}_4)$, and appropriately dimensioned matrices W_x and W_y, the subsequent LMI is satisfied:*

$$\begin{bmatrix} \Omega & \epsilon\Gamma_1 & \Gamma_2 \\ * & -\epsilon I_{8n\times 8n} & 0 \\ * & * & -\epsilon I_{8n\times 8n} \end{bmatrix} < 0, \quad (7)$$

where

$$\Omega = \begin{bmatrix} \Omega_{11} & \Omega_{12} \\ * & \Omega_{22} \end{bmatrix}, \quad \Gamma_1 = \begin{bmatrix} \Gamma_{11} & 0_{4n\times 4n} \\ 0_{4n\times 4n} & \Gamma_{12} \end{bmatrix}^T, \quad \Gamma_2 = \begin{bmatrix} \Gamma_{21} & 0_{4n\times 4n} \\ 0_{4n\times 4n} & \Gamma_{22} \end{bmatrix}$$

with

$$\Gamma_{11} = \text{diag}\{-N_{x_1}, 0, N_{x_2}, N_{x_3}\}, \quad \Gamma_{12} = \text{diag}\{-N_{y_1}, 0, N_{y_2}, N_{y_3}\},$$

$$\Gamma_{21} = \begin{bmatrix} P_xG & 0_{n\times n} & P_xG & P_xG \\ 0_{3n\times n} & 0_{3n\times n} & 0_{3n\times n} & 0_{3n\times n} \end{bmatrix}, \quad \Gamma_{22} = \begin{bmatrix} P_yG & 0_{n\times n} & P_yG & P_yG \\ 0_{3n\times n} & 0_{3n\times n} & 0_{3n\times n} & 0_{3n\times n} \end{bmatrix},$$

$$\Omega_{11} = \begin{bmatrix} \widehat{\Omega}_1 & 0 & P_x\widetilde{B}_x & P_x\widetilde{C}_x \\ * & \widehat{\Omega}_3 & 0 & 0 \\ * & * & -R_3 & 0 \\ * & * & * & -R_4 \end{bmatrix}, \quad \Omega_{22} = \begin{bmatrix} \widehat{\Omega}_2 & 0 & P_y\widetilde{B}_y & P_y\widetilde{C}_y \\ * & \widehat{\Omega}_4 & 0 & 0 \\ * & * & -R_1 & 0 \\ * & * & * & -R_2 \end{bmatrix},$$

$$\Omega_{12} = \begin{bmatrix} 0 & 0 & \frac{1}{2}R_1(L_{y_1}+L_{y_2}) & 0 \\ 0 & 0 & 0 & \frac{1}{2}R_2(L_{y_1}+L_{y_2}) \\ \frac{1}{2}R_3(L_{x_1}+L_{x_2}) & 0 & 0 & 0 \\ 0 & \frac{1}{2}R_4(L_{x_1}+L_{x_2}) & 0 & 0 \end{bmatrix}$$

in which $\widehat{\Omega}_1 = -P_x\widetilde{A}_x - \widetilde{A}_x^T P_x - L_{y_1}R_1L_{y_2} + \beta_x P_x - W_x - W_x^T$, $\widehat{\Omega}_2 = -P_y\widetilde{A}_y - \widetilde{A}_y^T P_y - L_{x_1}R_3L_{x_2} + \beta_y P_y - W_y - W_y^T$, $\widehat{\Omega}_3 = -L_{y_1}R_2L_{y_2} - \beta_x P_x$, $\widehat{\Omega}_4 = -L_{x_1}R_4 L_{x_2} - \beta_y P_y$. *Further, from the feasibility of the proposed LMI (7), the control gain matrices can be estimated as* $F_x = P_x^{-1}W_x$ *and* $F_y = P_y^{-1}W_y$.

Proof The proof of this theorem is given in the following two Steps 1 and 2. In Step 1, stability of the error system (4) without the effect of uncertainties has been given in the Mittag-Leffler sense. Followed by this, with the help of S-procedure Lemma in [24], a new synchronization criterion has been developed in the aspect of Mittag-Leffler stability for the considered error system (4) under the linear feedback controller.

Step 1: For this purpose, we first consider the error system in its nominal form as follows:

$$\begin{cases} D^\alpha e_x(t) = -(\widetilde{A}_x + F_x)e_x(t) + \widetilde{B}_x\hat{f}_x(e_y(t)) + \widetilde{C}_x\hat{f}_x(e_y(t-h_y(t))), \\ D^\alpha e_y(t) = -(\widetilde{A}_y + F_y)e_y(t) + \widetilde{B}_y\hat{f}_y(e_x(t)) + \widetilde{C}_y\hat{f}_y(e_x(t-h_x(t))). \end{cases} \quad (8)$$

In order to obtain the required results, the Lyapunov functional candidate can be considered as $V(t, e_x(t), e_y(t)) = V_x(t, e_x(t)) + V_y(t, e_y(t))$, where $V_x(t, e_x(t)) = e_x^T(t)P_xe_x(t)$, $V_y(t, e_y(t)) = e_y^T(t)P_ye_y(t)$.

Then, according to Lemma 2.14 in [19], we have

$$\begin{aligned} D^{\alpha}V(t, e_x(t), e_y(t)) &= D^{\alpha}\left(e_x^T(t)P_x e_x(t)\right) + D^{\alpha}\left(e_y^T(t)P_y e_y(t)\right) \\ &\leq e_x^T(t)P_x(D^{\alpha}e_x(t)) + (D^{\alpha}e_x(t))^T P_x e_x(t) \\ &\quad + e_y^T(t)P_y(D^{\alpha}e_y(t)) + (D^{\alpha}e_y(t))^T P_y e_y(t). \end{aligned} \tag{9}$$

Based on the Assumption 1, there exists diagonal matrices $R_i > 0\,(i \in \mathcal{I}_4)$ satisfying the following inequalities.

$$\begin{aligned} & e_x^T(t)L_{y_1}R_1L_{y_2}e_x(t) - e_x^T(t)R_1(L_{y_1}+L_{y_2})\hat{f}_y(e_x(t)) \\ & \quad + \hat{f}_y^T(e_x(t))R_1\hat{f}_y(e_x(t)) \leq 0, \end{aligned} \tag{10}$$

$$\begin{aligned} & e_y^T(t)L_{x_1}R_3L_{x_2}e_y(t) - e_y^T(t)R_3(L_{x_1}+L_{x_2})\hat{f}_x(e_y(t)) \\ & \quad + \hat{f}_x^T(e_y(t))R_3\hat{f}_x(e_y(t)) \leq 0, \end{aligned} \tag{11}$$

$$\begin{aligned} & e_x^T(t-h_x(t))L_{y_1}R_2L_{y_2}e_x(t-h_x(t)) - e_x^T(t-h_x(t))R_2(L_{y_1}+L_{y_2})\hat{f}_y(e_x(t-h_x(t))) \\ & \quad + \hat{f}_y^T(e_x(t-h_x(t)))R_2\hat{f}_y(e_x(t-h_x(t))) \leq 0, \end{aligned} \tag{12}$$

$$\begin{aligned} & e_y^T(t-h_y(t))L_{x_1}R_4L_{x_2}e_y(t-h_y(t)) - e_y^T(t-h_y(t))R_4(L_{x_1}+L_{x_2})\hat{f}_x(e_y(t-h_y(t))) \\ & \quad + \hat{f}_x^T(e_y(t-h_y(t)))R_4\hat{f}_x(e_y(t-h_y(t))) \leq 0, \end{aligned} \tag{13}$$

where $L_{x_1} = \text{diag}\left\{l_{x_1}^- l_{x_1}^+, l_{x_2}^- l_{x_2}^+, \ldots, l_{x_n}^- l_{x_n}^+\right\}$, $L_{y_1} = \text{diag}\left\{l_{y_1}^- l_{y_1}^+, l_{y_2}^- l_{y_2}^+, \ldots, l_{y_n}^- l_{y_n}^+\right\}$, $L_{x_2} = \text{diag}\left\{\frac{l_{x_1}^- + l_{x_1}^+}{2}, \frac{l_{x_2}^- + l_{x_2}^+}{2}, \ldots, \frac{l_{x_n}^- + l_{x_n}^+}{2}\right\}$, $L_{y_2} = \text{diag}\left\{\frac{l_{y_1}^- + l_{y_1}^+}{2}, \frac{l_{y_2}^- + l_{y_2}^+}{2}, \ldots, \frac{l_{y_n}^- + l_{y_n}^+}{2}\right\}$. From [25], it follows for some constants $p_x > 1$ and $p_y > 1$ that

$$\begin{aligned} & V_x(t+\theta_x, e_x(t+\theta_x)) < p_x V_x(t, e_x(t)), \quad \forall\, \theta_x \in [-h, 0] \\ \text{and } & V_y(t+\theta_y, e_y(t+\theta_y)) < p_y V_y(t, e_y(t)), \quad \forall\, \theta_y \in [-h, 0] \end{aligned}$$

and thus yields the following

$$\beta_x\left(e_x^T(t)p_xP_xe_x(t) - e_x^T(t-h_x(t))P_xe_x(t-h_x(t))\right) > 0, \tag{14}$$

$$\beta_y\left(e_y^T(t)p_yP_ye_y(t) - e_y^T(t-h_y(t))P_ye_y(t-h_y(t))\right) > 0, \tag{15}$$

for $\beta_x \geq 0$ and $\beta_y \geq 0$. From the inequalities (9)–(15), and as $p_x \to 1^+$ and $p_y \to 1^+$, we have

$$D^{\alpha}V(t, e_x(t), e_y(t)) \leq \zeta^T(t)\Omega\zeta(t), \tag{16}$$

where $\zeta(t) = [e_x(t)\; e_x(t-h_x(t))\; \hat{f}_x(e_y(t))\; \hat{f}_x(e_y(t-h_y(t)))\; e_y(t)\; e_y(t-h_y(t))\; \hat{f}_y(e_x(t))\; \hat{f}_y(e_x(t-h_x(t)))]^T$ and Ω is presented in the statement of the theorem.

Suppose $\Omega < 0$, then the following inequality is possible for some scalar $\delta_1 > 0$.

$$\Omega + \text{diag}\{\delta_1 I, 0, 0, 0, \delta_1 I, 0, 0, 0\} < 0.$$

Thus, we obtain

$$D^{\alpha}V(t, e_x(t), e_y(t)) < -\delta_1 \left(\|e_x(t)\|^2 + \|e_y(t)\|^2\right). \tag{17}$$

Further, the definition of the Lyapunov candidate $V(t, e_x(t), e_y(t))$ implies

$$\lambda_{\min}(P_x)\|e_x(t)\|^2 \leq V_x(t, e_x(t)) \leq \lambda_{\max}(P_x)\|e_x(t)\|^2,$$
$$\lambda_{\min}(P_y)\|e_y(t)\|^2 \leq V_y(t, e_y(t)) \leq \lambda_{\max}(P_y)\|e_y(t)\|^2.$$

From the above two inequalities, we get

$$\gamma_1 \left(\|e_x(t)\|^2 + \|e_y(t)\|^2\right) \leq V(t, e_x(t), e_y(t)) \leq \gamma_2 \left(\|e_x(t)\|^2 + \|e_y(t)\|^2\right), \tag{18}$$

where $\gamma_1 = \frac{1}{2}\min\{\lambda_{\min}(P_x),\ \lambda_{\min}(P_y)\}$ and $\gamma_2 = \frac{1}{2}\max\{\lambda_{\max}(P_x),\ \lambda_{\max}(P_y)\}$.

In view of Lemma 1, the inequalities (17) and (18) together yield that the error system (8) is Mittag-Leffler stable which in turn ensures the Mittag-Leffler synchronization for the master system (2) in respect of the slave system (3) in the absence of uncertain parameters. This completes the proof of Step 1.

Step 2: With reference to the system (2), the inequality (16) becomes,

$$D^{\alpha}V(t, e_x(t), e_y(t)) \leq \zeta^T(t)\begin{bmatrix} \Pi_1 & \Omega_{12} \\ * & \Pi_2 \end{bmatrix}\zeta(t), \tag{19}$$

where

$$\Pi_1 = \begin{bmatrix} \Pi_{11} & 0 & \Pi_{13} & \Pi_{14} \\ * & \widehat{\Omega}_3 & 0 & 0 \\ * & * & -R_3 & 0 \\ * & * & * & -R_4 \end{bmatrix},\ \Pi_2 = \begin{bmatrix} \Pi_{21} & 0 & \Pi_{23} & \Pi_{24} \\ * & \widehat{\Omega}_4 & 0 & 0 \\ * & * & -R_1 & 0 \\ * & * & * & -R_2 \end{bmatrix}$$

with $\Pi_{11} = -P_x(\widetilde{A}_x + GM_{x_1}(t)N_{x_1} + F_x) - (\widetilde{A}_x + GM_{x_1}(t)N_{x_1} + F_x)^T P_x - L_{y_1} R_1 L_{y_2} + \beta_x p_x P_x$, $\Pi_{13} = P_x(\widetilde{B}_x + GM_{x_2}(t)N_{x_2})$, $\Pi_{14} = P_x(\widetilde{C}_x + GM_{x_3}(t)N_{x_3})$, $\Pi_{21} = -P_y(\widetilde{A}_y + GM_{y_1}(t)N_{y_1} + F_y) - (\widetilde{A}_y + GM_{y_1}(t)N_{y_1} + F_y)^T P_y - L_{x_1} R_3 L_{x_2} + \beta_y p_y P_y$, $\Pi_{23} = P_y(\widetilde{B}_y + GM_{y_2}(t)N_{y_2})$, $\Pi_{24} = P_y(\widetilde{C}_y + GM_{y_3}(t)N_{y_3})$ and Ω_{12} is the same as in the statement of the theorem.

The inequality (19) can be rewritten as follows:

$$D^{\alpha}V(t,e_x(t), e_y(t)) \leq \zeta^T(t)\left(\Xi + \Gamma_1 M(t)\Gamma_2^T + \Gamma_2 M^T(t)\Gamma_1^T\right)\zeta(t), \tag{20}$$

where $\Xi = [\Xi_{ij}]_{8n\times 8n}$, $i, j \in \mathcal{I}_8$ with $\Xi_{11} = -P_x\widetilde{A}_x - \widetilde{A}_x^T P_x - L_{y_1} R_1 L_{y_2} + \beta_x p_x P_x - W_x - W_x^T$, $\Xi_{13} = P_x\widetilde{B}_x$, $\Xi_{14} = P_x\widetilde{C}_x$, $\Xi_{17} = \frac{1}{2}R_1(L_{y_1} + L_{y_2})$, $\Xi_{22} = -L_{y_1} R_2 L_{y_2} - \beta_x P_x$, $\Xi_{28} = \frac{1}{2}R_2(L_{y_1} + L_{y_2})$, $\Xi_{33} = -R_3$, $\Xi_{35} = \frac{1}{2}R_3(L_{x_1} + L_{x_2})$, $\Xi_{44} = -R_4$, $\Xi_{46} = \frac{1}{2}R_4(L_{x_1} + L_{x_2})$, $\Xi_{55} = -P_y\widetilde{A}_y - \widetilde{A}_y^T P_y - L_{x_1} R_3 L_{x_2} + \beta_y p_y P_y - W_y - W_y^T$, $\Xi_{57} = P_y\widetilde{B}_y$, $\Xi_{58} = P_y\widetilde{C}_y$, $\Xi_{66} = -L_{x_1} R_4 L_{x_2} - \beta_y P_y$, $\Xi_{77} = -R_1$, $\Xi_{88} =$

$-R_2$, Γ_1 and Γ_2 are given in (7), and $M(t) = \begin{bmatrix} M_1(t) & 0_{4n\times 4n} \\ 0_{4n\times 4n} & M_2(t) \end{bmatrix}$ with $M_1(t)=$ diag$\{M_{x_1}(t), 0, M_{x_2}(t), M_{x_3}(t)\}$ and $M_2(t)=$diag$\{M_{y_1}(t), 0, M_{y_2}(t), M_{y_3}(t)\}$. Taking $p_x \to 1^+$ and $p_y \to 1^+$, the inequality (20) becomes

$$D^\alpha V(t, e_x(t), e_y(t)) \leq \zeta^T(t)\left(\Omega + \Gamma_1 M(t)\Gamma_2^T + \Gamma_2 M^T(t)\Gamma_1^T\right)\zeta(t). \tag{21}$$

Suppose the following inequality holds:

$$\begin{bmatrix} \Omega & \epsilon\Gamma_1 & \Gamma_2 \\ * & -\epsilon I_{8n\times 8n} & 0 \\ * & * & -\epsilon I_{8n\times 8n} \end{bmatrix} < 0. \tag{22}$$

Then, from the Lemma 2 it follows that $\Omega + \epsilon\Gamma_1\Gamma_1^T + \epsilon^{-1}\Gamma_2\Gamma_2^T < 0$ for some scalar $\epsilon > 0$ which in turn implies that $\Omega + \Gamma_1 M(t)\Gamma_2^T + \Gamma_2 M^T(t)\Gamma_1^T < 0$ and hence from the inequality (21) we have

$$D^\alpha V(t, e_x(t), e_y(t)) < 0.$$

Therefore, a sufficiently small scalar $\delta_2 > 0$ can be obtained such that

$$\Omega + \Gamma_1 M(t)\Gamma_2^T + \Gamma_2 M^T(t)\Gamma_1^T \leq \text{diag}\{\delta_2 I, 0, 0, 0, \delta_2 I, 0, 0, 0\} < 0.$$

Thus, we obtain

$$D^\alpha V(t, e_x(t), e_y(t)) < -\delta_2\left(\|e_x(t)\|^2 + \|e_y(t)\|^2\right). \tag{23}$$

Further, the definition of the Lyapunov candidate $V(t, e_x(t), e_y(t))$ implies

$$\lambda_{\min}(P_x)\|e_x(t)\|^2 \leq V_x(t, e_x(t)) \leq \lambda_{\max}(P_x)\|e_x(t)\|^2,$$
$$\lambda_{\min}(P_y)\|e_y(t)\|^2 \leq V_y(t, e_y(t)) \leq \lambda_{\max}(P_y)\|e_y(t)\|^2.$$

From the above two inequalities, we get

$$\gamma_1\left(\|e_x(t)\|^2 + \|e_y(t)\|^2\right) \leq V(t, e_x(t), e_y(t)) \leq \gamma_2\left(\|e_x(t)\|^2 + \|e_y(t)\|^2\right), \tag{24}$$

where $\gamma_1 = \frac{1}{2}\min\{\lambda_{\min}(P_x),\ \lambda_{\min}(P_y)\}$ and $\gamma_2 = \frac{1}{2}\max\{\lambda_{\max}(P_x),\ \lambda_{\max}(P_y)\}$.

In view of Lemma 1, the inequalities (23) and (24) together yields that the error system (4) is Mittag-Leffler stable which in turn ensures the Mittag-Leffler synchronization for the master system (2) in respect of the slave system (3) when the uncertain parameters occur. This completes the proof.

Remark 4 From Lemma 1, it is evident that for systems involving fractional-order, the Mittag-Leffler stability implies asymptotic stability. In a similar manner, it can also be realized that the global asymptotic synchronization for the master system

(2) and the slave system (3) can be obtained by means of the global Mittag-Leffler synchronization condition established in Theorem 1.

Remark 5 When $\alpha = 1$, the considered class of FOMBAM NNs (1) is reduced to the integer-order memristor-based BAM NNs described by,

$$\begin{cases} \dot{x}(t) = -A_x x(t) + B_x f_x(y(t)) + C_x f_x(y(t - h_y(t))) + J_x(t), \\ \dot{y}(t) = -A_y y(t) + B_y f_y(x(t)) + C_y f_y(x(t - h_x(t))) + J_y(t), \end{cases} \tag{25}$$

where $t \geq 0$.

Based on the memristor characteristics, the system (25) can be transformed into

$$\begin{cases} \dot{x}(t) = -(\widetilde{A}_x + \Delta A_x(t))x(t) + (\widetilde{B}_x + \Delta B_x(t)) f_x(y(t)) + (\widetilde{C}_x + \Delta C_x(t)) \\ \qquad \times f_x(y(t - h_y(t))) + J_x(t), \\ \dot{y}(t) = -(\widetilde{A}_y + \Delta A_y(t))y(t) + (\widetilde{B}_y + \Delta B_y(t)) f_y(x(t)) + (\widetilde{C}_y + \Delta C_y(t)) \\ \qquad \times f_y(x(t - h_x(t))) + J_y(t). \end{cases} \tag{26}$$

For this system (26), the corresponding slave system and error system can be obtained from the systems (3) and (4) for $\alpha = 1$. Based on the Definition 3, the global Mittag-Leffler synchronization result in Theorem 1 can be reduced to the common global exponential synchronization of the integer-order NN (26).

Remark 6 Due to the memristor characteristics, in systems (2) and (3), the choice of the connection weights depends on their corresponding neuron states. Generally, the values of the connection weights are not unique. Due to this reason, the fractional-order systems (2) and (3) are known as the state-dependent switching systems. If the weight parameters, $\widehat{a}_{x_i}$, $\widetilde{a}_{x_i}$, $\widehat{a}_{y_i}$, $\widetilde{a}_{y_i}$, $\widehat{b}_{x_{ij}}$, $\widetilde{b}_{x_{ij}}$, $\widehat{b}_{y_{ij}}$, $\widetilde{b}_{y_{ij}}$, $\widehat{c}_{x_{ij}}$, $\widetilde{c}_{x_{ij}}$, $\widehat{c}_{y_{ij}}$, and $\widetilde{c}_{y_{ij}}$ are considered to be $\widehat{a}_{x_i} = \widetilde{a}_{x_i}$, $\widehat{a}_{y_i} = \widetilde{a}_{y_i}$,$\widehat{b}_{x_{ij}} = \widetilde{b}_{x_{ij}}$, $\widehat{b}_{y_{ij}} = \widetilde{b}_{y_{ij}}$, $\widehat{c}_{x_{ij}} = \widetilde{c}_{x_{ij}}$, and $\widehat{c}_{y_{ij}} = \widetilde{c}_{y_{ij}}$, then the fractional master system (2) and fractional slave system (3) are reduced to the FOBAM NNs and the established global Mittag-Leffler synchronical criteria can be specialized for the problem of global Mittag-Leffler synchronization of delayed FOBAM NNs.

4 Illustrative Example

In this section, one numerical example with simulation results is given to facilitate the effectiveness of the established theoretical results for the synchronization approach with prescribed control gain matrices F_x and F_y for the FOMBAM NNs discussed in the previous section.

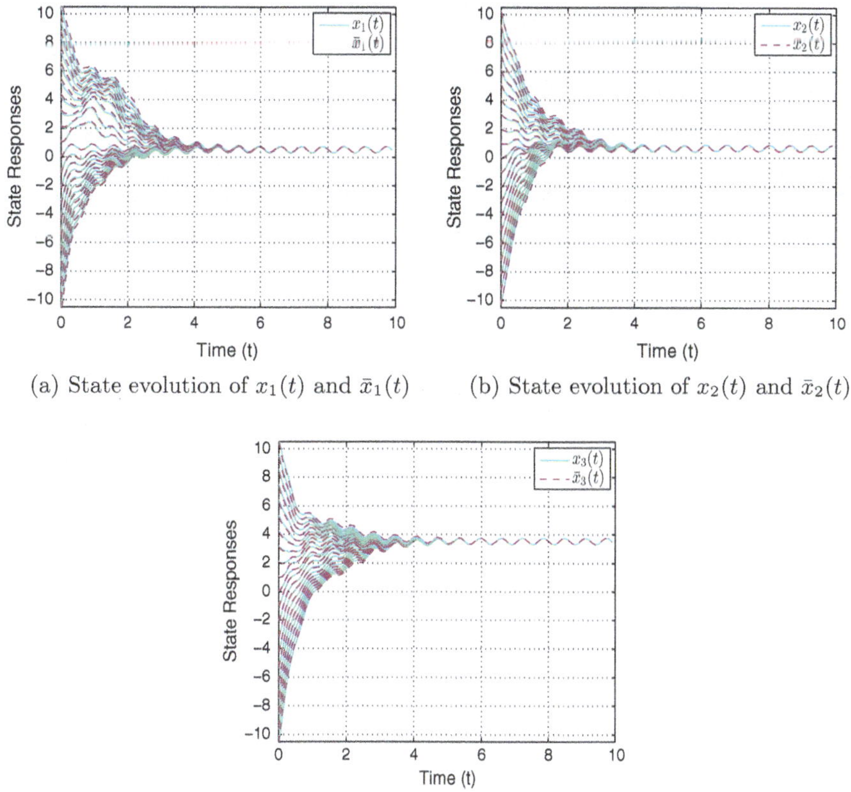

(a) State evolution of $x_1(t)$ and $\bar{x}_1(t)$

(b) State evolution of $x_2(t)$ and $\bar{x}_2(t)$

(c) State evolution of $x_3(t)$ and $\bar{x}_3(t)$

Fig. 1 Responses of state $x(t)$ and $\bar{x}(t)$

Example 1 Choose the three-dimensional FOMBAM NNs (2) with the following system parameters:

$$\widetilde{A}_x = \begin{bmatrix} 1 & 0 & 0 \\ 0 & 1 & 0 \\ 0 & 0 & 1 \end{bmatrix}, \quad \widetilde{B}_x = \begin{bmatrix} 1.75 & 0.25 & -0.22 \\ 0.8 & -1.1 & 0.55 \\ 0.26 & 0.12 & -1.7 \end{bmatrix}, \quad \widetilde{C}_x = \begin{bmatrix} -0.53 & -0.15 & 0.135 \\ 0.2 & 0.51 & 0.25 \\ 0.16 & -0.9 & -1 \end{bmatrix}$$

$$\widetilde{A}_y = \begin{bmatrix} 2.75 & 0 & 0 \\ 0 & 2.75 & 0 \\ 0 & 0 & 2.75 \end{bmatrix}, \quad \widetilde{B}_y = \begin{bmatrix} 2.5 & 1.2 & -2.2 \\ -2.6 & 2.3 & 3 \\ 1.4 & -1.1 & 1.8 \end{bmatrix}, \quad \widetilde{C}_y = \begin{bmatrix} 0.5 & 1.5 & -2.1 \\ -1.3 & 2.3 & 1.4 \\ 1 & -3 & 1.8 \end{bmatrix}.$$

The input parameters are taken as $J_x(t) = J_y(t) = [3 - 2\cos(t), 3 - 2\sin(t), 3 - 2\cos(t)]^T$. The activation functions are described by $f_x(y(t)) = f_y(x(t)) = [\tanh(t), \tanh(t), \tanh(t)]^T$.

For the above parameters, from Assumption 1, we obtain $L_{x_1} = L_{y_1} = \mathrm{diag}\{0, 0, 0\}$ and $L_{x_2} = L_{y_2} = \mathrm{diag}\{0.5, 0.5, 0.5\}$.

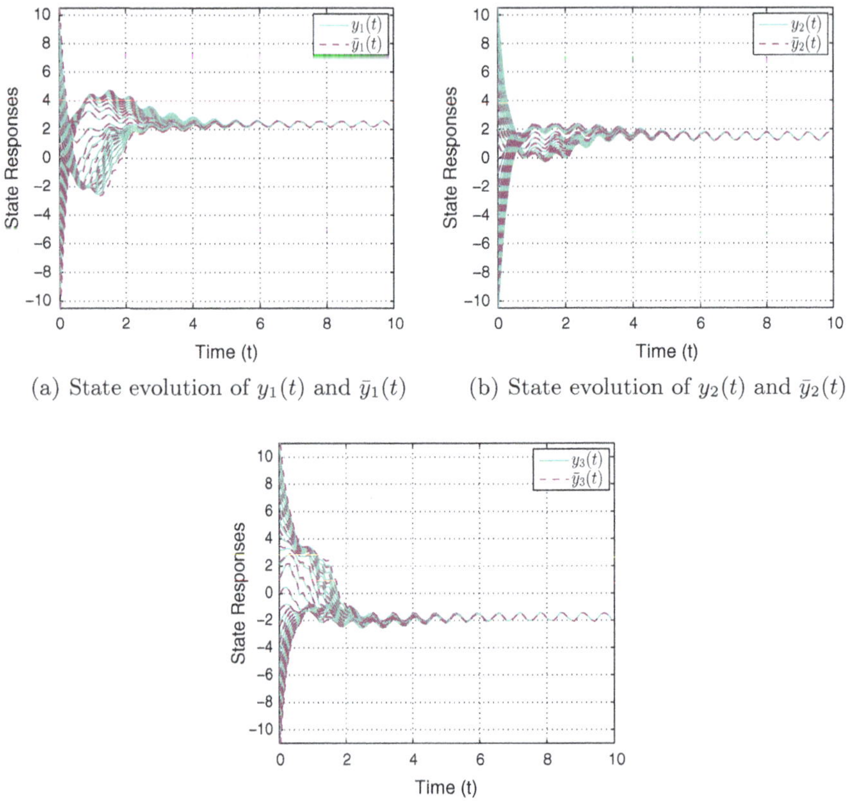

(a) State evolution of $y_1(t)$ and $\bar{y}_1(t)$

(b) State evolution of $y_2(t)$ and $\bar{y}_2(t)$

(c) State evolution of $y_3(t)$ and $\bar{y}_3(t)$

Fig. 2 Responses of state $y(t)$ and $\bar{y}(t)$

The uncertain parameters in the system (2) are considered as follows:

$$N_{x_1} = \begin{bmatrix} 0.1 & 0 & 0 \\ 0 & 0.15 & 0 \\ 0 & 0 & 0.2 \end{bmatrix}, \quad N_{x_2} = \begin{bmatrix} 1.2 & 0.1 & 0 \\ 0.1 & 0.2 & 0 \\ 0.1 & 0 & 0.2 \end{bmatrix}, \quad N_{x_3} = \begin{bmatrix} 0.15 & 0 & 0.4 \\ 0.05 & 0.15 & 0 \\ 0 & 0.1 & 0.2 \end{bmatrix},$$

$$N_{y_1} = \begin{bmatrix} 0.2 & 0 & 0 \\ 0 & 0.1 & 0 \\ 0 & 0 & 0.1 \end{bmatrix}, \quad N_{y_2} = \begin{bmatrix} 0.4 & 0.3 & 0.2 \\ 0.3 & 0.2 & 0.1 \\ 0.2 & 0.1 & 0 \end{bmatrix}, \quad N_{y_3} = \begin{bmatrix} 0.1 & 0.4 & 0.2 \\ 0.3 & 0 & 0.25 \\ 0 & 0 & 0.1 \end{bmatrix},$$

and $G = \mathrm{diag}\{1, 1, 1\}$.

By utilizing the MATLAB LMI toolbox to solve the LMI (7), we get the feasible solutions and the corresponding state feedback control gain matrices are calculated as

Fig. 3 Responses of error states of $x(t)$ and $y(t)$

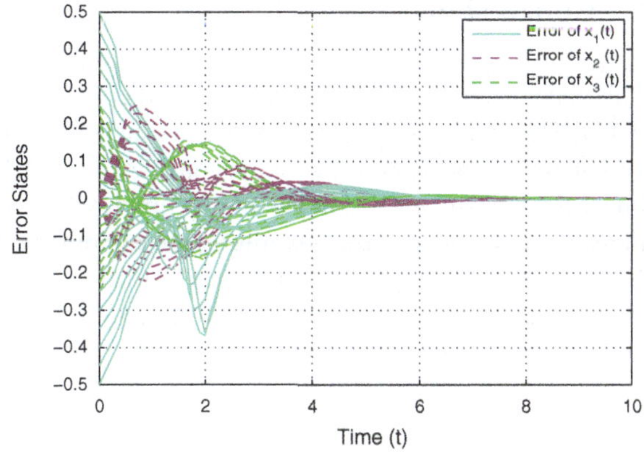

(a) State evolution of $e_{x1}(t), e_{x2}(t)$ and $e_{x3}(t)$

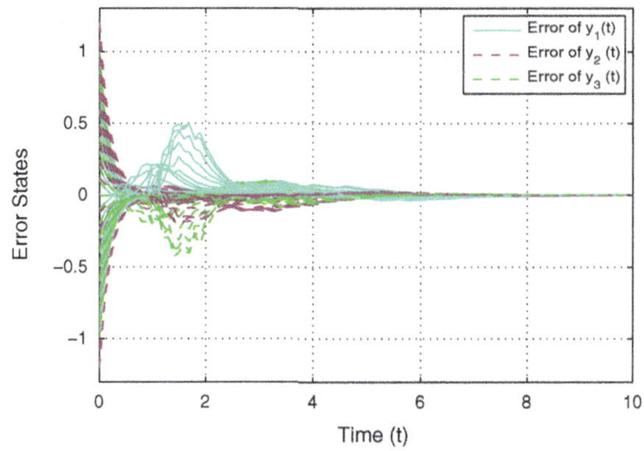

(b) State evolution of $e_{y1}(t), e_{y2}(t)$ and $e_{y3}(t)$

$$F_x = \begin{bmatrix} 5.6097 & -0.8228 & -2.0281 \\ -0.9288 & 4.0411 & -2.6552 \\ -1.9175 & -2.5096 & 5.0975 \end{bmatrix} \quad \text{and} \quad F_y = \begin{bmatrix} 4.5978 & -0.7757 & -2.9276 \\ -0.9741 & 5.3983 & -2.1997 \\ -2.9016 & -2.0010 & 5.7685 \end{bmatrix}.$$

Therefore, from Theorem 1, it follows that the systems (2) and (3) are globally synchronized in Mittag-Leffler sense with the effect of uncertainties which can further be verified by the simulation results in the following Figs. 1, 2 and 3. The simulation results for the state vectors along with its estimators and error states are depicted in Figs. 1, 2, and 3. Figure 1a–c depict the trajectories of the true state $x_1(t)$ with its estimator $\bar{x}_1(t)$, $x_2(t)$ with its estimator $\bar{x}_2(t)$ and $x_3(t)$ with its estimator $\bar{x}_3(t)$ with different initial conditions, respectively. Further, the trajectories of the true state $y_1(t)$ with its estimator $\bar{y}_1(t)$, $y_2(t)$ with its estimator $\bar{y}_2(t)$ and $y_3(t)$ with its estimator $\bar{y}_3(t)$ under different initial conditions are depicted in Figure 2a–c, respectively. Figure 3a, b show the dynamical evolution of the error states $e_{x_1}(t)$, $e_{x_2}(t)$, $e_{x_3}(t)$ and $e_{y_1}(t)$, $e_{y_2}(t)$, $e_{y_3}(t)$, respectively. These figures indicate that the states of the error system (4) converges to zero.

5 Conclusion

This paper dealt with the global Mittag-Leffler synchronization problem of the FOM-BAM NNs with time-varying delays through LMI framework. An efficient slave system has been constructed for the considered FOMBAM NNs. By considering the linear feedback control scheme and by employing a suitable Lyapunov functional, the sufficient synchronization criteria are established for the system of FOMBAM NNs to confirm the global stability in the Mittag-Leffler sense, which are derived in the form of LMIs. Finally, to confirm the efficiency of the acquired results, one numerical example along with its simulation results is explored.

References

1. Ali, M.S.: Stability analysis of Markovian jumping stochastic Cohen-Grossberg neural networks with discrete and distributed time varying delays. Chin. Phys. B **23**(6), 060702 (2014)
2. Ali, M.S., Gunasekaran, N., Rani, M.E.: Robust stability of Hopfield delayed neural networks via an augmented LK functional. Neurocomputing **234**, 198–204 (2017)
3. Ali, M.S., Balasubramaniam, P., Zhu, Q.: Stability of stochastic fuzzy BAM neural networks with discrete and distributed time-varying delays. Int. J. Mach. Learn. Cybernet. **8**(1), 263–273 (2017)
4. Kosko, B.: Adaptive bidirectional associative memories. Appl. Opt. **26**(23), 4947–4960 (1987)
5. Kosko, B.: Bidirectional associative memories. IEEE Trans. Syst. Man Cybernet. **18**(1), 49–60 (1988)
6. Chua, L.: Memristor-the missing circuit element. IEEE Trans. Circ. Theory **18**(5), 507–519 (1971)
7. Yang, S., Guo, Z., Wang, J.: Robust synchronization of multiple memristive neural networks with uncertain parameters via nonlinear coupling. IEEE Trans. Syst. Man Cybernet. Syst. **45**(7), 1077–1086 (2015)
8. Li, K.Z., Zhao, M.C., Fu, X.C.: Projective synchronization of driving-response systems and its application to secure communication. IEEE Trans. Circ. Syst. I: Reg. Pap. **56**(10), 2280–2291 (2009)
9. Kilbas, A.A., Srivastava, H.M., Trujillo, J.J.: Theory and Applications of Fractional Differential Equations, vol. 204, Elsevier (2006)
10. Podlubny, I.: Fractional Differential Equations. Academic Press, San Diego (1999)
11. Liu, Y., Li, C., Huang, T., Wang, X.: Robust adaptive lag synchronization of uncertain fuzzy memristive neural networks with time-varying delays. Neurocomputing **190**, 188–196 (2016)
12. Pecora, L.M., Carroll, T.L.: Synchronization in chaotic systems. Phys. Rev. Lett. **64**(8), 821 (1990)
13. Fan, Y., Huang, X., Wang, Z., Li, Y.: Nonlinear dynamics and chaos in a simplified memristor-based fractional-order neural network with discontinuous memductance function. Nonlinear Dyn. **93**(2), 611–627 (2018)
14. Kong, F., Zhu, Q.: New fixed-time synchronization control of discontinuous inertial neural networks via indefinite Lyapunov-Krasovskii functional method. Int. J. Robust Nonlinear Control **31**(2), 471–495 (2021)
15. Kong, F., Zhu, Q., Sakthivel, R., Mohammadzadeh, A.: Fixed-time synchronization analysis for discontinuous fuzzy inertial neural networks with parameter uncertainties. Neurocomputing **422**, 295–313 (2021)
16. Bao, H.B., Cao, J.D.: Projective synchronization of fractional-order memristor-based neural networks. Neural Netw. **63**, 1–9 (2015)

17. Bao, H., Park, J.H., Cao, J.: Adaptive synchronization of fractional-order memristor-based neural networks with time delay. Nonlinear Dyn. **82**(3), 1343–1354 (2015)
18. Aadhithiyan, S., Raja, R., Zhu, Q., Alzabut, J., Niezabitowski, M., Lim, C.P.: Exponential synchronization of nonlinear multi-weighted complex dynamic networks with hybrid time varying delays. Neural Process. Lett., 1–29 (2021)
19. Nagamani, G., Shafiya, M., Soundararajan, G.: An LMI based state estimation for fractional-order memristive neural networks with leakage and time delays. Neural Process. Lett. **52**(3), 2089–2108 (2020)
20. Liu, Y., Wang, Z., Liu, X.: Global exponential stability of generalized recurrent neural networks with discrete and distributed delays. Neural Netw. **19**(5), 667–675 (2006)
21. Xu, J., Cao, Y.Y., Sun, Y., Tang, J.: Absolute exponential stability of recurrent neural networks with generalized activation function. IEEE Trans. Neural Netw. **19**(6), 1075–1089 (2008)
22. Li, Y., Chen, Y., Podlubny, I.: Stability of fractional-order nonlinear dynamic systems: Lyapunov direct method and generalized Mittag-Leffler stability. Comput. Math. Appl. **59**(5), 1810–1821 (2010)
23. Nagamani, G., Shafiya, M., Soundararajan, G., Prakash, M.: Robust state estimation for fractional-order delayed BAM neural networks via LMI approach. J. Franklin Inst. (2020)
24. Boyd, S., El Ghaoui, L., Feron, E., Balakrishnan, V.: Linear matrix inequalities in system and control theory. In: Society for Industrial and Applied Mathematics (1994)
25. Yang, D., Hu, C., Wang, Z., Liang, X.: New global stability criteria of neural networks with time delays. In: 2008 7th World Congress on Intelligent Control and Automation, pp. 5317–5320. IEEE (2008)

Graphs and Networks

r-Dynamic Chromatic Number of Extended Neighborhood Corona of Complete Graph with Some Graphs

V. Aparna and N. Mohanapriya

Abstract The graphs $H = (V, E)$ taken into account here are undirected, simple, finite, and connected. The neighborhood corona of the graphs H_1 and H_2 is constructed by first taking a copy of the graph H_1 and $|H_1|$ copies of the graph H_2 and then by connecting each neighbor of the ith vertex of H_1 to every vertex of the ith copy of H_2. The extended neighborhood corona of two graphs H_1 and H_2, $H_1 * H_2$ is obtained by taking the neighborhood corona of H_1 and H_2 and then connecting each vertex of the ith copy of H_2 to every vertex of the jth copy of H_2 whenever the vertices v_i and v_j are adjacent in the graph H_1. In this paper, we have attained the lower bound for the r-dynamic chromatic number of the extended neighborhood corona of the complete graph $K_q, q \geq 2$ with any simple, connected and finite graph H with at least 2 vertices and the exact r-dynamic chromatic number of the extended neighborhood corona of complete graph K_q with path P_t, cycle C_t, complete graph K_t, star graph $K_{1,t}$ and wheel graph W_t.

Keywords r-dynamic coloring · Neighborhood corona · Extended neighborhood corona

1 Introduction

The graphs that we consider throughout this paper are undirected, simple, finite, and connected. The concept of r-dynamic coloring was put forward by Bruce Montgomery in [11]. Since proper graph coloring describes the allocation of vertices to categories, it is ideal in many applications that vertices have neighbors in several categories. The number of colors needed could be increased by this. By proper vertex coloring of a graph H, we mean a function $f : V(H) \to A$, where A is a set of distinct colors such that any two adjacent vertices are allocated different colors. For a natural number r, the proper k-coloring of the vertices of the graph H where in

V. Aparna · N. Mohanapriya (✉)
PG and Research Department of Mathematics, Kongunadu Arts and Science College (Autonomous), Coimbatore, Tamil Nadu 641 029, India

P. Balasubramaniam et al. (eds.), *Mathematical Modelling and Computational Intelligence Techniques*, Springer Proceedings in Mathematics & Statistics 376,
https://doi.org/10.1007/978-981-16-6018-4_15

$|f(\mathcal{N}(b))| \geq \min\{r, d(b)\}$ for each $b \in V(H)$ is said to be r-dynamic coloring of a graph H, i.e., the neighbors of each vertex b receives at least $\min\{r, d(b)\}$ distinct colors. Here $\mathcal{N}(b)$ and $d(b)$ denotes the neighborhood set of vertex b and number of vertices adjacent to b, respectively. The lowest k that enables H to have an r-dynamic coloring with k colors is known as the r-dynamic chromatic number and it is represented as $\chi_r(H)$.

When $r = 1$ it is just the chromatic number $\chi(H)$ and the 2-dynamic chromatic number is referred as the dynamic chromatic number of H and is labeled as $\chi_d(H)$. Dynamic coloring has been studied in detail in the following papers [1, 2, 7, 9, 10].

There are numerous lower and upper bounds for dynamic and r-dynamic chromatic number in terms of graph parameters. Here are some of them: For regular graphs, Montgomery proposed the conjecture $\chi_r(H) \leq \chi(H) + 2$. Lai et al. in [7] proved that $\chi_2(H) \leq \Delta(H) + 1$ for all graphs with $\Delta \geq 3$ except for cycle C_5. An upper bound for dynamic chromatic number in terms of minimum degree δ, maximum degree Δ and chromatic number $\chi(H)$ is $\chi_2(H) - \chi(H) \leq \lceil (\Delta e)/\delta log$ $(2e(\Delta^2 + 1)) \rceil$ was put forward by Taherkhani in [12] where H is a k-regular graph. The problem of finding the value for $\chi_r(H)$ for planar bipartite graphs with maximum degree at most 3 and high girth is an NP-hard problem was proved by Li et al. in [8]. Also $\chi_2(H) \leq 4$ for planar graphs which has no component of C_5 and $\chi_d(H) \leq 5$ when H is planar was proved in [6] by Kim et al. $\chi_r(H) \geq min\{r, \Delta(H)\} + 1$ is one of the most familiar lower bound for $\chi_r(H)$ and it was put forward by Montgomery and Lai in [7]. Networking usually deals with fully connected networks(complete graph) which forms the motivation for this work. The r-dynamic coloring [3, 5, 12] of different graphs is being studied extensively by many nowadays.

2 Preliminaries

Let H_1 and H_2 be graphs with vertex disjointed sets of h_1 and h_2 vertices. The neighborhood corona of the graphs H_1 and H_2 is constructed by first taking a copy of the graph H_1 and h_1 copies of the graph H_2 and then by connecting each neighbor of the ith vertex of H_1 to every vertex of the ith copy of H_2. The extended neighborhood corona [4] of two graphs H_1 and H_2 is obtained by taking the neighborhood corona of H_1 and H_2 and then connecting each vertex of the ith copy of H_2 to every vertex of the jth copy of H_2 whenever the vertices v_i and v_j are adjacent in the graph H_1. It is denoted as $H_1 * H_2$. The extended neighborhood corona of two graphs H_1 and H_2 is a graph with $h_1h_2 + h_1$ vertices where $|V(H_1)| = h_1$ and $|V(H_2)| = h_2$. Let the vertex set of H_1 and H_2 be defined as $V(H_1) = \{v_1, v_2, \ldots, v_q\}$ and $V(H_2) = \{u_1, u_2, \ldots, u_t\}$ then the vertex set of $H_1 * H_2$ be defined as $V(H_1 * H_2) = \{v_1, v_2, \ldots, v_q\} \cup \{u_{i1}, u_{i2}, \ldots, u_{it} : 1 \leq i \leq q\}$ where $u_{i1}, u_{i2}, \ldots, u_{it}$ denote the t vertices of the ith copy of H_2. The degree of the vertex v_i of $H_1 * H_2$ is $d_{H_1 * H_2}(v_i) = d_{H_1}(v_i)[|H_2| + 1] = d_{H_1}(v_i)[h_2 + 1]$, where $|H_2|$ denotes the number of vertices in

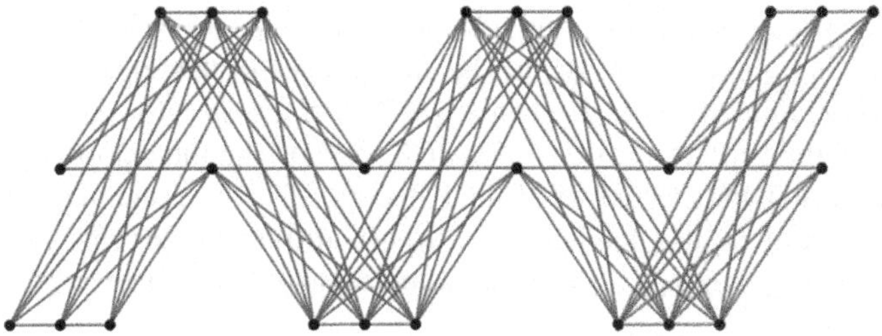

Fig. 1 Extended neighborhood corona of path P_6 with path P_3

the graph H_2 and degree of the vertex u_{ij} of $H_1 * H_2$ is $d_{H_1*H_2}(u_{ij}) = d_{H_2}(u_j) + d_{H_1}(v_i)[|H_2| + 1] = \deg_{H_2}(u_j) + \deg_{H_1}(v_i)[h_2 + 1]$. Figure 1 depicts the extended neighborhood corona of path P_6 with path P_3.

3 Results

Lemma 1 *Let K_q be the complete graph on q vertices with $q \geq 2$ and H be any simple, connected, and finite graph with at least t vertices with $t \geq 2$ then the lower bound for the r-dynamic chromatic number of the extended neighborhood corona of the complete graph K_q with H is*

$$\chi_r(K_q * H) \geq \begin{cases} 2q & \text{for } 1 \leq r \leq 2(q-1) \\ r+3 & \text{for } 2q-1 \leq r \leq 3(q-1) \\ r+4 & \text{for } 3q-2 \leq r \leq 4(q-1) \\ \vdots & \\ r+t+1 & \text{for } t(q-1)+1 \leq r \leq (t+1)(q-1) \\ qt+q & \text{for } r \geq (t+1)(q-1)+1 \end{cases}$$

$$= \begin{cases} 2q & \text{for } 1 \leq r \leq 2(q-1) \\ r+i & \text{for } (i-1)(q-1)+1 \leq r \leq i(q-1), \\ & \quad 3 \leq i \leq (t+1) \\ qt+q & \text{for } r \geq (t+1)(q-1)+1 \end{cases}$$

Proof Each vertex v_i of complete graph K_q is adjacent to all other vertices v_j : $i \neq j$ of K_q and all the t vertices of the $(q-1)$ copies of H except the ith copy of H by the elucidation of extended neighborhood corona of graphs. Thus, each v_i has degree $(t+1)(q-1)$ which is the minimum degree of the graph $K_q * H$, i.e., $\delta(K_q * H) = (t+1)(q-1)$.
Case 1: When $1 \leq r \leq 2(q-1)$.
Since the graph H is simple, connected graph, there exists at least one edge e, say between the vertices u and v in H. Now from the definition of extended neighborhood

corona of graphs every vertex in the ith copy of H is adjacent to all the t vertices in the remaining copies of H. Hence the end vertices of the edge e in every copy of H induce a clique of order $2q$, hence $\chi_r(K_q * H) \geq 2q$. Now assign the colors $1, 2, \ldots, q$ to the vertices $v_1, v_2, \ldots, v_q$ of K_q and assign the colors $i, q+i$ to the end vertices u, v in the ith copy of H for $1 \leq i \leq q$. By this assignment, we can see that each vertex of K_q has neighbors in $2(q-1)$ different color classes. Hence we have the lower bound $\chi_r(K_q * H) \geq 2q$ when $1 \leq r \leq 2(q-1)$.
Case 2: When $2q-1 \leq r \leq 3(q-1)$.
First, let us consider the case $r = 2q-1$. First allot the color i to the vertex v_i for $1 \leq i \leq q$ of K_q. By this assignment, each v_i of K_q has neighbors in $q-1$ different color classes and it further requires neighbors in q more different color classes. Consider the vertex v_1, first assign two new colors $q+1, q+2$ to any two vertices in the ith copy of H where $i \geq 2$. Suppose that it is assigned to any two vertices in the 2nd copy of H. And allot one new color to any one vertex in each of $3^{rd}, 4^{th}, \ldots, q^{th}$ copy of H. By this, the vertex v_1 has attained neighbors in $2q-1$ different color classes. Now moving to the next vertex v_2, by the earlier assignment of colors it has $q-1+q-2 = 2q-3$ differently colored neighbors and it requires two more colors to satisfy its r-adjacency condition. Hence we can assign those two colors to any two vertices of the 1st copy of H or one new color to the any vertex of 1st copy of H and other new color to any one vertex which has not been colored before of 3rd or 4th or $\cdots$ q^{th} copy of H. Assume the former, that is assigning two new colors to any two vertices of the 1st copy of H and the latter case is also similar. By this we can observe that all vertices v_i of K_q have neighbors in at least $2q-1$ different color classes and we have used $q+2+q-2+2 = 2q+2$ colors. Therefore $\chi_r(K_q * H) \geq 2q+2 = r+3$. Now consider the case when $r = 2q$ and consider the vertex v_1, to satisfy $2q$-adjacency condition it requires one more new color $2q+3$. Now we can assign this new color to any one vertex which has not been colored before in the 2nd or 3rd or $\cdots$ q^{th} copy of H. Assign that color to any one vertex which has not been colored before of 3rd or 4th or $\cdots$ qth copy of H for convenience say it is assigned in the 3rd copy of H. The cases in which it is assigned to 4th or 5th or $\cdots$ q^{th} copy of H is similar. By this assignment, the $2q$-adjacency of all v_i will be satisfied and we have used $2q+3$ colors. Suppose the case in which the new color is assigned to any one vertex which has not been colored before of 2nd copy of H. Then when we consider the vertex v_2 it now has neighbors only in $2q-1$ different color classes and for satisfying its r-adjacency condition we require one more additional color $2q+4$. By this the $2q$-adjacency of all v_i will be satisfied, and we have used a total of $2q+4$ colors. Since our coloring is minimum coloring, we should not consider this and we follow the assignment provided before. Hence $\chi_r(K_q * H) \geq 2q+3 = r+3$. Now consider the case when we assign three or more colors to the 2nd copy of H when $r = 2q-1$ then it will result in the usage of more colors other than the previously used $2q+2 = r+3$ colors, and hence, this case is not considered. When $r = 2q+1$, consider the vertex v_1 in order to satisfy its r-adjacency condition we require one new color $2q+4$ which can assigned to any one vertex which has not been colored before of 2nd or 3rd or 4th or $\cdots$ q^{th} copy of H. For convenience let it be assigned in the 4th copy of H. The case in which it is

assigned to any one vertex which has not been colored before of 5th or 6th or $\cdots$ q^{th} copy of H is similar. Thus, the r-adjacency condition of all v_i of K_q is satisfied, and we have used a total of $2q + 4$ colors. But if it is assigned to any one vertex which has not been colored before of 2nd or 3rd copy of H we will require more colors than the previously used $2q + 4$ colors and hence this case is neglected. Hence when $r = 2q + 1$ we have $\chi_r(K_q * H) \geq 2q + 4 = r + 3$. Similarly proceeding as above we require one new color at each stage of r other than the previously used colors of the $(r - 1)^{\text{th}}$ stage. This process continues till all copies of H has exactly two different colors which occurs when $r = 3q - 3 = 3(q - 1)$. Thus, in this case we can conclude that the lower bound is $\chi_r(K_q * H) \geq r + 3$.

Case 3: When $3q - 2 \leq r \leq 4(q - 1)$.

At the end of Case 2, the vertex v_1 has neighbors in $3(q - 1)$ different color classes. Hence for its $r = (3q - 2)$-adjacency condition, we require a new color $3q + 1$ which can be assigned to any one vertex which has not been colored before of 2nd or 3rd or 4th or $\cdots$ q^{th} copy of H let us assume it is assigned in the 2nd copy of H. The cases of assignment of that color in the 3rd or 4th or $\cdots$ q^{th} copy of H is similar. Moving on next to the vertex v_2 for its r-adjacency we require a new color $3q + 2$ which can be assigned to any one vertex which has not been colored before of 1st or 3rd or 4th or $\cdots$ q^{th} copy of H. Assume that it is assigned in the 1st copy of H for convenience. By this the r-adjacency condition of all v_i will be satisfied. Hence we require at least $3q + 2$ colors in this case, i.e., $\chi_r(K_q * H) \geq 3q + 2 = r + 4$. When $r = 3q - 1$, consider the vertex v_1 which requires a new color $3q + 3$ other than the previously used $3q + 2$ colors in order to satisfy its $r = (3q - 1)$-adjacency condition. This new color can be assigned to any one vertex which has not been colored before of 2nd or 3rd or 4th or $\cdots$ q^{th} copy of H. Assume it is assigned in the 3rd or 4th or $\cdots$ q^{th} copy of H for convenience let us assign it in 3rd copy of H and the assignment in other cases is also similar then by this the $r = (3q - 1)$-adjacency of all v_i will be satisfied. But if it is assigned in the 2nd copy of H, we will require more colors to satisfy the r-adjacency of other $v_i's$ hence this case is omitted. Therefore we require at least $3q + 3$ colors in this case, i.e., $\chi_r(K_q * H) \geq 3q + 3 = r + 4$. When $r = 3q$ we assign the new color to any one vertex which has not been colored before in the 4th copy of H and in the subsequent cases of r we keep on assigning the new color in the 5th, 6th, $\cdots$ till the qth copy of H which occurs when $r = 4(q - 1)$. That is, this case stops when each copy of H has exactly 3 different colors. Thus, we can conclude that the lower bound is $\chi_r(K_q * H) \geq r + 4$.

Case i: When $(i - 1)(q - 1) + 1 \leq r \leq i(q - 1), 3 \leq i \leq (t + 1)$.

We can observe from Case 2, Case 3 and deduce easily that we require $r + i$ colors when $(i - 1)(q - 1) + 1 \leq r \leq i(q - 1), 3 \leq i \leq (t + 1)$. And these cases end when all the t vertices in each copy of the graph H have received exactly t different colors and this occurs when $r = (t + 1)(q - 1)$, i.e., when $i = t + 1$. Therefore we can conclude from this that $\chi_r(K_q * H) \geq r + i$ when $(i - 1)(q - 1) + 1 \leq r \leq i(q - 1)$ and $3 \leq i \leq (t + 1)$.

Final Case: When $r \geq (t+1)(q-1)+1$.
By $r = (t+1)(q-1)$ all the $qt+q$ vertices of the graph $K_q * H$ has received different colors and we no longer require new colors hence $\chi_r(K_q * H) \geq qt+q$ when $r \geq (t+1)(q-1)+1$.

Let the vertex set for the extended neighborhood corona of complete graph K_q with any graph H with t vertices be defined as $V(K_q * H) = \{v_i : 1 \leq i \leq q\} \cup \{u_{ik} : 1 \leq i \leq q, 1 \leq k \leq t\}$.

Theorem 1 *For positive integers $q \geq 2, t \geq 3$, the r-dynamic chromatic number of the extended neighborhood corona of complete graph K_q with path P_t is*

$$\chi_r(K_q * P_t) = \begin{cases} 2q & \text{for } 1 \leq r \leq 2(q-1) \\ r+i & \text{for } (i-1)(q-1)+1 \leq r \leq i(q-1), 3 \leq i \leq (t+1) \\ qt+q & \text{for } r \geq (t+1)(q-1)+1 \end{cases}$$

Proof The minimum degree, $\delta(K_q * P_t) = (t+1)(q-1)$ and maximum degree, $\Delta(K_q * P_t) = (t+1)(q-1)+2$.
Case 1: When $1 \leq r \leq 2(q-1)$.
By Lemma 1 we have the lower bound $\chi_r(K_q * P_t) \geq 2q$. The function $f : V(K_q * P_t) \rightarrow \{1, 2, \ldots, 2q\}$ described as below provides the upper bound.
$f(v_i) = i$ for $1 \leq i \leq q$

$$f(u_{ik}) = \begin{cases} i, & \text{when } k \text{ is odd} \\ q+i, & \text{when } k \text{ is even} \end{cases}$$

Hence $\chi_r(K_q * P_t) = 2q$.
Case 2: When $(i-1)(q-1)+1 \leq r \leq i(q-1), 3 \leq i \leq (t+1)$.
By the above lemma we have the lower bound as $\chi_r(K_q * P_t) \geq r+i$. The coloring for each $r = (i-1)(q-1)+x$ where $1 \leq x \leq (q-1)$ is given as separate cases as below using the coloring function $f : V(K_q * P_t) \rightarrow \{1, 2, \ldots, r+i\}$.
The coloring for the vertices v_i is same as given in Case 1.
For $3 \leq i \leq t$.
Subcase: When $r = (i-1)(q-1)+1$.

$$\begin{aligned} f(u_{11}, u_{12}, \ldots, u_{1t}) &= \{1, q+(q-1)i-(2q-4), \ldots, (i-1)q+2, 1, q \\ &\quad +(q-1)i-(2q-4), \ldots, (i-1)q+2, \ldots\} \\ f(u_{21}, u_{22}, \ldots, u_{2t}) &= \{2, q+1, \ldots, q+i-1, 2, q+1, \ldots, q+i-1, \ldots\} \\ f(u_{31}, u_{32}, \ldots, u_{3t}) &= \{3, q+i, \ldots, q+2i-3, 3, q+i, \ldots, q+2i-3, \ldots\} \\ f(u_{41}, u_{42}, \ldots, u_{4t}) &= \{4, q+2i-2, \ldots, q+3i-5, 4, q+2i-2, \ldots, q \\ &\quad +3i-5, \ldots\} \\ f(u_{51}, u_{52}, \ldots, u_{5t}) &= \{5, q+3i-4, \ldots, q+4i-7, 5, q+3i-4, \ldots, q \\ &\quad +4i-7, \ldots\} \end{aligned}$$

$$\vdots$$

$$f(u_{q1}, u_{q2}, \ldots, u_{qt}) = \{q, q + (q-2)i - (2q-6), \ldots, q + (q-1)i - (2q-3), q, q + (q-2)i - (2q-6), \ldots, q + (q-1)i - (2q-3), \ldots\}$$

That is, in this case first $(i-1)$ new colors are added in the 2nd copy of H, $(i-2)$ new colors are added in each of 3rd, 4th, $\cdots$, q^{th} copy of H, and finally $(i-1)$ new colors are added in the 1st copy of H.

Subcase: When $r = (i-1)(q-1) + 2$.

$$f(u_{11}, u_{12}, \ldots, u_{1t}) = \{1, q + (q-1)i - (2q-5), \ldots, (i-1)q + 3, 1, q + (q-1)i - (2q-3), \ldots, (i-1)q + 3, \ldots\}$$

$$f(u_{21}, u_{22}, \ldots, u_{2t}) = \{2, q+1, \ldots, q+i-1, 2, q+1, \ldots, q+i-1, \ldots\}$$

$$f(u_{31}, u_{32}, \ldots, u_{3t}) = \{3, q+i, \ldots, q+2i-2, 3, q+i, \ldots, q+2i-2, \ldots\}$$

$$f(u_{41}, u_{42}, \ldots, u_{4t}) = \{4, q+2i-1, \ldots, q+3i-4, 4, q+2i-1, \ldots, q+3i-4, \ldots\}$$

$$f(u_{51}, u_{52}, \ldots, u_{5t}) = \{5, q+3i-3, \ldots, q+4i-6, 5, q+3i-3, \ldots, q+4i-6, \ldots\}$$

$$\vdots$$

$$f(u_{q1}, u_{q2}, \ldots, u_{qt}) = \{q, q + (q-2)i - (2q-7), \ldots, q + (q-1)i - (2q-4), q, q + (q-2)i - (2q-7), \ldots, q + (q-1)i - (2q-4), \ldots\}$$

That is, in this case first $(i-1)$ new colors are added in each of 2nd and 3rd copy of H, $(i-2)$ new colors are added in each of 4th, 5th, $\cdots$, q^{th} copy of H and finally $(i-1)$ new colors are added in the 1st copy of H.

Subcase: When $r = (i-1)(q-1) + 3$.

$$f(u_{11}, u_{12}, \ldots, u_{1t}) = \{1, q + (q-1)i - (2q-6), \ldots, (i-1)q + 4, 1, q + (q-1)i - (2q-4), \ldots, (i-1)q + 4, \ldots\}$$

$$f(u_{21}, u_{22}, \ldots, u_{2t}) = \{2, q+1, \ldots, q+i-1, 2, q+1, \ldots, q+i-1, \ldots\}$$

$$f(u_{31}, u_{32}, \ldots, u_{3t}) = \{3, q+i, \ldots, q+2i-2, 3, q+i, \ldots, q+2i-2, \ldots\}$$

$$f(u_{41}, u_{42}, \ldots, u_{4t}) = \{4, q+2i-1, \ldots, q+3i-3, 4, q+2i-1, \ldots, q+3i-3, \ldots\}$$

$$f(u_{51}, u_{52}, \ldots, u_{5t}) = \{5, q+3i-2, \ldots, q+4i-5, 5, q+3i-2, \ldots, q+4i-5, \ldots\}$$

$$f(u_{61}, u_{62}, \ldots, u_{6t}) = \{6, q+4i-4, \ldots, q+5i-7, 6, q+4i-4, \ldots, q+5i-7, \ldots\}$$

$$\vdots$$

$$f(u_{q1}, u_{q2}, \ldots, u_{qt}) = \{q, q + (q-2)i - (2q-8), \ldots, q + (q-1)i - (2q-5), q, q + (q-2)i - (2q-8), \ldots, q + (q-1)i - (2q-5), \ldots\}$$

That is, in this case first $(i-1)$ new colors are added in each of 2nd, 3rd, and 4th copy of H, $(i-2)$ new colors are added in each of 5th, 6th, ..., qth copy of H and finally $(i-1)$ new colors are added in the 1st copy of H.
In general, we can see that when $r = (i-1)(q-1) + x$ where $1 \le x \le (q-1)$ first $(i-1)$ new colors are added in each of 2nd, ..., $(x+1)$th copy of H then $(i-2)$ new colors are added in each of $(x+2)$th, ..., qth copy of H and finally $(i-1)$ new colors in the 1st copy of H. This same pattern is followed for coloring all graphs H in $K_q * H$, but the coloring sequence varies for different graphs H.
Subcase: When $r = (i-1)(q-1) + (q-1) = i(q-1)$ the coloring is as below:

$$f(u_{11}, u_{12}, \ldots, u_{1t}) = \{1, q + (q-1)i - (q-2), \ldots, iq, 1, q + (q-1)i - (q-2), \ldots, iq, \ldots\}$$
$$f(u_{21}, u_{22}, \ldots, u_{2t}) = \{2, q+1, \ldots, q+i-1, 2, q+1, \ldots, q+i-1, \ldots\}$$
$$f(u_{31}, u_{32}, \ldots, u_{3t}) = \{3, q+i, \ldots, q+2i-2, 3, q+i, \ldots, q+2i-2, \ldots\}$$
$$f(u_{41}, u_{42}, \ldots, u_{4t}) = \{4, q+2i-1, \ldots, q+3i-3, 4, q+2i-1, \ldots, q+3i-3, \ldots\}$$
$$f(u_{51}, u_{52}, \ldots, u_{5t}) = \{5, q+3i-2, \ldots, q+4i-4, 5, q+3i-2, \ldots, q+4i-4, \ldots\}$$
$$f(u_{61}, u_{62}, \ldots, u_{6t}) = \{6, q+4i-3, \ldots, q+5i-5, 6, q+4i-3, \ldots, q+5i-5, \ldots\}$$
$$\vdots$$
$$f(u_{q1}, u_{q2}, \ldots, u_{qt}) = \{q, q + (q-2)i - (q-3), \ldots, q + (q-1)i - (q-1), q, q + (q-2)i - (q-3), \ldots, q + (q-1)i - (q-1), \ldots\}$$

For $i = t + 1$.
Subcase: When $r = (i-1)(q-1) + 1 = t(q-1) + 1$.

$$f(u_{11}, u_{12}, \ldots, u_{1t}) = \{(q-1)t + 3, (q-1)t + 4, \ldots, qt + 2\}$$
$$f(u_{21}, u_{22}, \ldots, u_{2t}) = \{q+1, q+2, \ldots, q+t\}$$
$$f(u_{31}, u_{32}, \ldots, u_{3t}) = \{3, q+t+1, \ldots, q+2t-1\}$$
$$f(u_{41}, u_{42}, \ldots, u_{4t}) = \{4, q+2t, \ldots, q+3t-2\}$$
$$f(u_{51}, u_{52}, \ldots, u_{5t}) = \{5, q+3t-1, \ldots, q+4t-4\}$$
$$\vdots$$
$$f(u_{q1}, u_{q2}, \ldots, u_{qt}) = \{q, q + (q-2)t - (q-4), \ldots, (q-1)t + 2\}$$

That is, in this case first t new colors are added in the 2nd copy of H, $t-1$ new colors are added in each of 3rd, 4th, $\cdots$, q^{th} copy of H, and finally t new colors are added in the 1st copy of H.
Subcase: When $r = t(q-1)+2$.

$$\begin{aligned}
f(u_{11}, u_{12}, \ldots, u_{1t}) &= \{(q-1)t+4, (q-1)t+5, \ldots, qt+3\}\\
f(u_{21}, u_{22}, \ldots, u_{2t}) &= \{q+1, q+2, \ldots, q+t\}\\
f(u_{31}, u_{32}, \ldots, u_{3t}) &= \{q+t+1, q+t+2, \ldots, q+2t\}\\
f(u_{41}, u_{42}, \ldots, u_{4t}) &= \{4, q+2t+1, \ldots, q+3t-1\}\\
f(u_{51}, u_{52}, \ldots, u_{5t}) &= \{5, q+3t, \ldots, q+4t-2\}\\
f(u_{61}, u_{62}, \ldots, u_{6t}) &= \{6, q+4t-1, \ldots, q+5t-3\}\\
&\vdots\\
f(u_{q1}, u_{q2}, \ldots, u_{qt}) &= \{q, q+(q-2)t-(q-5), \ldots, (q-1)t+3\}
\end{aligned}$$

That is, in this case first t new colors are added in each of 2nd and 3rd copy of H, $t-1$ new colors are added in each of 4th, $\cdots$, q^{th} copy of H, and finally t new colors are added in the 1st copy of H.
Subcase: When $r = t(q-1)+3$.

$$\begin{aligned}
f(u_{11}, u_{12}, \ldots, u_{1t}) &= \{(q-1)t+5, (q-1)t+6, \ldots, qt+4\}\\
f(u_{21}, u_{22}, \ldots, u_{2t}) &= \{q+1, q+2, \ldots, q+t\}\\
f(u_{31}, u_{32}, \ldots, u_{3t}) &= \{q+t+1, q+t+2, \ldots, q+2t\}\\
f(u_{41}, u_{42}, \ldots, u_{4t}) &= \{q+2t+1, \ldots, q+3t\}\\
f(u_{51}, u_{52}, \ldots, u_{5t}) &= \{5, q+3t+1, \ldots, q+4t-1\}\\
f(u_{61}, u_{62}, \ldots, u_{6t}) &= \{6, q+4t, \ldots, q+5t-2\}\\
&\vdots\\
f(u_{q1}, u_{q2}, \ldots, u_{qt}) &= \{q, q+(q-2)t-(q-6), \ldots, (q-1)t+4\}
\end{aligned}$$

That is, in this case first t new colors are added in each of 2nd, 3rd, and 4th copy of H, $t-1$ new colors are added in each of 5th, $\cdots$, q^{th} copy of H, and finally t new colors are added in the 1st copy of H. Thus, we can conclude that when $r = (i-1)(q-1)+x$ where $1 \leq x \leq (q-1)$ first t new colors are added in each of $2^{\text{nd}}, \ldots, (x+1)^{\text{th}}$ copy of H then $t-1$ new colors are added in each of $(x+2)^{\text{th}}$, $\ldots$, q^{th} copy of H and finally t new colors are added in the 1st copy of H.
Subcase: At last when $r = (t+1)(q-1)$.

$$\begin{aligned}
f(u_{11}, u_{12}, \ldots, u_{1t}) &= \{(q-1)t+q+1, \ldots, qt+q\}\\
f(u_{21}, u_{22}, \ldots, u_{2t}) &= \{q+1, q+2, \ldots, q+t\}\\
f(u_{31}, u_{32}, \ldots, u_{3t}) &= \{q+t+1, q+t+2, \ldots, q+2t\}\\
f(u_{41}, u_{42}, \ldots, u_{4t}) &= \{q+2t+1, \ldots, q+3t\}
\end{aligned}$$

$$\vdots$$

$$f(u_{q1}, u_{q2}, \ldots, u_{qt}) = \{q + (q-2)t + 1, \ldots, (q-1)t + q\}$$

Thus, we have the upper bound $\chi_r(K_q * P_t) \leq r + i$. Hence $\chi_r(K_q * P_t) = r + i$ when $(i-1)(q-1) + 1 \leq r \leq i(q-1), 3 \leq i \leq (t+1)$.
Case 3: When $r \geq (t+1)(q-1) + 1$.
By $r = (t+1)(q-1)$ all $qt + q$ vertices of $K_q * P_t$ will be colored differently and hence we no longer need any extra colors. Thus $\chi_r(K_q * P_t) = qt + q$.

Theorem 2 *For positive integers $q \geq 2, t \geq 3$, the r-dynamic chromatic number of the extended neighborhood corona of complete graph K_q with cycle C_t is*

$$\chi_r(K_q * C_t) = \begin{cases} 2q & \text{for } 1 \leq r \leq 2(q-1), t \text{ is even} \\ 3q & \text{for } 1 \leq r \leq 3(q-1), t \text{ is odd} \\ r + i & \text{for } (i-1)(q-1) + 1 \leq r \leq i(q-1) \\ & \quad \text{where } i = 3 \text{ for } t \text{ is even and } 4 \leq i \leq (t+1) \\ qt + q & \text{for } r \geq (t+1)(q-1) + 1 \end{cases}$$

Proof The minimum degree, $\delta(K_q * C_t) = (t+1)(q-1)$ and maximum degree, $\Delta(K_q * C_t) = (t+1)(q-1) + 2$.
Case 1: When $1 \leq r \leq 2(q-1)$, t is even.
By Lemma 1 we have the lower bound $\chi_r(K_q * C_t) \geq 2q$. The function $f : V(K_q * P_t) \to \{1, 2, \ldots, 2q\}$ described as below provides the upper bound.
$f(v_i) = i$ for $1 \leq i \leq q$

$$f(u_{ik}) = \begin{cases} i, & \text{when } k \text{ is odd} \\ q + i, & \text{when } k \text{ is even} \end{cases}$$

Hence $\chi_r(K_q * C_t) = 2q$.
Case 2: When $1 \leq r \leq 3(q-1)$, t is odd.
Using Lemma 1 we get $\chi_r(K_q * C_t) \geq 2q$, but here the cycle C_t is an odd cycle so $\chi(C_t) = 3$, i.e., we require at least 3 different colors and by the definition of the extended corona each copy of H is adjacent to all the remaining copies of H. Thus we require 3 colors for each copy of H that is we require a minimum of $3q$ colors. Hence the lower bound transforms into $\chi_r(K_q * C_t) \geq 3q$ when t is odd. Consider the coloring function $f : V(K_q * C_t) \to \{1, 2, \ldots, 3q\}$ which yields the upper bound $\chi_r(K_q * C_t) \leq 3q$.

The coloring for the vertices v_i is same as given in Case 1.

$$f(u_{11}, u_{12}, \ldots, u_{1t}) = \{1, q+1, q+2, q+1, q+2, q+1, q+2, \ldots\}$$
$$f(u_{21}, u_{22}, \ldots, u_{2t}) = \{2, q+3, q+4, q+3, q+4, q+3, q+4, \ldots\}$$
$$f(u_{31}, u_{32}, \ldots, u_{3t}) = \{3, q+5, q+6, q+5, q+6, q+5, q+6, \ldots\}$$
$$\vdots$$
$$f(u_{q1}, u_{q2}, \ldots, u_{qt}) = \{q, 3q-1, 3q, 3q-1, 3q, 3q-1, 3q, \ldots\}$$

Therefore we have $\chi_r(K_q * C_t) = 3q$ when t is odd.
Case 3: When $(i-1)(q-1)+1 \le r \le i(q-1)$ where $i = 3$ for t is even and $4 \le i \le (t+1)$ for all t.
By Lemma 1 we have the lower bound $\chi_r(K_q * C_t) \ge r+3$ when $2(q-1)+1 \le r \le 3(q-1)$ for t is even, i.e., $i = 3$ in $(i-1)(q-1)+1 \le r \le i(q-1)$. Also by the same lemma, we have the lower bound $\chi_r(K_q * C_t) \ge r+i$ when $(i-1)(q-1)+1 \le r \le i(q-1), 4 \le i \le t+1$ for all t. The coloring for each $r = (i-1)(q-1)+x$ where $1 \le x \le (q-1)$ is given as separate cases as below using the coloring function $f : V(K_q * C_t) \to \{1, 2, \ldots, r+i\}$. The coloring for the vertices v_i is same as given in Case 1.
For $3 \le i \le t$.
Subcase: When $r = (i-1)(q-1)+1$.

$$\begin{aligned}
f(u_{11}, u_{12}, \ldots, u_{1t}) &= \{1, q+(q-1)i-(2q-4), \ldots, (i-1)q+2, q \\
&\quad +(q-1)i-(2q-4), \ldots, (i-1)q+2, \ldots\} \\
f(u_{21}, u_{22}, \ldots, u_{2t}) &= \{2, q+1, \ldots, q+i-1, q+1, \ldots, q+i-1, \ldots\} \\
f(u_{31}, u_{32}, \ldots, u_{3t}) &= \{3, q+i, \ldots, q+2i-3, q+i, \ldots, q+2i-3, \ldots\} \\
f(u_{41}, u_{42}, \ldots, u_{4t}) &= \{4, q+2i-2, \ldots, q+3i-5, q+2i-2, \ldots, q \\
&\quad +3i-5, \ldots\} \\
f(u_{51}, u_{52}, \ldots, u_{5t}) &= \{5, q+3i-4, \ldots, q+4i-7, q+3i-4, \ldots, q \\
&\quad +4i-7, \ldots\} \\
&\vdots \\
f(u_{q1}, u_{q2}, \ldots, u_{qt}) &= \{q, q+(q-2)i-(2q-6), \ldots, q+(q-1)i \\
&\quad -(2q-3), q+(q-2)i-(2q-6), \ldots, q+(q-1)i-(2q-3), \ldots\}
\end{aligned}$$

Subcase: When $r = (i-1)(q-1)+2$.

$$\begin{aligned}
f(u_{11}, u_{12}, \ldots, u_{1t}) &= \{1, q+(q-1)i-(2q-5), \ldots, (i-1)q+3, q \\
&\quad +(q-1)i-(2q-5), \ldots, (i-1)q+3, \ldots\} \\
f(u_{21}, u_{22}, \ldots, u_{2t}) &= \{2, q+1, \ldots, q+i-1, q+1, \ldots, q+i-1, \ldots\} \\
f(u_{31}, u_{32}, \ldots, u_{3t}) &= \{3, q+i, \ldots, q+2i-2, q+i, \ldots, q+2i-2, \ldots\} \\
f(u_{41}, u_{42}, \ldots, u_{4t}) &= \{4, q+2i-1, \ldots, q+3i-4, q+2i-1, \ldots, q \\
&\quad +3i-4, \ldots\} \\
f(u_{51}, u_{52}, \ldots, u_{5t}) &= \{5, q+3i-3, \ldots, q+4i-6, q+3i-3, \ldots, q \\
&\quad +4i-6, \ldots\} \\
&\vdots \\
f(u_{q1}, u_{q2}, \ldots, u_{qt}) &= \{q, q+(q-2)i-(2q-7), \ldots, q+(q-1)i \\
&\quad -(2q-4), q+(q-2)i-(2q-7), \ldots, q+(q-1)i-(2q-4), \ldots\}
\end{aligned}$$

Proceeding in the similar pattern when $r = i(q-1)$ we have the coloring as follows:

Subcase : When $r = i(q-1)$.

$$f(u_{11}, u_{12}, \ldots, u_{1t}) = \{1, q+(q-1)i-(q-2), \ldots, iq, q+(q-1)i -(q-2), \ldots, iq, \ldots\}$$
$$f(u_{21}, u_{22}, \ldots, u_{2t}) = \{2, q+1, \ldots, q+i-1, q+1, \ldots, q+i-1, \ldots\}$$
$$f(u_{31}, u_{32}, \ldots, u_{3t}) = \{3, q+i, \ldots, q+2i-2, q+i, \ldots, q+2i-2, \ldots\}$$
$$f(u_{41}, u_{42}, \ldots, u_{4t}) = \{4, q+2i-1, \ldots, q+3i-3, q+2i-1, \ldots, q +3i-3, \ldots\}$$
$$f(u_{51}, u_{52}, \ldots, u_{5t}) = \{5, q+3i-2, \ldots, q+4i-4, q+3i-2, \ldots, q +4i-4, \ldots\}$$
$$f(u_{61}, u_{62}, \ldots, u_{6t}) = \{6, q+4i-3, \ldots, q+5i-5, q+4i-3, \ldots, q +5i-5, \ldots\}$$
$$\vdots$$
$$f(u_{q1}, u_{q2}, \ldots, u_{qt}) = \{q, q+(q-2)i-(q-3), \ldots, q+(q-1)i -(q-1), q+(q-2)i-(q-3), \ldots, q+(q-1)i-(q-1), \ldots\}$$

When $i = t+1$ the coloring is same as the one given in Case 2 of Theorem 1. Therefore $\chi_r(K_q * C_t) = r+i$ when $(i-1)(q-1)+1 \le r \le i(q-1)$ where $i = 3$ for t is even and $4 \le i \le (t+1)$ for all t.
Case 4: When $r \ge (t+1)(q-1)+1$.
By $r = (t+1)(q-1)$ all $qt+q$ vertices of $K_q * C_t$ will be colored differently, and hence, we no longer need any extra colors. Thus $\chi_r(K_q * C_t) = qt+q$.

Theorem 3 *For positive integers $q \ge 2, t \ge 2$, the r-dynamic chromatic number of the extended neighborhood corona of complete graph K_q with complete graph K_t is*

$$\chi_r(K_q * K_t) = \begin{cases} qt & \text{for } 1 \le r \le t(q-1) \\ r+t+1 & \text{for } t(q-1)+1 \le r \le (t+1)(q-1) \\ qt+q & \text{for } r \ge (t+1)(q-1)+1 \end{cases}$$

Proof The minimum degree, $\delta(K_q * K_t) = (t+1)(q-1)$ and maximum degree, $\Delta(K_q * K_t) = (t+1)(q-1)+t-1$.
Case 1: When $1 \le r \le t(q-1)$.
Since the complete graph K_t requires t different colors and each copy of K_t is adjacent to all the remaining copies of K_t. There are q such copies of K_t we require at least qt different colors in this case, and thus, the lower bound is $\chi_r(K_q * K_t) \ge qt$. The upper bound is provided using the function $f : V(K_q * K_t) \to \{1, 2, \ldots, qt\}$ as below:
$f(v_i) = i$

$$
\begin{aligned}
&f(u_{11}, u_{12}, \ldots, u_{1t}) - \{1, q+1, q+2, \ldots, q+t-1\}\\
&f(u_{21}, u_{22}, \ldots, u_{2t}) = \{2, q+t, q+t+1, \ldots, q+2t-2\}\\
&f(u_{31}, u_{32}, \ldots, u_{3t}) = \{3, q+2t-1, q+2t, \ldots, q+3t-3\}\\
&f(u_{41}, u_{42}, \ldots, u_{4t}) = \{4, q+3t-2, q+3t-1, \ldots, q+4t-4\}\\
&\vdots\\
&f(u_{q1}, u_{q2}, \ldots, u_{qt}) = \{q, q+(q-1)t-(q-2), \ldots, qt\}
\end{aligned}
$$

Therefore $\chi_r(K_q * K_t) = qt$.
Case 2: When $t(q-1)+1 \le r \le (t+1)(q-1)$.
By Lemma 1 we have the lower bound $\chi_r(K_q * K_t) \ge r+i$ when $(i-1)(q-1)+1 \le r \le i(q-1)$. When $i = t+1$ in $(i-1)(q-1)+1 \le r \le i(q-1)$, we have the values of r as in our case. Thus, we have the lower bound in this case as $\chi_r(K_q * K_t) \ge r+i = r+t+1$. The coloring in this case is same the one for $i = t+1$ of Case 2 in Theorem 1. Therefore $\chi_r(K_q * K_t) = r+t+1$.
Case 3: When $r \ge (t+1)(q-1)+1$.
By $r = (t+1)(q-1)$ all $qt+q$ vertices of $K_q * K_t$ will be colored differently and hence we no longer need any extra colors. Thus $\chi_r(K_q * K_t) = qt+q$.

Theorem 4 *For positive integers $q \ge 2, t \ge 2$, the r-dynamic chromatic number of the extended neighborhood corona of complete graph K_q with star graph $K_{1,t}$ is*

$$
\chi_r(K_q * K_{1,t}) = \begin{cases} 2q & \text{for } 1 \le r \le 2(q-1)\\ r+i & \text{for } (i-1)(q-1)+1 \le r \le i(q-1), 3 \le i \le (t+2)\\ qt+2q & \text{for } r \ge (t+2)(q-1)+1 \end{cases}
$$

Proof Here the graph $K_{1,t}$ is a graph with $t+1$ vertices. The vertex set of $K_q * K_{1,t}$ is $V(K_q * K_{1,t}) = \{v_i : 1 \le i \le q\} \cup \{u_{ik} : 1 \le i \le q, 1 \le k \le t+1\}$ where u_{i1}'s are the central vertex of $K_{1,t}$ which is adjacent with the t vertices $u_{i2}, u_{i3}, \ldots, u_{i(t+1)}$. The minimum degree, $\delta(K_q * K_{1,t}) = (t+2)(q-1)$ and maximum degree, $\Delta(K_q * K_{1,t}) = (t+2)(q-1)+t$.
Case 1: When $1 \le r \le 2(q-1)$.
By Lemma 1 we have the lower bound $\chi_r(K_q * K_{1,t}) \ge 2q$. We provide the upper bound in this case using the map $f : V(K_q * K_{1,t}) \to \{1, 2, \ldots, 2q\}$ as follows:
$f(v_i) = i$ for $1 \le i \le q$

$$
f(u_{ik}) = \begin{cases} i, & \text{when } k = 1\\ q+i, & \text{when } 2 \le k \le t+1 \end{cases}
$$

Thus, we have the upper bound $\chi_r(K_q * K_{1,t}) \le 2q$. Hence $\chi_r(K_q * K_{1,t}) = 2q$.
Case 2: When $(i-1)(q-1)+1 \le r \le i(q-1), 3 \le i \le (t+2)$.
By Lemma 1 we have the lower bound $\chi_r(K_q * H) \ge r+i$ when $(i-1)(q-1)+1 \le r \le i(q-1), 3 \le i \le (t+1)$ and we apply this lower bound here to get $\chi_r(K_q * K_{1,t}) \ge r+i$ when $(i-1)(q-1)+1 \le r \le i(q-1), 3 \le i \le (t+2)$.

The coloring for each $r = (i-1)(q-1) + x$ where $1 \leq x \leq (q-1)$ is given as separate cases as below using the coloring function $f : V(K_q * K_{1,t}) \rightarrow \{1, 2, \ldots, r + i\}$. The coloring for the vertices v_i is same as given in Case 1.
For $3 \leq i \leq t + 1$.
Subcase: When $r = (i-1)(q-1) + 1$.

$$f(u_{11}, u_{12}, \ldots, u_{1(t+1)}) = \{1, q + (q-1)i - (2q-4), \ldots, (i-1)q + 2, q$$
$$+(q-1)i - (2q-4), q + (q-1)i - (2q-4), q + (q-1)i - (2q-4), \ldots\}$$
$$f(u_{21}, u_{22}, \ldots, u_{2(t+1)}) = \{2, q+1, \ldots, q+i-1, q+1, q+1, q+1, \ldots\}$$
$$f(u_{31}, u_{32}, \ldots, u_{3(t+1)}) = \{3, q+i, \ldots, q+2i-3, q+i, q+i, q+i, \ldots\}$$
$$f(u_{41}, u_{42}, \ldots, u_{4(t+1)}) = \{4, q+2i-2, \ldots, q+3i-5, q+2i-2, q$$
$$+2i-2, q+2i-2, \ldots\}$$
$$\vdots$$
$$f(u_{q1}, u_{q2}, \ldots, u_{q(t+1)}) = \{q, q + (q-2)i - (2q-6), \ldots, q + (q-1)i$$
$$-(2q-3), q + (q-2)i - (2q-6), q + (q-2)i - (2q-6), q + (q-2)i$$
$$-(2q-6), \ldots\}$$

Subcase: When $r = (i-1)(q-1) + 2$.

$$f(u_{11}, u_{12}, \ldots, u_{1(t+1)}) = \{1, q + (q-1)i - (2q-5), \ldots, (i-1)q + 3, q$$
$$+(q-1)i - (2q-5), q + (q-1)i - (2q-5), q + (q-1)i - (2q-5), \ldots\}$$
$$f(u_{21}, u_{22}, \ldots, u_{2(t+1)}) = \{2, q+1, \ldots, q+i-1, q+1, q+1, q+1, \ldots\}$$
$$f(u_{31}, u_{32}, \ldots, u_{3(t+1)}) = \{3, q+i, \ldots, q+2i-2, q+i, q+i, q+i, \ldots\}$$
$$f(u_{41}, u_{42}, \ldots, u_{4(t+1)}) = \{4, q+2i-1, \ldots, q+3i-4, q+2i-1, q$$
$$+2i-1, q+2i-1, \ldots\}$$
$$f(u_{51}, u_{52}, \ldots, u_{5(t+1)}) = \{5, q+3i-3, \ldots, q+4i-6, q+3i-3, q$$
$$+3i-3, q+3i-3, \ldots\}$$
$$\vdots$$
$$f(u_{q1}, u_{q2}, \ldots, u_{qv}) = \{q, q + (q-2)i - (2q-7), \ldots, q + (q-1)i$$
$$-(2q-4), q + (q-2)i - (2q-7), q + (q-2)i - (2q-7), q + (q-2)i$$
$$-(2q-7), \ldots\}$$

Proceeding in the similar pattern when $r = i(q-1)$ we have the coloring as follows:

Subcase: When $r = i(q-1)$.

$$f(u_{11}, u_{12}, \ldots, u_{1(t+1)}) = \{1, q+(q-1)i-(q-2), \ldots, iq, q+(q-1)i -(q-2), q+(q-1)i-(q-2), q+(q-1)i-(q-2), \ldots\}$$
$$f(u_{21}, u_{22}, \ldots, u_{2(t+1)}) = \{2, q+1, \ldots, q+i-1, q+1, q+1, q+1, \ldots\}$$
$$f(u_{31}, u_{32}, \ldots, u_{3(t+1)}) = \{3, q+i, \ldots, q+2i-2, q+i, q+i, q+i, \ldots\}$$
$$f(u_{41}, u_{42}, \ldots, u_{4(t+1)}) = \{4, q+2i-1, \ldots, q+3i-3, q+2i-1, q +2i-1, q+2i-1, \ldots\}$$
$$f(u_{51}, u_{52}, \ldots, u_{5(t+1)}) = \{5, q+3i-2, \ldots, q+4i-4, q+3i-2, q +3i-2, q+3i-2, \ldots\}$$
$$f(u_{61}, u_{62}, \ldots, u_{6(t+1)}) = \{6, q+4i-3, \ldots, q+5i-5, q+4i-3, q +4i-3, q+4i-3, \ldots\}$$
$$\vdots$$
$$f(u_{q1}, u_{q2}, \ldots, u_{q(t+1)}) = \{q, q+(q-2)i-(q-3), \ldots, q+(q-1)i -(q-1), q+(q-2)i-(q-3), q+(q-2)i-(q-3), q+(q-2)i -(q-3), \ldots\}$$

For $i = t+2$.
Subcase: When $r = (i-1)(q-1)+1 = q(t+1)-t$.

$$f(u_{11}, u_{12}, \ldots, u_{1(t+1)}) = \{q+(q-1)t+2, \ldots, q(t+1)+2\}$$
$$f(u_{21}, u_{22}, \ldots, u_{2(t+1)}) = \{q+1, q+2, \ldots, q+t+1\}$$
$$f(u_{31}, u_{32}, \ldots, u_{3(t+1)}) = \{3, q+t+2, \ldots, q+2t+1\}$$
$$f(u_{41}, u_{42}, \ldots, u_{4(t+1)}) = \{4, q+2t+2, \ldots, q+3t+1\}$$
$$f(u_{51}, u_{52}, \ldots, u_{5(t+1)}) = \{5, q+3t+2, \ldots, q+4t+1\}$$
$$\vdots$$
$$f(u_{q1}, u_{q2}, \ldots, u_{q(t+1)}) = \{q, q+(q-2)t+2, \ldots, q+(q-1)t+1\}$$

Subcase: When $r = q(t+1)-t+1$.

$$f(u_{11}, u_{12}, \ldots, u_{1(t+1)}) = \{q+(q-1)t+3, \ldots, q(t+1)+3\}$$
$$f(u_{21}, u_{22}, \ldots, u_{2(t+1)}) = \{q+1, q+2, \ldots, q+t+1\}$$
$$f(u_{31}, u_{32}, \ldots, u_{3(t+1)}) = \{q+t+2, \ldots, q+2t+2\}$$
$$f(u_{41}, u_{42}, \ldots, u_{4(t+1)}) = \{4, q+2t+3, \ldots, q+3t+2\}$$
$$f(u_{51}, u_{52}, \ldots, u_{5(t+1)}) = \{5, q+3t+3, \ldots, q+4t+2\}$$
$$\vdots$$
$$f(u_{q1}, u_{q2}, \ldots, u_{q(t+1)}) = \{q, q+(q-2)t+3, \ldots, q+(q-1)t+2\}$$

Proceeding similarly at last when $r = (t+2)(q-1)$ we get the coloring as below;
Subcase: When $r = (t+2)(q-1)$.

$$
\begin{aligned}
f(u_{11}, u_{12}, \ldots, u_{1(t+1)}) &= \{2q + (q-1)t, \ldots, q(t+2)\} \\
f(u_{21}, u_{22}, \ldots, u_{2(t+1)}) &= \{q+1, q+2, \ldots, q+t+1\} \\
f(u_{31}, u_{32}, \ldots, u_{3(t+1)}) &= \{q+t+2, \ldots, q+2t+2\} \\
f(u_{41}, u_{42}, \ldots, u_{4(t+1)}) &= \{q+2t+3, \ldots, q+3t+3\} \\
&\vdots \\
f(u_{q1}, u_{q2}, \ldots, u_{q(t+1)}) &= \{q + (q-2)t + (q-1), \ldots, 2q + (q-1)t - 1)\}
\end{aligned}
$$

Therefore $\chi_r(K_q * K_{1,t}) = r + i$ when $(i-1)(q-1)+1 \le r \le i(q-1), 3 \le i \le (t+2)$.
Case 3: When $r \ge (t+2)(q-1)+1$.
By $r = (t+1)(q-1)$ all $qt + 2q$ vertices of $K_q * K_{1,t}$ will be colored differently and hence we no longer need any extra colors. Thus $\chi_r(K_q * K_{1,t}) = qt + 2q$.

Theorem 5 *For positive integers $q \ge 2, t \ge 4$, the r-dynamic chromatic number of the extended neighborhood corona of complete graph K_q with wheel graph W_t is*

$$
\chi_r(K_q * W_t) = \begin{cases} 3q & \text{for } 1 \le r \le 3(q-1), t \text{ is odd} \\ 4q & \text{for } 1 \le r \le 4(q-1), t \text{ is even} \\ r+i & \text{for } (i-1)(q-1)+1 \le r \le i(q-1) \\ & \quad \text{where } i = 4 \text{ for } t \text{ is odd and } 5 \le i \le (t+1) \\ qt+q & \text{for } r \ge (t+1)(q-1)+1 \end{cases}
$$

Proof The minimum degree, $\delta(K_q * W_t) = (t+1)(q-1)$ and maximum degree, $\Delta(K_q * W_t) = (t+1)(q-1) + t - 1$. Here u_{i1} is the hub vertex of the wheel graph W_t which is adjacent to the $t-1$ vertices $\{u_{i2}, u_{i3}, \ldots, u_{it}\}$ in the respective copies.
Case 1: When $1 \le r \le 3(q-1)$, t is odd.
Since we know that the wheel graph W_t of odd order requires 3 different colors, and each copy of W_t is adjacent to all other copies of W_t in $K_q * W_t$ we require at least $3q$ different colors. Thus the lower bound is $\chi_r(K_q * W_t) \ge 3q$. Now we provide the upper bound $\chi_r(K_q * W_t) \le 3q$ using the coloring map $f : V(K_q * W_t) \to \{1, 2, \ldots, 3q\}$. The coloring for this case is same as the one given in Case 2 of Theorem 2. Hence we have $\chi_r(K_q * W_t) = 3q$.
Case 2: When $1 \le r \le 4(q-1)$, t is even.
Observe the fact that the wheel graph W_t of even order requires 4 different colors, i.e., $\chi(W_t) = 4$ and each copy of W_t is adjacent to all other copies of W_t in $K_q * W_t$ we require at least $4q$ different colors. Thus the lower bound is $\chi_r(K_q * W_t) \ge 4q$. Now we provide the upper bound $\chi_r(K_q * W_t) \le 4q$ using the coloring map $f : V(K_q * W_t) \to \{1, 2, \ldots, 4q\}$ as follows:
$f(v_i) = i$

$$f(u_{11}, u_{12}, \ldots, u_{1t}) = \{1, q+1, q+2, q+3, q+2, q+3, q+2, q+3, \ldots\}$$
$$f(u_{21}, u_{22}, \ldots, u_{2t}) = \{2, q+4, q+5, q+6, q+5, q+6, q+5, q+6, \ldots\}$$
$$f(u_{31}, u_{32}, \ldots, u_{3t}) = \{3, q+7, q+8, q+9, q+8, q+9, q+8, q+9, \ldots\}$$
$$\vdots$$
$$f(u_{q1}, u_{q2}, \ldots, u_{qt}) = \{q, 4q-2, 4q-1, 4q, 4q-1, 4q, 4q-1, 4q, \ldots\}$$

Therefore we have $\chi_r(K_q * C_t) = 4q$ when t is even.
Case 3: When $(i-1)(q-1)+1 \le r \le i(q-1)$ where $i = 4$ for t is odd and $5 \le i \le (t+1)$ for all t.
By Lemma 1 we have the lower bound $\chi_r(K_q * W_t) \ge r+4$ when $3(q-1)+1 \le r \le 4(q-1)$ for t is odd i.e., $i = 4$ in $(i-1)(q-1)+1 \le r \le i(q-1)$. Also by using the same lemma we have the lower bound $\chi_r(K_q * W_t) \ge r+i$ when $(i-1)(q-1)+1 \le r \le i(q-1), 5 \le i \le t+1$ for all t. The coloring for each $r = (i-1)(q-1)+x$ where $1 \le x \le (q-1)$ is given as separate cases as below using the coloring function $f : V(K_q * W_t) \to \{1, 2, \ldots, r+i\}$. The coloring for the vertices v_i is same as given in Case 1.
For $3 \le i \le t$.
Subcase: When $r = (i-1)(q-1)+1$.

$$f(u_{11}, u_{12}, \ldots, u_{1t}) = \{1, q+(q-1)i-(2q-4), q+(q-1)i -(2q-5), \ldots, (i-1)q+2, q+(q-1)i-(2q-5), \ldots, (i-1)q+2, q +(q-1)i-(2q-5), \ldots, (i-1)q+2, \ldots\}$$
$$f(u_{21}, u_{22}, \ldots, u_{2t}) = \{2, q+1, q+2, \ldots, q+i-1, q+2, \ldots, q +i-1, q+2, \ldots, q+i-1, \ldots\}$$
$$f(u_{31}, u_{32}, \ldots, u_{3t}) = \{3, q+i, q+i+1, \ldots, q+2i-3, q+i+1, \ldots, q +2i-3, q+i+1, \ldots, q+2i-3, \ldots\}$$
$$f(u_{41}, u_{42}, \ldots, u_{4t}) = \{4, q+2i-2, q+2i-1, \ldots, q+3i-5, q +2i-1, \ldots, q+3i-5, q+2i-1, \ldots, q+3i-5, \ldots\}$$
$$f(u_{51}, u_{52}, \ldots, u_{5t}) = \{5, q+3i-4, q+3i-3, \ldots, q+4i-7, q +3i-3, \ldots, q+4i-7, q+3i-3, \ldots, q+4i-7, \ldots\}$$
$$\vdots$$
$$f(u_{q1}, u_{q2}, \ldots, u_{qt}) = \{q, q+(q-2)i-(2q-6), q+(q-2)i -(2q-7), \ldots, q+(q-1)i-(2q-3), q+(q-2)i-(2q-7), \ldots, q +(q-1)i-(2q-3), q+(q-2)i-(2q-7), \ldots, q+(q-1)i -(2q-3), \ldots\}$$

Subcase: When $r = (i-1)(q-1)+2$.

$$f(u_{11}, u_{12}, \ldots, u_{1t}) = \{1, q+(q-1)i-(2q-5), q+(q-1)i$$
$$-(2q-6), \ldots, (i-1)q+3, q+(q-1)i-(2q-6), \ldots, (i-1)q+3, q$$
$$+(q-1)i-(2q-6), \ldots, (i-1)q+3, \ldots\}$$
$$f(u_{21}, u_{22}, \ldots, u_{2t}) = \{2, q+1, q+2, \ldots, q+i-1, q+2, \ldots, q$$
$$+i-1, q+2, \ldots, q+i-1, \ldots\}$$
$$f(u_{31}, u_{32}, \ldots, u_{3t}) = \{3, q+i, q+i+1, \ldots, q+2i-2, q+i+1, \ldots, q$$
$$+2i-2, q+i+1, \ldots, q+2i-2, \ldots\}$$
$$f(u_{41}, u_{42}, \ldots, u_{4t}) = \{4, q+2i-1, q+2i, \ldots, q+3i-4, q+2i, \ldots, q$$
$$+3i-4, q+2i, \ldots, q+3i-4, \ldots\}$$
$$f(u_{51}, u_{52}, \ldots, u_{5t}) = \{5, q+3i-3, q+3i-2, \ldots, q+4i-6, q$$
$$+3i-2, \ldots, q+4i-6, q+3i-2, \ldots, q+4i-6, \ldots\}$$
$$\vdots$$
$$f(u_{q1}, u_{q2}, \ldots, u_{qt}) = \{q, q+(q-2)i-(2q-7), q+(q-2)i$$
$$-(2q-8), \ldots, q+(q-1)i-(2q-4), q+(q-2)i-(2q-8), \ldots, q$$
$$+(q-1)i-(2q-4), q+(q-2)i-(2q-8), \ldots, q+(q-1)i$$
$$-(2q-4), \ldots\}$$

Proceeding in the similar pattern when $r = i(q-1)$ we have the coloring as follows:
Subcase: When $r = i(q-1)$.

$$f(u_{11}, u_{12}, \ldots, u_{1t}) = \{1, q+(q-1)i-(q-2), q+(q-1)i$$
$$-(q-3), \ldots, iq, q+(q-1)i-(q-3), \ldots, iq, q+(q-1)i$$
$$-(q-3), \ldots, iq, \ldots\}$$
$$f(u_{21}, u_{22}, \ldots, u_{2t}) = \{2, q+1, q+2, \ldots, q+i-1, q+2, \ldots, q$$
$$+i-1, q+2, \ldots, q+i-1, \ldots\}$$
$$f(u_{31}, u_{32}, \ldots, u_{3t}) = \{3, q+i, q+i+1, \ldots, q$$
$$+2i-2, q+i+1, \ldots, q+2i-2, q+i+1, \ldots, q+2i-2, \ldots\}$$
$$f(u_{41}, u_{42}, \ldots, u_{4t}) = \{4, q+2i-1, q+2i, \ldots, q$$
$$+3i-3, q+2i, \ldots, q+3i-3, q+2i, \ldots, q+3i-3, \ldots\}$$
$$f(u_{51}, u_{52}, \ldots, u_{5t}) = \{5, q+3i-2, q+3i-1, \ldots, q+4i-4, q$$
$$+3i-1, \ldots, q+4i-4, q+3i-1, \ldots, q+4i-4, \ldots\}$$
$$f(u_{61}, u_{62}, \ldots, u_{6t}) = \{6, q+4i-3, q+4i-2, \ldots, q+5i-5, q$$
$$+4i-2, \ldots, q+5i-5, q+4i-2, \ldots, q+5i-5, \ldots\}$$
$$\vdots$$
$$f(u_{q1}, u_{q2}, \ldots, u_{qt}) = \{q, q+(q-2)i-(q-3), q+(q-2)i$$

$$-(q-4),\ldots,q+(q-1)i-(q-1),q+(q-2)i-(q-4),\ldots,q$$
$$+(q-1)i-(q-1),q+(q-2)i-(q-4),\ldots,q+(q-1)i-(q-1),\ldots\}$$

When $i = t + 1$ the coloring is same as the one given in Case 2 of Theorem 1. Thus we have the upper bound $\chi_r(K_q * W_t) \leq r + i$. Therefore $\chi_r(K_q * W_t) = r + i$ when $(i-1)(q-1)+1 \leq r \leq i(q-1)$ where $i = 4$ for t is odd and $5 \leq i \leq (t+1)$ for all t.

Case 4: When $r \geq (t+1)(q-1)+1$.
By $r = (t+1)(q-1)$ all $qt + q$ vertices of $K_q * W_t$ will be colored differently and hence we no longer need any extra colors. Thus $\chi_r(K_q * W_t) = qt + q$.

4 Conclusion

Here we have determined the lower bound for the r-dynamic chromatic number of the extended neighborhood corona of the complete graph $K_q, q \geq 2$ with any simple, connected and finite graph H with at least 2 vertices and further the exact r-dynamic chromatic number of the extended neighborhood corona of complete graph with path, cycle, complete graph, star graph, and wheel graph. We can broaden this work for various other graphs also. However, determining the exact value of the r-dynamic chromatic number of certain graphs is a time-consuming operation, leaving several problems unresolved.

Acknowledgements With due respect, the writers express their gratitude to the referees for their thorough reading, insightful remarks, and useful suggestions that have improved the content of this manuscript.

References

1. Ahadi, A., Akbari, S., Dehghana, A., Ghanbari, M.: On the difference between chromatic and dynamic chromatic number of graphs. Discr. Math. **312**, 2579–2583 (2012)
2. Akbari, S., Ghanbari, M., Jahanbakam, S.: On the dynamic chromatic number of graphs, combinatorics and graphs. Contemp. Math. Am. Math. Soc. **531**, 11–18 (2010)
3. Aparna, V., Mohanapriya, N.: On r-dynamic coloring of neighborhood corona of path with some graphs. J. Phys. Conf. Ser. **1523**(012001), 1–11 (2020)
4. Adiga, C., Rakshith, B.R., Subba Krishna, K.N.: Spectra of extended neighborhood corona and extended corona of two graphs. Electron. J. Graph Theory Appl. **4**(1), 101–110 (2016)
5. Augestina, I.H., Dafik, D., Harsyaa, A.Y.: On r-dynamic coloring of some graph operations. Indonesian J. Combin. **1**(1), 22–30 (2016)
6. Kim, S.J., Lee, S.J., Park, W.J.: Dynamic coloring and list dynamic coloring of planar graphs. Discr. Appl. Math. **161**, 2207–2212 (2013)
7. Lai, H.J., Montgomery, B., Poon, H.: Upper bounds of dynamic chromatic number. Ars Combin. **68**, 193–201 (2003)

8. Li, X., Yao, X., Zhou, W., Broersma, H.: Complexity of conditional colorability of graphs. Appl. Math. Lett. **22**, 320–324 (2009)
9. Mohanapriya, N., Vernold, V.J., Venkatachalam, M.: On dynamic coloring of Fan graphs. Int. J. Pure Appl. Math. **106**(8), 169–174 (2016)
10. Mohanapriya, N.: A study on dynamic coloring of graphs. Ph.D. thesis. Bharathiar University, Coimbatore, India (2017)
11. Montgomery, B.: Dynamic coloring of graphs. Ph.D. thesis. ProQuest LLC, Ann Arbor, MI, West Virginia University (2001)
12. Taherkhani, A.: r-dynamic chromatic number of graphs. Discr. Appl. Math. **201**, 222–227 (2016)

Corona Domination Number of Graphs

G. Mahadevan, M. Vimala Suganthi, and C. Sivagnanam

Abstract We initiate a study of the domination parameter for graph which is defined as "a dominating set S of a graph G is said to be a corona dominating set (CD-set), if every vertex in $< S >$ is either a pendent vertex or a support vertex. The minimum cardinality of a CD-set is called as corona domination number and is denoted by $\gamma_{\text{CD}}(G)$". In this paper, we initiate a study of this parameter and obtained its results for some derived graphs of path and cycle.

Keywords Paired domination number · Total domination number · Corona domination number · Pendent vertex · Isolated vertex

1 Introduction

The initiation of paired domination number was done by Studer et al. [7] "a paired dominating set [5] S is an induced-paired dominating set if the edges of the matching are the induced edges of $< S >$. The minimum cardinality of an induced-paired dominating set is called the induced-paired domination number and is notated as $\gamma_{ip}(G)$". The initiation of total domination number was done by Cockayne et al. [2] defined as "a dominating set S is a total dominating set if the induced subgraph $< S >$ has no isolated vertices. The total domination number $\gamma_t(G)$ of a graph G is the minimum cardinality of a total dominating set". If a vertex is adjacent to a pendent vertex, we call that vertex to be support vertex. Having above as a motivation the authors added a constrain to $< S >$ and introduce a new parameter called corona domination of graphs and it is named so as the structure of $< S >$ resembles the structure of corona cells.

G. Mahadevan (✉) · M. Vimala Suganthi
Department of Mathematics, Gandhigram Rural Institute-Deemed to be University, Gandhigram, India

C. Sivagnanam
Department of General Requirments, University of Technology and Applied Sciences-Sur, Sur, Sultanate of Oman

P. Balasubramaniam et al. (eds.), *Mathematical Modelling and Computational Intelligence Techniques*, Springer Proceedings in Mathematics & Statistics 376,
https://doi.org/10.1007/978-981-16-6018-4_16

All the basic graph notation and definitions are referred from [1], and all the domination definitions and notations are referred from [3, 4]. The graph G obtained by taking one copy of G_1 of order r and r copies of G_2 and then joining the ith vertex of G_1 to every vertex in the ith copy of G_2 is called as the corona product of $G_1 \odot G_2$. Given any graph G, its square graph G^2 with $V(G^2) = V(G)$ and two vertices are connected in G^2 if they are at length 1 or 2. The graph $l_n = P_2 \times P_n$ is called the ladder graph. The splitting graph of G, denoted by $S(G)$ and is obtained from G by adding for each vertex v of G a new vertex v' so that v' is adjacent to every vertex that is adjacent to v. The vertex v' is called duplication vertex of v. For a given graph G, the graph obtained by subdividing edge and joining all the non-adjacent vertices of G is called the central graph of G, and it is denoted by $C(G)$. The middle graph of G is denoted by $M(G)$ and is defined as, $M(G)$ is a graph whose vertex set is $V(G) \cup E(G)$ and in which two vertices are adjacent if and only if either they are adjacent edges of G or one is a vertex of G and the other an edge incident with it. The total graph [6] $T(G)$ of a graph G is a graph such that the vertex set of $T(G)$ corresponds to the vertices and edges of G and two vertices are adjacent in $T(G)$ if and only if their corresponding elements are either adjacent or incident in G. In this paper, we have investigate this number for some standard and some derived graphs.

2 Characterization of Corona Domination Number of a Graph

Definition 1 *A dominating set S of a graph G is said to be a corona dominating set (CD-set) if every vertex in $< S >$ is either a pendent vertex or a support vertex. The minimum cardinality of a CD-set is called the corona domination number and is denoted by $\gamma_{CD}(G)$.*

In Fig. 1, $S = \{v_2, v_5, v_6\}$ is a corona dominating set, and it is minimum, whose cardinality is 3. Hence, $\gamma_{CD} = 3$.

Observation 1 Since every CD-set is a dominating set as well as a total dominating set, we have $\gamma(G) \leq \gamma_t(G) \leq \gamma_{CD}(G)$.

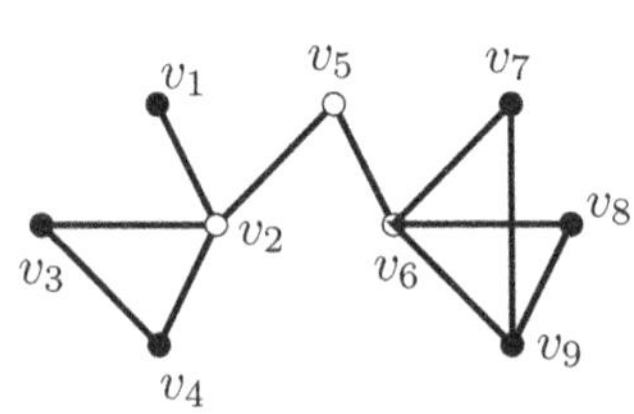

Fig. 1

Example In Fig. 2, $S_1 = \{v_1, v_3, v_4, v_5\}$ is the $\gamma-$set whose cardinality is 4. $S_2 = \{v_1, v_3, v_4, v_5, v_6\}$ is the γ_t- set whose cardinality is 5. $S_3 = \{v_1, v_3, v_4, v_5, v_6, v_8, v_9, v_{10}\}$ is a γ_{CD}-set whose cardinality is 8. Hence, $\gamma(G) \leq \gamma_t(G) \leq \gamma_{CD}(G)$.

Observation 2 Since every induced paired dominating set is a γ_{CD}-set, we have $\gamma_{CD}(G) \leq \gamma_{ip}(G)$.

Example In Fig. 3, $S_1 = \{v_1, v_2, v_5\}$ is the CD-set whose cardinality is 3. $S_2 = \{v_1, v_4, v_5, v_7\}$ is a γ_{ip}-set whose cardinality is 4. Hence, $\gamma_{CD}(G) \leq \gamma_{ip}(G)$.

Observation 3 Since every CD-set is a dominating set and $\lceil \frac{n}{\Delta+1} \rceil \leq \gamma(G)$, we have $\gamma_{CD}(G) \geq \lceil \frac{n}{\Delta+1} \rceil$.

Theorem 1 *For a nontrivial path* P_n, $\gamma_{CD}(P_n) = \begin{cases} \frac{n}{2}+1 & if\, n \equiv 2\ (mod\ 4) \\ \lceil \frac{n}{2} \rceil & otherwise. \end{cases}$

Proof Let $(v_1, v_2, \ldots, v_n)$, $n \geq 2$ be the path P_n and let $S_1 = \{v_i\ :\ i\ \equiv 2\ or\ 3\ (\text{mod}\ 4)\}$. Now let $S = \begin{cases} S_1 \cup \{v_{n-1}\} & \text{if}\ n\ \equiv\ 1\ or\ 2\ (\text{mod}\ 4) \\ S_1 & \text{otherwise.} \end{cases}$ Then S is a CD-set of P_n and hence $\gamma_{CD} \leq |S| = \begin{cases} \frac{n}{2}+1 & \text{if}\ n\ \equiv 2\ (\text{mod}\ 4) \\ \lceil \frac{n}{2} \rceil & \text{otherwise.} \end{cases}$ Let S be any CD-set of P_n then every component of $< S >$ contains at least two vertices. Suppose $n \equiv 2 (\text{mod}\ 4)$ then any dominating set D of cardinality $\frac{n}{2}$ contains at least one isolated vertex in

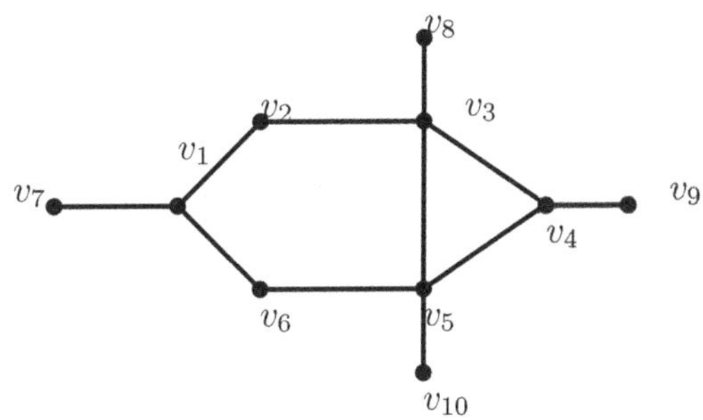

Fig. 2

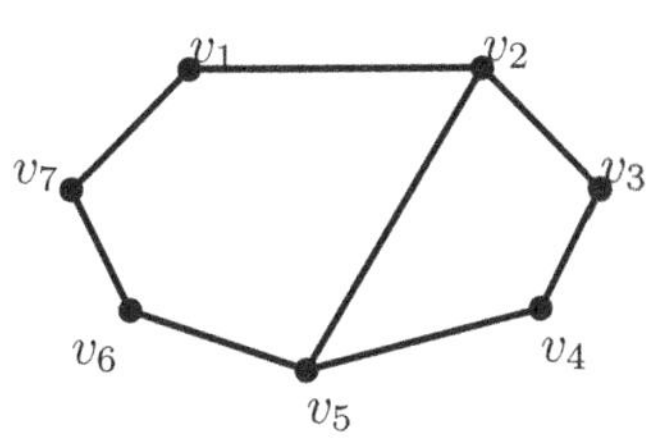

Fig. 3

$< D >$. Also, if $n \not\equiv 2 (\mathrm{mod}\ 4)$, then any dominating set D of cardinality $\lceil \frac{n}{2} \rceil - 1$ contains at least one isolated vertex in $< D >$. Thus $|S| \geq \begin{cases} \frac{n}{2} + 1 & \text{if } n \equiv 2\ (\mathrm{mod}\ 4) \\ \lceil \frac{n}{2} \rceil & \text{otherwise.} \end{cases}$
Hence, the result follows.

Theorem 2 *For a cycle* C_n, $\gamma_{CD}(C_n) = \begin{cases} \frac{n}{2} + 1 & if\ n \equiv 2\ (mod\ 4) \\ \lceil \frac{n}{2} \rceil & otherwise. \end{cases}$

Proof Let $(v_1, v_2, \ldots, v_n, v_1)$ be a cycle C_n and let $S_1 = \{v_i : i \equiv 2\ or\ 3\ (\mathrm{mod}\ 4)\}$.
Now let $S = \begin{cases} S_1 \cup \{v_{n-1}\} & \text{if } n \equiv 1\ or\ 2\ (\mathrm{mod}\ 4) \\ S_1 & \text{otherwise.} \end{cases}$ Then S is a CD-set of C_n and
hence $\gamma_{CD} \leq |S| = \begin{cases} \frac{n}{2} + 1 & \text{if } n \equiv 2\ (\mathrm{mod}\ 4) \\ \lceil \frac{n}{2} \rceil & \text{otherwise.} \end{cases}$
Since $\gamma_t \leq \gamma_{CD}$ and $\gamma_t(C_n) = \begin{cases} \frac{n}{2} + 1 & \text{if } n \equiv 2\ (\mathrm{mod}\ 4) \\ \lceil \frac{n}{2} \rceil & \text{otherwise.} \end{cases}$
We have $\gamma_{CD}(C_n) \geq \begin{cases} \frac{n}{2} + 1 & \text{if } n \equiv 2\ (\mathrm{mod}\ 4) \\ \lceil \frac{n}{2} \rceil & \text{otherwise.} \end{cases}$ Hence the result follows.

Observation 4 1. $\gamma_{CD}(K_n) = 2$
2. $\gamma_{CD}(K_{r,s}) = 2$
3. $\gamma_{CD}(K_{1,n-1}) = 2$
4. $\gamma_{CD}(K_{1,n-1} \odot K_n) = n + 1$
5. $\gamma_{CD}(B_{r,s}) = 2$
6. $\gamma_{CD}(W_r) = 2$.

3 Corona Domination Number for Some Standard Graphs

Theorem 3 *For a non trivial path* P_r, $\gamma_{CD}(P_2 \times P_r) = 2\lceil \frac{r}{3} \rceil$.

Proof Let $V = \{u_1, u_2, \ldots, u_r, v_1, \ldots, v_r\}$ and $E = \{u_i v_i, u_j u_{j+1}, v_j v_{j+1} : 1 \leq i \leq r, 1 \leq j \leq r - 1\}$. Now, assume $S_1 = \{u_i, v_i : i \equiv 2(\mathrm{mod}\ 3)\}$, then $S = \begin{cases} S_1 \cup \{u_r, v_r\} & if\ r \equiv 1(\mathrm{mod}\ 3) \\ S_1 & \text{otherwise} \end{cases}$ is a CD-set of $P_2 \times P_r$. Hence, $\gamma_{CD}(P_2 \times P_r) \leq |S| = 2\lceil \frac{r}{3} \rceil$. Now, since $\gamma_t \leq \gamma_{CD}$ and $\gamma_t(P_2 \times P_r) = 2\lceil \frac{r}{3} \rceil$. We have $\gamma_{CD}(P_2 \times P_r) \geq 2\lceil \frac{r}{3} \rceil$. Hence, the result follows.

Example In Fig. 4, collection of lightened vertices is a CD-set. For the graph $P_2 \times P_4$, the set $S_1 = \{v_2, v_4, u_2, u_4\}$ is a minimum CD-set and hence $\gamma_{CD}(P_2 \times P_4) = 4$. For the graph $P_2 \times P_5$ the set $S_2 = \{v_2, v_5, u_2, u_5\}$ is a minimum CD-set and hence $\gamma_{CD}(P_2 \times P_5) = 4$.

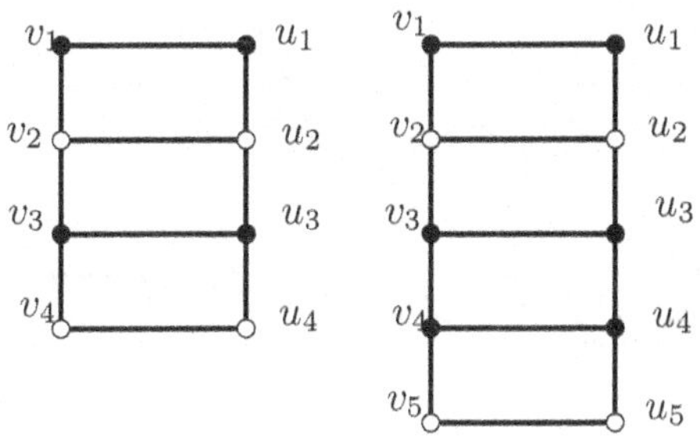

Fig. 4

Theorem 4 *For a cycle C_r, $\gamma_{CD}(P_2 \times C_r) = 2\lceil \frac{r}{3} \rceil$.*

Proof Let $V = \{u_1, u_2, \ldots, u_r, v_1, \ldots, v_r\}$ and $E = \{u_i v_i, u_j u_{j+1}, v_j v_{j+1} : 1 \leq i \leq r, 1 \leq j \leq r-1\}$. Now, assume $S_1 = \{u_i, v_i : i \equiv 2(\text{mod}\, 3)\}$, then $S = \begin{cases} S_1 \cup \{u_r, v_r\}\ if\ r \equiv 1(\text{mod}\, 3) \\ S_1 \quad \text{otherwise} \end{cases}$ is a CD-set of $P_2 \times C_r$. Hence, $\gamma_{\text{CD}}(P_2 \times C_r) \leq |S| = 2\lceil \frac{r}{3} \rceil$. Now, since $\gamma_t \leq \gamma_{\text{CD}}$ and $\gamma_t(P_2 \times C_r) = 2\lceil \frac{r}{3} \rceil$, we have $\gamma_{\text{CD}}(P_2 \times C_r) \geq 2\lceil \frac{r}{3} \rceil$. Hence the theorem.

Theorem 5 *For a path P_n, $n \geq 3$, $\gamma_{CD}(P_n^2) = \begin{cases} \lceil \frac{2n}{7} \rceil + 1\ if\, n \equiv 3\ (mod 7) \\ \lceil \frac{2n}{7} \rceil \quad otherwise. \end{cases}$*

Proof Let $(v_1, v_2, \ldots, v_n)$ be a path P_n. It is clear that $E(P_n^2) = \{v_i v_{i+1}, v_i v_{i+2}, v_{n-1} v_n : 1 \leq i \leq n-2\}$. Let $S_1 = \{v_i : i \equiv 3\ or\ 5\ (\text{mod}\, 7)\}$.

Now, let $S = \begin{cases} S_1 & \text{if } n \equiv 0\ or\ 5\ or\ 6\ (\text{mod}\, 7) \\ S_1 \cup \{v_{n-1}\} & \text{if } n \equiv 1\ or\ 3\ (\text{mod}\, 7) \\ S_1 \cup \{v_{n-2}\} & \text{if } n \equiv 2\ (\text{mod}\, 7) \\ S_1 \cup \{v_n\} & \text{if } n \equiv 4\ (\text{mod}\, 7) \end{cases}$ then S is a CD-set of P_n^2.

Hence $\gamma_{\text{CD}}(P_n^2) \leq |S| = \begin{cases} \lceil \frac{2n}{7} \rceil + 1 \text{ if } n \equiv 3\ (\text{mod}\, 7) \\ \lceil \frac{2n}{7} \rceil \quad \text{otherwise.} \end{cases}$ Suppose there exist $S' \subset V$ which is a CD-set of P_n^2. As a dominating set D of cardinality $k = \begin{cases} \lceil \frac{2n}{7} \rceil \text{ if } n \equiv 3\ (\text{mod}\, 7) \\ \lceil \frac{2n}{7} \rceil - 1 \text{ otherwise} \end{cases}$ contains at least one isolated vertex in $< D >$, we have

$|S'| \geq k + 1 = \begin{cases} \lceil \frac{2n}{7} \rceil + 1 \text{ if } n \equiv 3\ (\text{mod}\, 7) \\ \lceil \frac{2n}{7} \rceil \quad \text{otherwise.} \end{cases}$

Hence the theorem.

Example In Fig. 5, collection of lightened vertices is a CD-set. For the graph P_8^2, the set $S_1 = \{v_3, v_5, v_7\}$ is a minimum CD-set and, hence, $\gamma_{\text{CD}}(P_8^2) = 3$. For the graph P_{10}^2, the set $S_2 = \{v_3, v_5, v_{10}\}$ is a minimum CD-set and, hence, $\gamma_{\text{CD}}(P_{10}^2) = 3$.

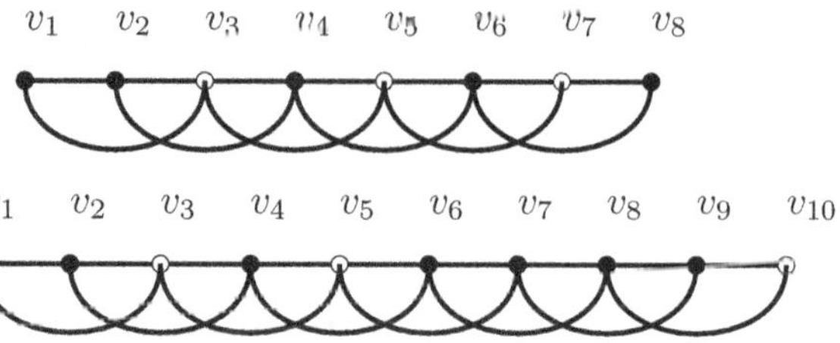

Fig. 5

Theorem 6 *For a cycle C_n, $n \geq 3$,* $\gamma_{CD}(C_n^2) = \begin{cases} \lceil \frac{2n}{7} \rceil + 1 & if\, n \equiv 3\ (mod\, 7) \\ \lceil \frac{2n}{7} \rceil & otherwise. \end{cases}$

Proof Let $(v_1, v_2, \ldots, v_n, v_1)$ be the cycle C_n. It is clear that $E(C_n^2) = \{v_i v_{i+1}, v_i v_{i+2}, v_{n-1} v_n : 1 \leq i \leq n-2\}$. Let $S_1 = \{v_i : i \equiv 3\ or\ 5\ (\text{mod}\ 7)\}$.

Now, let $S = \begin{cases} S_1 & \text{if } n \equiv 0\ or\ 5\ or\ 6\ (\text{mod}\ 7) \\ S_1 \cup \{v_{n-1}\} & \text{if } n \equiv 1\ or\ 3\ (\text{mod}\ 7) \\ S_1 \cup \{v_{n-2}\} & \text{if } n \equiv 2\ (\text{mod}\ 7) \\ S_1 \cup \{v_n\} & if\ n \equiv 4\ (\text{mod}\ 7) \end{cases}$ then S is a CD-set of C_n^2.

Hence $\gamma_{CD}(C_n^2) \leq |S| = \begin{cases} \lceil \frac{2n}{7} \rceil + 1 & \text{if } n \equiv 3\ (\text{mod}\ 7) \\ \lceil \frac{2n}{7} \rceil & \text{otherwise.} \end{cases}$ Suppose, there exist $S' \subset V$ which is a CD-set of C_n^2. As a dominating set D of cardinality $k = \begin{cases} \lceil \frac{2n}{7} \rceil & \text{if } n \equiv 3\ (\text{mod}\ 7) \\ \lceil \frac{2n}{7} \rceil - 1 & \text{otherwise} \end{cases}$ contains at least one isolated vertex in $< D >$, we have $|S'| \geq k + 1 = \begin{cases} \lceil \frac{2n}{7} \rceil + 1 & \text{if } n \equiv 3\ (\text{mod}\ 7) \\ \lceil \frac{2n}{7} \rceil & \text{otherwise.} \end{cases}$ Hence the theorem.

Theorem 7 *For a path $P_n, n \geq 3$,*

$$\gamma_{CD}(S(P_n)) = \begin{cases} \lceil \frac{n}{2} \rceil + 1 & if\, n \equiv 2\ (mod\, 4) \\ \lceil \frac{n}{2} \rceil & otherwise. \end{cases}$$

Proof Let $P_n = (v_1, v_2, \ldots, v_n)$ be a path P_n and let v_i' be the duplication vertex of v_i, $1 \leq i \leq n$. It is clear that $E(S(P_n)) = \{v_i v_{i+1}, v_i' v_{i+1}, v_i v_{i+1}' : 1 \leq i \leq n-1\}$. Let $S_1 = \{v_i, v_{i+1} : i \equiv 2(\text{mod}\ 4)\}$.

Let $S = \begin{cases} S_1 & if\ n \equiv 0, 3\ (\text{mod}\ 4) \\ S_1 \cup \{v_{n-1}\} & \text{otherwise.} \end{cases}$ Then S is a CD set of $S(P_n)$.

Hence, $\gamma_{CD}(S(P_n)) \leq |S| = \begin{cases} \lceil \frac{n}{2} \rceil + 1 & \text{if } n \equiv 2\ (\text{mod}\ 4) \\ \lceil \frac{n}{2} \rceil & \text{otherwise.} \end{cases}$ Suppose there exist $S' \subset V$ which is a CD-set of $S(P_n)$. As a dominating set D of cardinality $k = \begin{cases} \lceil \frac{n}{2} \rceil & if\ n \equiv 2\ (\text{mod}\ 4) \\ \lceil \frac{n}{2} \rceil - 1 & \text{otherwise} \end{cases}$ contains at least one isolated vertex in $< D >$.

We have $|S'| \geq k + 1 = \begin{cases} \lceil \frac{n}{2} \rceil + 1 & if\ n \equiv 2\ (\text{mod}\ 4) \\ \lceil \frac{n}{2} \rceil & \text{otherwise.} \end{cases}$ Hence the theorem.

Example In Fig. 6, collection of lightened vertices is a CD-set. For the graph $S(P_9)$, the set $S_1 = \{v_2, v_3, v_6, v_7, v_8\}$ is a minimum CD-set and hence $\gamma_{CD}(S(P_9)) = 5$. For the graph $S(P_{10})$, the set $S_2 = \{v_2, v_3, v_6, v_7, v_9, v_{10}\}$ is a minimum CD-set and hence $\gamma_{CD}(S(P_{10})) = 6$.

Theorem 8 *For a cycle $C_n, n \geq 3$,*

$$\gamma_{CD}(S(C_n)) = \begin{cases} \lceil \frac{n}{2} \rceil + 1 & if\ n \equiv 2\ (mod\,4) \\ \lceil \frac{n}{2} \rceil & otherwise. \end{cases}$$

Proof Let $C_n = (v_1, v_2, \ldots, v_n, v_1), n \geq 3$ and let v_i' be the duplication vertex $v_i, 1 \leq i \leq n$. It is clear that $E(S(C_n)) = \{v_i v_{i+1}, v_i' v_{i+1}, v_i v_{i+1}' : 1 \leq i \leq n\}$. Let $S_1 = \{v_i, v_{i+1} : i \equiv 2(\text{mod}\,4)\}$.

Let $S = \begin{cases} S_1 & if\ n \equiv 0, 3\ (\text{mod}\,4) \\ S_1 \cup \{v_{n-1}\} & \text{otherwise}, \end{cases}$

then S is a CD set of $S(C_n)$.

Hence, $\gamma_{CD}(S(C_n)) \leq |S| = \begin{cases} \lceil \frac{n}{2} \rceil + 1 & if\ n \equiv 2\ (\text{mod}\,4) \\ \lceil \frac{n}{2} \rceil & \text{otherwise}. \end{cases}$

Suppose there exist $S' \subset V$ which is a CD-set of $S(C_n)$ As a dominating set D of cardinality $k = \begin{cases} \lceil \frac{n}{2} \rceil & if\ n \equiv 2\ (\text{mod}\,4) \\ \lceil \frac{n}{2} \rceil - 1 & \text{otherwise} \end{cases}$

contains at least one isolated vertex in $< D >$,

we have $|S'| \geq k + 1 = \begin{cases} \lceil \frac{n}{2} \rceil + 1 & if\ n \equiv 2\ (\text{mod}\,4) \\ \lceil \frac{n}{2} \rceil & \text{otherwise}. \end{cases}$ Hence the theorem.

Theorem 9 *For a path $P_n, n \geq 3$*

$$\gamma_{CD}(M(P_n)) = \begin{cases} 2\lfloor \frac{n}{3} \rfloor + 1 & if\ n \equiv 1\ (mod\,3) \\ 2\lceil \frac{n}{3} \rceil & otherwise. \end{cases}$$

Proof Let $(v_1, v_2, \ldots, v_n)$ be the path P_n and Let $e_i = v_i v_{i+1}, 1 \leq i \leq n-1$. Let u_i be the vertex in $M(P_n)$ corresponding to the edge $e_i, 1 \leq i \leq n-1$. Let $S_1 = \{u_i u_{i+1} : i \equiv 1\ (\text{mod}\,3)\}$,

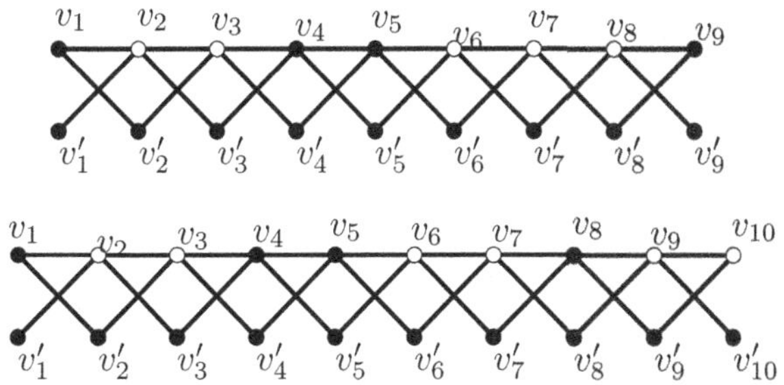

Fig. 6

then $S = \begin{cases} S_1 & if\ n \equiv 0(\bmod 3) \\ S_1 \cup \{u_{n-1}\} & \text{otherwise} \end{cases}$ is a CD-set for the graph $M(P_n)$. Hence,
$\gamma_{CD}(M(P_n)) \leq |S| = \begin{cases} 2\lfloor \frac{n}{3} \rfloor + 1 & if\ n \equiv 1 \ (\bmod 3) \\ 2\lceil \frac{n}{3} \rceil & \text{otherwise.} \end{cases}$
Let S' be a CD-set of $M(P_n)$. Since any dominating set D of cardinality $k = \begin{cases} 2\lfloor \frac{n}{3} \rfloor & if\ n \equiv 1 \ (\bmod 3) \\ 2\lceil \frac{n}{3} - 1 \rceil & \text{otherwise} \end{cases}$ contains at least one isolated vertex in $< D >$, we have $|S'| \geq k + 1 = \begin{cases} 2\lfloor \frac{n}{3} \rfloor + 1 & if\ n \equiv 1 \ (\bmod 3) \\ 2\lceil \frac{n}{3} \rceil & \text{otherwise} \end{cases}$, and hence the result follows.

Example In Fig. 7, collection of lightened vertices is a CD-set. For the graph $M(P_6)$, the set $S_1 = \{u_1, u_2, u_4, u_5\}$ is a minimum CD-set and hence $\gamma_{CD}(M(P_6)) = 4$. For the graph $M(P_7)$, the set $S_2 = \{u_1, u_2, u_4, u_5, u_6\}$ is a minimum CD-set, and hence, $\gamma_{CD}(M(P_7)) = 5$.

Theorem 10 *For a cycle* C_n,
$\gamma_{CD}(M(C_n)) = \begin{cases} 2\lfloor \frac{n}{3} \rfloor + 1 & if\ n \equiv 1 \ (mod\ 3) \\ 2\lceil \frac{n}{3} \rceil & otherwise. \end{cases}$

Proof Let $(v_1, v_2, \ldots, v_n, v_1)$ be a cycle C_n and let $e_i = v_i v_{i+1}, 1 \leq i \leq n-1, e_n = v_n v_1$. Let u_i be the vertex in $M(C_n)$ corresponding to the edge $e_i, 1 \leq i \leq n-1$. Let $S_1 = \{u_i u_{i+1} : i \equiv 1 \ (\bmod 3)\}$.
Then $S = \begin{cases} S_1 & n \equiv 0(\bmod 3) \\ S_1 \cup \{u_{n-1}\} & \text{otherwise} \end{cases}$ is a CD-set for the graph $M(C_n)$. Hence,
$\gamma_{CD}(M(C_n)) \leq |S| = \begin{cases} 2\lfloor \frac{n}{3} \rfloor + 1 & if\ n \equiv 1 \ (\bmod 3) \\ 2\lceil \frac{n}{3} \rceil & \text{otherwise.} \end{cases}$
Suppose there exist $S' \subset V$ which is a CD-set of $M(C_n)$. As a dominating set D of cardinality $k = \begin{cases} 2\lfloor \frac{n}{3} \rfloor & n \equiv 1 \ (\bmod 3) \\ 2\lceil \frac{n}{3} - 1 \rceil & \text{otherwise} \end{cases}$ contains at least one isolated vertex in $< D >$, we have $|S'| \geq k + 1 = \begin{cases} 2\lfloor \frac{n}{3} \rfloor + 1 & n \equiv 1 \ (\bmod 3) \\ 2\lceil \frac{n}{3} \rceil & \text{otherwise} \end{cases}$, and hence the result follows.

Theorem 11 *For a path* $P_n, n \geq 3$
$\gamma_{CD}(C(P_n)) = \begin{cases} 2\lfloor \frac{n}{3} \rfloor + 1 & if\ n \equiv 2 \ (mod\ 3) \\ 2\lfloor \frac{n}{3} \rfloor & otherwise. \end{cases}$

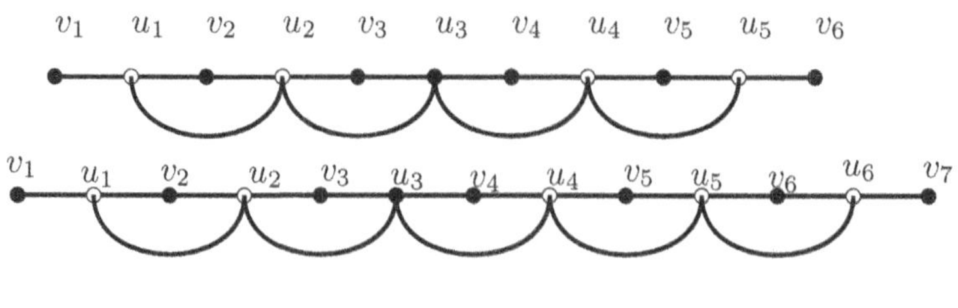

Fig. 7

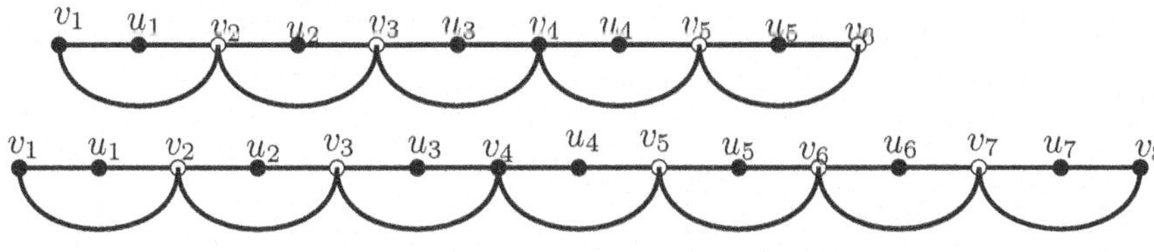

Fig. 8

Proof Let $(v_1, v_2, \ldots, v_n)$ be a path P_n and let $e_i = v_i v_{i+1}$, $1 \le i \le n-1$. Let u_i be the vertex in $C(P_n)$ corresponding to the edge e_i, $1 \le i \le n-1$. Let $S_1 = \{v_i v_{i+1} : i \equiv 2 \pmod 3\}$.

Then $S = \begin{cases} S_1 \cup \{v_{n-1}\} & if\ n \equiv 2 \pmod 3 \\ S_1 & \text{otherwise} \end{cases}$ is a CD-set for the graph $C(P_n)$. Hence,

$\gamma_{CD}(C(P_n)) \le |S| = \begin{cases} 2\lfloor \frac{n}{3} \rfloor + 1 & if\ n \equiv 2 \pmod 3 \\ 2\lfloor \frac{n}{3} \rfloor & \text{otherwise.} \end{cases}$

Suppose there exist $S' \subset V$ which is a CD-set of $C(P_n)$. As a dominating set D of cardinality $k = \begin{cases} 2\lfloor \frac{n}{3} \rfloor & if\ n \equiv 2 \pmod 3 \\ 2\lfloor \frac{n}{3} \rfloor - 1 & \text{otherwise} \end{cases}$ contains at least one isolated vertex in $< D >$, we have $|S'| \ge k + 1 = \begin{cases} 2\lfloor \frac{n}{3} \rfloor + 1 & if\ n \equiv 2 \pmod 3 \\ 2\lfloor \frac{n}{3} \rfloor & \text{otherwise} \end{cases}$, and hence the result follows.

Example In Fig. 8, collection of lightened vertices is a CD-set. For the graph $C(P_6)$, the set $S_1 = \{v_2, v_3, v_5, v_6\}$ is a minimum CD-set, and hence, $\gamma_{CD}(C(P_6)) = 4$. For the graph $C(C_7)$, the set $S_2 = \{v_2, v_3, v_5, v_6, v_7\}$ is a minimum CD-set, and hence, $\gamma_{CD}(C(P_7)) = 5$.

Theorem 12 *For a cycle C_n,*

$$\gamma_{CD}(C(C_n)) = \begin{cases} 2\lfloor \frac{n}{3} \rfloor + 1 & if\ n \equiv 2 \pmod 3 \\ 2\lceil \frac{n}{3} \rceil & otherwise. \end{cases}$$

Proof Let $(v_1, v_2, \ldots, v_n, v_1)$ be a cycle C_n and let $e_i = v_i v_{i+1}$, $1 \le i \le n-1$, $e_n = v_n v_1$. Let u_i be the vertex in $C(C_n)$ corresponding to the edge e_i, $1 \le i \le n-1$. Let $S_1 = \{v_i v_{i+1} : i \equiv 2 \pmod 3\}$.

Then $S = \begin{cases} S_1 & if\ n \equiv 0 \pmod 3 \\ S_1 \cup \{v_n\} & if\ n \equiv 1 \pmod 3 \\ S_1 \cup \{v_{n-1}\} & if\ n \equiv 2 \pmod 3 \end{cases}$ is a CD-set for the graph $C(C_n)$.

Hence, $\gamma_{CD}(C(C_n)) \le |S| = \begin{cases} 2\lfloor \frac{n}{3} \rfloor + 1 & if\ n \equiv 2 \pmod 3 \\ 2\lceil \frac{n}{3} \rceil & \text{otherwise.} \end{cases}$

Suppose there exist $S' \subset V$ which is a CD-set of $C(C_n)$. As a dominating set D of cardinality $k = \begin{cases} 2\lfloor \frac{n}{3} \rfloor & if\ n \equiv 2 \pmod 3 \\ 2\lceil \frac{n}{3} - 1 \rceil & \text{otherwise} \end{cases}$ contains at least one isolated vertex in $< D >$, we have $|S'| \geq k+1 = \begin{cases} 2\lfloor \frac{n}{3} \rfloor + 1 & n \equiv 2 \pmod 3 \\ 2\lceil \frac{n}{3} \rceil & \text{otherwise} \end{cases}$, and hence the result follows.

Theorem 13 *For a path P_n, $n \geq 3$*
$\gamma_{CD}(T(P_n)) = \begin{cases} 2\lfloor \frac{n}{3} \rfloor + 1 & if\ n \equiv 2\ (mod\ 3) \\ 2\lfloor \frac{n}{3} \rfloor & otherwise. \end{cases}$

Proof Let $(v_1, v_2, \ldots, v_n)$ be a path P_n and let $e_i = v_i v_{i+1}, 1 \leq i \leq n-1$. Let $u_i \in V(T(P_n))$ corresponding to the edge $e_i, 1 \leq i \leq n-1$. Let $S_1 = \{v_i v_{i+1} : i \equiv 2 \pmod 3\}$,
then $S = \begin{cases} S_1 \cup \{v_{n-1}\} & \text{if } n \equiv 2 \pmod 3 \\ S_1 & \text{otherwise} \end{cases}$ is a CD-set for the graph $T(P_n)$. Hence,
$\gamma_{CD}(T(P_n)) \leq |S| = \begin{cases} 2\lfloor \frac{n}{3} \rfloor + 1 & if\ n \equiv 2 \pmod 3 \\ 2\lfloor \frac{n}{3} \rfloor & \text{otherwise.} \end{cases}$
Suppose there exist $S' \subset V$ which is a CD-set of $T(P_n)$. As a dominating set D of cardinality $k = \begin{cases} 2\lfloor \frac{n}{3} \rfloor & if\ n \equiv 2 \pmod 3 \\ 2\lfloor \frac{n}{3} \rfloor - 1 & \text{otherwise} \end{cases}$ contains atleast one isolated vertex in $< D >$, we have $|S'| \geq k+1 = \begin{cases} 2\lfloor \frac{n}{3} \rfloor + 1 & if\ n \equiv 2 \pmod 3 \\ 2\lfloor \frac{n}{3} \rfloor & \text{otherwise} \end{cases}$ and hence the result follows.

Example In Fig. 9, collection of lightened vertices is a CD-set. For the graph $T(P_6)$ the set $S_1 = \{v_2, v_3, v_5, v_6\}$ is a minimum CD-set and hence $\gamma_{CD}(T(P_6)) = 4$. For the graph $T(P_7)$, the set $S_2 = \{v_2, v_3, v_5, v_6, v_7\}$ is a minimum CD-set and hence $\gamma_{CD}(T(P_7)) = 5$.

Theorem 14 *For a cycle C_n,*

$$\gamma_{CD}(T(C_n)) = \begin{cases} 2\lfloor \frac{n}{3} \rfloor + 1 & if\ n \equiv 2\ (mod\ 3) \\ 2\lceil \frac{n}{3} \rceil & otherwise. \end{cases}$$

Proof Let $(v_1, v_2, \ldots, v_n, v_1)$ be a cycle C_n and Let $e_i = v_i v_{i+1}, 1 \leq i \leq n-1, e_n = v_n v_1$. Let $u_i \in V(T(C_n))$ corresponding to the edge $e_i, 1 \leq i \leq n-1$. Let

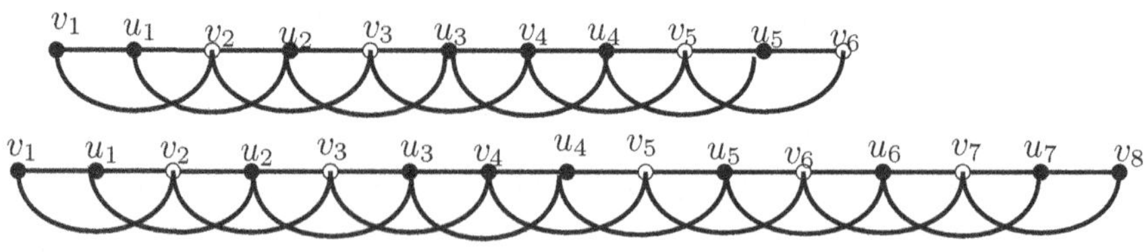

Fig. 9

$S_1 = \{v_i v_{i+1} : i \equiv 2 \pmod 3\}$,

then $S = \begin{cases} S_1 & \text{if } n \equiv 0 \pmod 3 \\ S_1 \cup \{v_n\} & if\ n \equiv 1 \pmod 3 \\ S_1 \cup \{v_{n-1}\} & if\ n \equiv 2 \pmod 3 \end{cases}$ is a CD-set for the graph $T(C_n)$. Hence

$\gamma_{CD}(T(C_n)) \leq |S| = \begin{cases} 2\lfloor \frac{n}{3} \rfloor + 1 & if\ n \equiv 2 \pmod 3 \\ 2\lceil \frac{n}{3} \rceil & \text{otherwise.} \end{cases}$

Suppose there exist $S' \subset V$ which is a CD-set of $T(C_n)$. As a dominating set D of cardinality $k = \begin{cases} 2\lfloor \frac{n}{3} \rfloor - 1 & if\ n \equiv 2 \pmod 3 \\ 2\lceil \frac{n}{3} \rceil & \text{otherwise} \end{cases}$ contains at least one isolated vertex in $< D >$, we have $|S'| \geq k + 1 = \begin{cases} 2\lfloor \frac{n}{3} \rfloor + 1 & if\ n \equiv 2 \pmod 3 \\ 2\lceil \frac{n}{3} \rceil & \text{otherwise} \end{cases}$ and hence the result follows.

4 Conclusion

In this paper, we have initiated a study of a new domination parameter called corona domination of graphs and discussed its nature. Also, we have found this number for some standard graphs. We have also obtained the results for some tree graphs which will be in our subsequent papers.

Acknowledgements The research work was supported by the Research seed money for the project proposal of the Department of Mathematics, Gandhigram Rural Institute-Deemed to be University, Gandhigram.

References

1. Chartarnd, G., Lesniak, L.: Graphs and Digraphs. CRC (2005)
2. Cockayne, E.J., Dawes, R.M., Hedetniemi, S.T.: Total domination in graphs. Network **10**, 211–219 (1980)
3. Haynes, T.W., Hedetniemi, S.T., Slater, P.J.: Fundamentals of Domination in Graphs. Marcel Dekker Inc., New York (1997)
4. Haynes, T.W., Hedetniemi, S.T., Slater, P.J.: Domination in Graphs Advanced Topics. Marcel Dekker Inc., New York (1997)
5. Haynes, T.W., Slater, P.J.: Paired domination in graphs. Networks **32**, 199–209 (1998)
6. Sivagnanam, C.: Neighborhood connected domination number of total graphs. Gen. Math. Notes **25**(1), 27–32 (2014)
7. Studer, D.S., Haynes, T.W., Lawson, L.M.: Induce-paired domination in graphs. Ars. Comb. **57**, 111–128 (2000)

An AHP-Based Unmanned Aerial Vehicle Selection for Data Collection in Wireless Sensor Networks

Immanuel Johnraja Jebadurai, Getzi Jeba Leelipushpam Paulraj, Jebaveerasingh Jebadurai, and Nancy Emymal Samuel

Abstract Wireless sensor networks (WSN) consist of sparse sensor nodes that cooperatively monitor the environment and communicate the information to a base station for computation, analysis, and storage. The sensor nodes communicate the data to the sink node which in turn relay them to the base station. However, the static sink node causes an energy-hole problem. This depletes the energy of the node and pushes the node to an inactive state. Hence, researchers have introduced a mobile sink that collects data by moving around in the WSN. The objective of the proposed work is to reduce energy consumption by utilizing unmanned aerial vehicle (UAV) as the mobile sink in collecting data from WSN. Various researchers have selected the mobile sink based on their distance from the sensor node. However, other parameters such as speed of the mobile sink and energy were not considered. This paper proposes an analytical hierarchy process-based unmanned aerial vehicle (UAV) mobile sink selection considering three criteria, viz. distance, energy, and speed of the sink node for data collection. Simulation results show that the proposed technique has outperformed the state-of-the-art techniques available in the literature.

Keywords Wireless sensor networks · Mobile sinks · Analytical hierarchy process · Unmanned aerial vehicle

---

I. J. Jebadurai · G. J. L. Paulraj (✉) · J. Jebadurai
Karunya Institute of Technology and Sciences, Coimbatore, India
e-mail: getzi@karunya.edu

J. Jebadurai
e-mail: jebaveerasingh@karunya.edu

N. E. Samuel
Sam Salt Works, Tuticorin, India

P. Balasubramaniam et al. (eds.), *Mathematical Modelling and Computational Intelligence Techniques*, Springer Proceedings in Mathematics & Statistics 376,
https://doi.org/10.1007/978-981-16-6018-4_17

1 Introduction

WSN has a wide spectrum of applications that include agriculture, environmental monitoring, managing urban areas [1–4]. It has sparsely distributed sensor nodes consisting of a sensing module, power module, and transceiver module. The sensing module collects the environment data where the sensors are deployed. Using the transceiver module, the sensor node transmits the collected data to the sink node. The sink node relays the data to the infrastructure-based base station for computation, storage, and analysis [5].

2 Motivation

Sink nodes can be deployed in fixed positions in the sensor network for collecting information and relaying the data to the base station through other sink nodes. This could result in energy depletion of nodes near the sink node commonly called an energy-hole problem or hotspot problem [5]. Hence, the concept of mobile sink has been introduced in many research contributions. The network has mobile sinks that periodically traverse across the network collecting information and transmitting it to the base station. Introducing mobile sinks improves the performance of the WSN in terms of energy efficiency and packet delivery ratio [6]. However, introducing mobile sinks has a lot of open issues and challenges [6]. It is evident from the literature [7] that selection of a mobile sink is really important to disseminate the data from sensor nodes to the base station. Works of literature have proposed papers in which mobile sink has been selected for data dissemination using the distance between the sensor node and sink node [8, 9]. When the mobile sink is implemented using UAVs especially in applications such as disaster management, selecting a mobile sink using distance parameter alone is not sufficient. Hence, an analytical hierarchy process (AHP)-based mobile sink selection technique for data collection has been proposed.

In this technique, the ranking module ranks the sink node using three criteria, viz. the distance between the sensor node and sink node, speed of the UAV sink node, and the energy of the UAV sink node. Then, the selection module selects the sink node for data transfer. The proposed technique has been simulated using NS-2 under various test cases. The vital parameters such as packet delivery ratio and energy efficiency have been measured. Comparative study shows that the proposed technique outperforms the existing technique.

The remainder of the paper is organized as follows: Sect. 2 surveys various papers that introduce mobile sink for data collection in WSN. Section 3 explains the proposed AHP-based data collection technique for WSN, and Sect. 4 analyzes the performance of the proposed technique with the existing techniques from the literature. Section 5 concludes the paper with the future research challenges.

3 Related Works

Mobile sink has been a proven method for improving the performance of WSN, and many researchers have contributed to improving the application using mobile sinks. This section reviews various pieces of literature related to mobile sinks. In [10], the authors have proposed a technique to reduce the delivery latency using mobile sinks. The objective of this paper is to reduce the data delivery latency in a mobile sink environment. The static sensors are randomly deployed in the sensor network. Initially, the mobile sink traverses the sensor network to identify the location of sensor nodes. Then, the mobile sink forms an anchor point. The anchor point forms the traverse route for the sink node to move in the network. The anchor points are formed based on the distance between the mobile sink and the sensor node. However, forming a fixed anchor point in a dynamic topology environment does introduce more dropping of packets. The authors have proposed a rechargeable sensor network in which the mobile sink travels to collect data in a predefined path [11]. A distributed data gathering approach has been proposed to ensure optimal data gathering from the sensor nodes to the mobile sink. The system has been proved to be energy-efficient using simulation results.

Comprehensive data gathering using graphing technique has been proposed in [12]. A genetic algorithm has been proposed to equalize the total energy consumption in the network. Route discovery protocol with improved shortest path tree enables energy-efficient data gathering for grid-based sensor environment. Simulation in NS-3 shows that the graphing-based method improves the lifetime and energy efficiency in the sensor environment. In [7], the authors have proposed opportunistic data collection using a mobile sink. In this technique, packet delivery delay is estimated using contact aware technique. Link quality between sensors and mobile sinks is also estimated. These estimations are combined with Lyapunov optimization theory to offer packet transmission delay, reliability, scalability, communication overhead, and storage overheads. Time-efficient data collection using iMIMO has been proposed in [13]. In this technique, compatible pairs are formed to collect data from an uncovered area of the mobile sink. The sensor node discovers the neighbor nodes in the uncovered area and forms pairs. The pairing allows the data to transmit to the sensor node which relays the data to the mobile sink.

A reliable technique to improve the availability of mobile sink has been proposed in [14]. Mobile sinks are set with a failure probability rate. When the mobile sink is about to fail, it does not move or collect data in the network. These mobile sinks are serviced by the base station and then resume their data collection process. Energy consumption and delay in WSN are improved by introducing virtual grid infrastructure-based mobile sink data collection in [9]. The sensor network is divided into a virtual grid-based structure. The sensor nodes in the intersection points select the nearest approaching mobile sink. The virtual grid infrastructure and the mobile sink reduce the energy consumption and delay in the network. The mobile sink's location is updated to the sensor nodes using messages. This causes overhead in the network. To reduce such message, overhead grid cycle routing protocol has been proposed. The WSN is divided into small grids. A grid cell head is elected from every grid. The mobile sink is assumed to move over the periphery of the grids and updates its location to the nearest grid cell head. This grid cell head communicates this information to the other heads using virtual cycles. The sensor nodes send the data to the nearest head which in turn relay the data to the neighboring head to reach the mobile sink.

Modes of the visit of the mobile sink to collect data are addressed in [15]. One is the direct mode where the mobile sink moves to every node and collects data. The other is the rendezvous mode where the mobile sink moves to certain nodes and collects data. In this paper, rendezvous mode is used. The rendezvous points are selected using the unsupervised agglomerative clustering technique. A technique to determine optimal data collection points for mobile sink has been proposed in [16]. Virtual grid architecture is defined to select the rendezvous node for data collection. In both the techniques, packet delivery ratio and energy efficiency are improved and control overhead and delay are reduced. However, implementing mobile selection technique-based AHP will improve the performance still further.

From the literature review, it is understood that the performance of the WSN is improved in terms of energy efficiency and packet delivery ratio by introducing a mobile sink. However, in every technique, the nearest mobile sink is selected for data collection by the sensor node. This could result in packet drop and data transmission discontinuity in the UAV sink layer environment. Hence, the technique is considering the distance, the energy of the sink node, and the speed of the UAV sink node. This paper proposed an AHP-based ranking and selection technique for data collection in a mobile sink environment.

4 Proposed Technique

Figure 1 shows the block diagram for the WSN network with UAVs serving as mobile sinks.

In this network, the sensor nodes are deployed in the area of application. The UAVs serve as mobile sinks collecting data from the sensor nodes. The UAV communicates the data to the base station in its reachability, and the data can be transmitted to the cloud for storage and analysis. The mobile sink has to be carefully selected by the sensor node to prevent delay and data loss. Hence, every sensor node selects its mobile sink based on three criteria, namely distance, the energy of the sink node, and speed of the sink node. This multi-criteria decision-making is solved using the analytical hierarchy process technique.

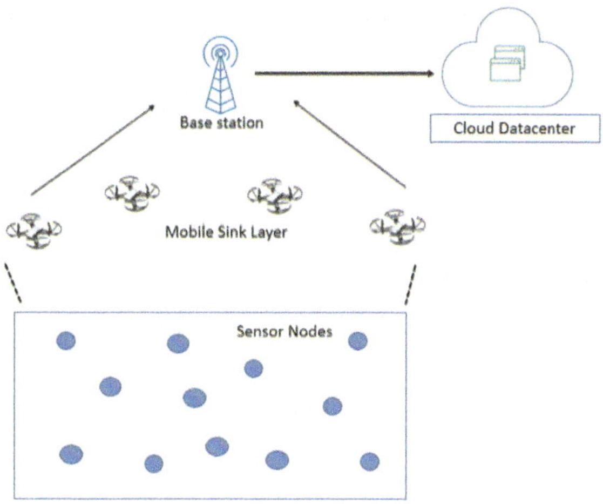

Fig. 1 WSN network with UAV serving as mobile sinks

4.1 AHP-Based Mobile Sink Selection Technique

The AHP-based mobile sink selection is a mathematical model that prioritizes the sink UAVs based on the distance between the sink and the sensor node, the energy of the sink node, and the speed of the sink node [17].

Distance between the sink and the sensor node: The distance between the sink and the sensor nodes is calculated as follows: Let N be the WSN network. S_i is the static sensor installed in network N. Let MS_n be the mobile sinks under the coverage of any sensor node S_i. Let h be the height of the UAV sink node from the ground. Let r be the distance between the sensor node and the foot of the UAV sink node. The distance D between the sensor and the sink node is calculated using Eq. 1. [18]

$$D = \sqrt{h^2 + r^2} \tag{1}$$

Energy of the UAV Sink Node Every UAV sink node sends a beacon signal periodically. Energy information E is communicated to the sensor node through the beacon signal sent from the sink node as it enters the coverage of the sink node.

Speed of the UAV Sink Node Speed of the sink node SP is measured using Eq. 2.

$$\mathrm{SP} = D * \mathrm{delay}_{\mathrm{RTT}} \tag{2}$$

D is the distance measured using Eq. 1. $\mathrm{delay}_{\mathrm{RTT}}$ is the round trip time measured between the sensor node and sink node. Based on these three criteria, the UAV sink nodes are ranked using the AHP technique.

Initially, the UAV sink introduces its presence to the sensor node using a beacon signal. The sensor nodes rank the UAV sink in its visibility using the AHP technique. The first step is to create a comparison matrix between the criteria. The comparison

Table 1 Rating scale for criteria

Rating number	Preferences
1	Equally preferred
2	Strongly preferred
3	Very strongly preferred
4	Extremely preferred

matrix R is given by Eq. 3.

$$R = \begin{array}{c} \\ \text{Speed} \\ \text{Energy} \\ \text{Distance} \end{array} \begin{array}{c} \text{Speed Energy Distance} \\ \begin{bmatrix} 1 & 1/2 & 3 \\ 2 & 1 & 4 \\ 1/3 & 1/4 & 1 \end{bmatrix} \end{array} \tag{3}$$

The values of the ranking matrix are based on the rating scale that is shown in Table 1

In Eq. 3, value of 1 has been assigned for any criterion compared against itself. The criterion energy is extremely preferred when compared with distance. Hence, a rating number of four is assigned. However, when distance is compared against energy, the value of ¼ is assigned.

$$\text{Normalized Ranking Matrix, } R_N = \begin{bmatrix} 0.30 & 0.29 & 0.38 \\ 0.60 & 0.57 & 0.50 \\ 0.10 & 0.14 & 0.13 \end{bmatrix}$$
$$\begin{matrix} 1.00 & 1.00 & 1.00 \end{matrix}$$

This assignment proves that an energy depriving UAV sink node very close to the sensor node is of less use for data collection. After the formulation of the ranking matrix, the priority vector for the criteria is calculated. The ranking matrix R is normalized by calculating the column sum and dividing each element in the matrix by the corresponding column sum as given below. Averaging each element of the normalized matrix R_N gives the priority vector PV_{MS} as given by Eq. 4.

$$\text{Priority vector, } PV_{MS} = \begin{bmatrix} 0.30 \\ 0.60 \\ 0.10 \end{bmatrix} \tag{4}$$

The priority value of 0.30, 0.60, and 0.10 has been assigned to the criteria speed, energy, and distance. The consistency of the priority vector has been checked for consistency using mathematical calculations, and it is observed that the consistency 0.03 which is less than 0.1. Hence, the priority vector for speed, energy, and distance can be assigned as 0.30, 0.60, and 0.10, respectively. The next step is to form a

Table 2 Rating scale for UAV sink node—speed criteria

Rating numbers	Preferences	Remarks
−8	Strongly not preferred	The UAV moves away from the sensor node with the speed of 30 m/s
−6	Not preferred as two resources are high compared with the source	The UAV moves away from the sensor node with the speed of 20 m/s
− 4	Can be preferred as one resource is high compared with the source	The UAV moves away from the sensor node with the speed of 10 m/s
−2	Equally preferred	The UAV moves away from the sensor node with the speed of 5 m/s
2	Strongly preferred	The UAV move toward the sensor node with the speed of m/s
4	Very strongly preferred	The UAV move toward the sensor node with the speed of 10 m/s
6	Extremely preferred	The UAV move toward the sensor node with the speed of 20 m/s
8	Very extremely preferred	The UAV move toward the sensor node with the speed of 30 m/s

comparison matrix for every UAV sink node against speed, energy, and distance criteria. The first step is to form a comparison matrix for all UAV sink nodes under the range of the sensor node using the speed criteria.

To form a comparison matrix, the speed of the UAV node is considered in two directions. A positive rating is allotted for the UAV moving toward the sensor node, and a negative rating is allotted for the UAV moving away from the sensor node. The speed range is assumed from 5 to 30 m/s. The rating scale of the speed criteria is given in Table 2.

For instance, let us assume that a sensor node S receives a beacon from UAV sink nodes MS_1, $MS_2 ... MS_{n.}$ Considering the values in Table 2, the comparison matrix of UAV sink nodes R_{SP} is formed. The next step is to form the comparison matrix for the UAV sink node against criteria energy criteria. The energy of the sink node is assumed to be from 500 to 1000 J. The rating scale for energy criteria is shown in Table 3.

As per the rating scale shown in Table 3, the comparison matrix of UAV sink nodes R_{EN} is formulated. Similarly, considering the rating scale in Table 4, the comparison matrix for UAV sink nodes R_{DIS} for distance criteria is calculated. As the communication technology is assumed to be Wi-Fi, the distance range is assumed to be 10 to 50 m.

After forming the comparison matrix, the priority vector matrix $PV_{SP}, PV_{EN}, PV_{DIS}$ is formed. The single-dimensional ranking matrix R_{SIN} is calculated using Eq. 5.

$$R_{SIN} = (PV_{SP} * 0.30) + (PV_{EN} * 0.60) + (PV_{DIS} * 0.10) \quad (5)$$

Table 3 Rating scale for UAV sink node—energy criteria

Rating numbers	Preferences	Remarks
1	Strongly not preferred	Energy of sink node = 100–300 J
2	Not preferred as two resources are high compared with the source	Energy of sink node = 400 J
3	Can be preferred as one resource is high compared with the source	Energy of sink node = 500 J
4	Equally preferred	Energy of sink node = 600 J
5	Strongly preferred	Energy of sink node = 700 J
6	Very strongly preferred	Energy of sink node = 800 J
7	Extremely preferred	Energy of sink node = 900 J
8	Very extremely preferred	Energy of sink node = 1000 J

Table 4 Rating scale for UAV sink node—distance criteria

Rating numbers	Preferences	Remarks
1	Strongly not preferred	Distance = 45–50 m
2	Not preferred as two resources are high compared with the source	Distance = 40 m
3	Can be preferred as one resource is high compared with the source	Distance = 35 m
4	Equally preferred	Distance = 30 m
5	Strongly preferred	Distance = 25 m
6	Very strongly preferred	Distance = 20 m
7	Extremely preferred	Distance = 15 m
8	Very extremely preferred	Distance = 10 m

Every sensor node maintains the priority vector matrix R_{SIN}. Whenever the sensor node has a packet to transmit, they select the UAV sink node of high priority and start transmission. The ranking table is updated for every period of T seconds.

5 Implementation Details

Simulation has been carried out using the NS-2 tool for the proposed technique. Energy consumption, average delay, and network lifetime are measured for the proposed AHP-based mobile sink selection technique using virtual grid-based geographic routing (VGBGR-AHP) and multiple ring-based nested routing (MRNR-AHP) considering three criteria such as speed, energy, and distance. The simulation parameters are listed in Table 5.

Table 5 Simulation parameters

Simulation parameters	Values
Number of sink nodes	1–10
Number of sensor nodes	100–400
Topography	500 m × 500 m
Data packet size	4000 bits
Beacon message size	100 bits
Communication range	90 m
Energy of the sensor nodes	2 J
Speed of each sink node	5–30 m/s

5.1 Virtual Grid-Based Geographic Routing-AHP Technique

Initially, in the VGBGR-AHP technique, depending on the number of sensor nodes, the network is divided into square-shaped grids. The closest nodes to the intersection points have been identified as intersection nodes. The sink nodes approaching the intersection nodes send beacon signals communicating their speed, remaining energy. The intersection nodes perform AHP-based ranking of the mobile sink nodes and maintain the ranking table. When the sensor nodes have data to transmit, it sends REQ_TRANSMIT, a request message to the nearby intersection node. The intersection node responds with CLR_TRANSMIT, clear to send message to the sensor nodes. This message has the sink node details on that sensor node. The sensor node then communicates the data to the sink node.

5.2 Multiple Ring-Based Nested Routing-AHP Technique

Similarly, in the MRNR-AHP technique, rings are formed with the distance of $L/(2*r)$, where L stands for the length of the topographical region and r stands for the radio range of the sensor nodes. The virtual infrastructure is constructed as given by Tang et al. [10]. The sink node sends the MSG_Update position message to one node of every ring. The node that receives the MSG_Update message calculates the AHP and ranks all the sink nodes from whom it receives the update message. The rank table is communicated to the sensor nodes of its ring. Whenever the sensor node wants to send some data, it sends the data to the suitable sink node referring to the rank table.

6 Performance Analysis

The experiments were conducted, and the performance of VGBGR-AHP and MRNR-AHP has been compared with the existing virtual grid-based dynamic routing (VGDRA) [19], virtual grid-based technique (VGB) [9], and ring routing [10] in terms of energy consumption, average delay, and packet delivery ratio [19].

6.1 Performance Analysis on Energy Consumption

Energy consumption is measured in terms of total energy consumed in the network and the average energy consumption of each node by varying the simulation time. The performance comparison of VGBGR-AHP and MRNR-AHP has been performed with that of the existing VGDRA, VGB, and nested routing technique. Figures 2 and 3 show the average energy consumed and total energy consumption of each node, respectively.

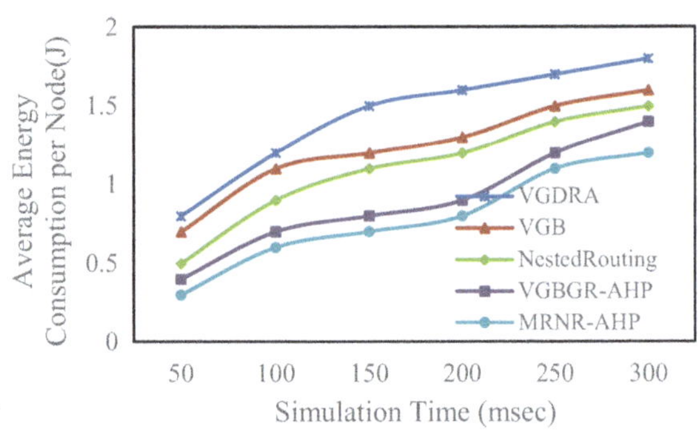

Fig. 2 Average energy consumption per node (J) versus simulation time (ms)

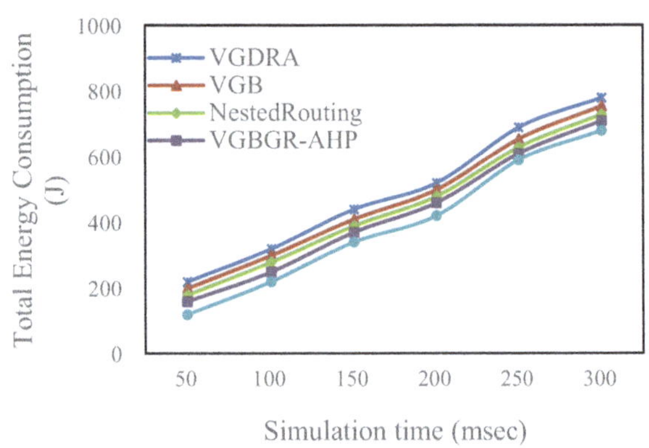

Fig. 3 Total energy consumption (J) versus simulation time (ms)

The average energy consumption per node in Joules by varying simulation time is shown in Fig. 2. It is observed that as the simulation time increases, the energy consumption is increased in VGDRA and higher than VGB, ring routing, VGBGR-AHP, and MRNR-AHP. This is because, as the sink nodes move outside the topography, the sensor nodes search for new sink nodes which incur more energy consumption.

However, in VGB and nested routing techniques, the sink node position is communicated to the sensor nodes through intersection nodes and router nodes, respectively. This enables the sensor nodes to select proper sink nodes leading to consistent data delivery. However, crucial parameters such as the direction and speed of the sink node and the energy of the sink node are not considered which lead to the election of the sink node which consumes some energy. But in the VGBGR-AHP and MRNR-AHP technique, the sink selection occurs after careful consideration of speed, energy, and distance parameters.

This leads to selection and suitable mobile sink and leads to more consistent data collection which saves energy. However, the VGBGR-AHP has higher energy consumption than the MRNR-AHP due to hotspot problems near the intersection points [10]. Similarly, in Fig. 3, the total energy consumption measured by varying simulation time is shown. The total energy consumption of each node in the proposed VGBGR-AHP and MRNR-AHP technique is lesser than the VGDRA, VGB, and nested routing techniques.

As the intersection nodes rank the UAV mobile sink nodes based on the speed, energy, and distance parameter, the sensor nodes function is to deliver the data, and the sink node selection rarely happens. This improves the energy consumption of the proposed technique compared with that of the VGDRA, VGB, and nested routing technique.

6.2 *Performance Analysis on Average Delay*

Average delay is measured as the average time taken for all the sensor nodes to reach the sink node for the simulation time. The average delay is measured by varying the number of sensor nodes from 100 to 400.

Figure 4 shows the average delay measured for the proposed VGBGR-AHP and MRNR-AHP technique compared with that of the VGDRA, VGB, and nested routing. It is observed that when the sensor nodes are less in the number of 100, the delay is minimal in VGB, nested routing, VGBGR-AHP, and MRNR-AHP technique and higher in VGDRA technique. This is because the sink position is not known to the sensor nodes, and the sensor nodes waste most of their time in searching for the sink node.

VGB and nested routing have less delay compared to the VGDRA technique but higher delay than the VGBGR-AHP and MRNR-AHP. This is because in VGB and nested routing technique, it calculates the sink position only considering the distance.

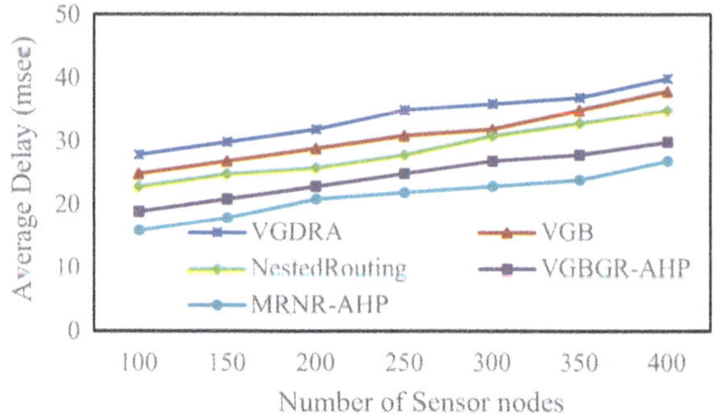

Fig. 4 Average delay (ms) versus number of sensor nodes

This leads to the recalculation of sink position when it moves away from the topology or dies due to energy depreciation which incurs delay in the packet transmission.

As the sensor nodes increase gradually, delay incurred is also higher as there is more need for recalculation in more sensor nodes. But, the AHP-based technique enables the sensor nodes to select the sink node based on speed, energy, and distance criteria. This enables the sensor nodes to transmit data for more time as the time required for recalculation is used for data transmission. This proves the higher scalability of the proposed technique.

6.3 *Performance Analysis on Packet Delivery Ratio*

Packet delivery ratio is a measure of successful packet delivery from the source sensor node to the destination node through the UAV sink node. The packet delivery ratio is measured by varying the number of sensor nodes. Figure 5 shows the packet delivery ratio of VGBGR-AHP and MRNR-AHP compared with that of VGDRA, VGB, and nested routing measured by varying the sensor nodes from 100 to 400.

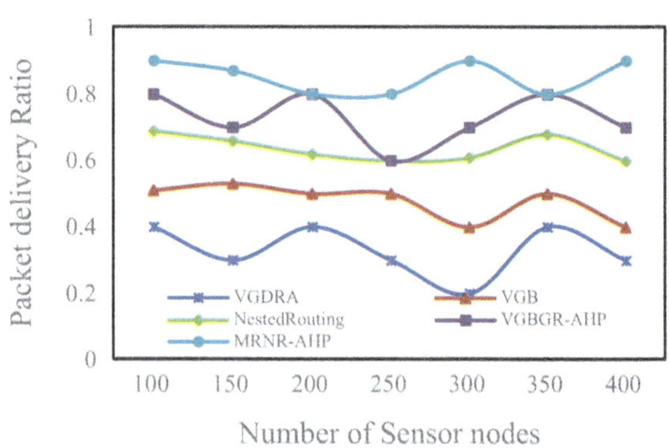

Fig. 5 Packet delivery ratio versus number of sensor nodes

From Fig. 5, it is observed that the packet delivery ratio of VGDRA, VGB, and nested routing technique is less compared with the AHP-based technique. This is due to the loss of packets when the sink node moves away from the sensor node or the energy of the UAV sink node reduces. However, in the AHP-based technique, the UAV mobile sink node is selected considering the speed, energy, and distance which assure the consistent connection with the sink node that improves the packet delivery ratio. As the number of sensor nodes increases, a trench and peak in packet delivery ratio are observed in the techniques due to the recalculation strategy followed as the mobile sink moves. The fluctuation was observed severely in VGDRA and VGBGR-AHP technique. In the MRNR-AHP technique, the fluctuation in packet delivery ratio is observed only when the number of nodes increases above 350 nodes. This technique has a higher packet delivery ratio compared to other techniques as the time is allotted for packet delivery, and time taken for recalculation is very less.

7 Statistical Analysis of the Proposed Techniques

The statistical test helps to analyze the obtained results and arrive at comparative conclusions and suitable decision-making. In this paper, a two-tailed test has been carried out to compare the performance of VGDRA, VGB and nested routing, VGBGR-AHP, and MRNR-AHP protocols.

7.1 Two-Tailed Test

The two-tailed test is performed to compare the performance of the proposed technique with that of the existing techniques. The steps involved in the two-tailed test are as follows:

Step 1: Statement of Hypothesis
In the statement of hypothesis, a null hypothesis (H_0) has to be proposed, and an alternate hypothesis (H_1) has also to be proposed. The hypothesis is proposed for the VGDRA, VGB and nested routing, VGBGR-AHP, and MRNR-AHP protocols for the measurement of average delay and packet delivery ratio by varying number of sensor nodes from 100 to 400.

Formulation of Null Hypothesis (H_0)
There is no difference in the performance of VGDRA, VGB, and nested routing compared with that of VGBGR-AHP and MRNR-AHP protocols in terms of packet delivery ratio and average delay by a varying number of sensor nodes from 100 to 400.

Formulation of Alternate Hypothesis (H_1)

There is a difference in the performance of VGDRA, VGB, and nested routing compared with that of VGBGR-AHP and MRNR-AHP protocols in terms of packet delivery ratio and average delay by a varying number of sensor nodes from 100 to 400.

Step2: Formulate the Analysis Scheme

The analysis is performed by taking packet delivery ratio and average delay values of all the protocols by varying the number of sensor nodes from 100 to 400. The t-test value is calculated and presented in Step 3.

Step3: Analysis of Sampled Data

The analysis of the sampled data is presented in Table 6.

Step 4: Interpretation of Results

From the table, it is understood that the "*t*" score for packet delivery ratio of the proposed VGBGR-AHP and MRNR-AHP protocols is higher than VGDRA, VGB, and nested routing. This clearly shows that the performance of the proposed protocol is superior to the existing protocols. Similarly, when checking the "*t*" score for average delay, the proposed protocols have outperformed the existing protocols.

Table 6 Two-tailed test report for VGDRA, VGB and nested routing, VGBGR-AHP, and MRNR-AHP protocols (number of sensor node varies from 100 to 400)

	Mean	Standard deviation	"*t*" score
Virtual grid-based geographic routing (VGBGR-AHP)			
Packet delivery ratio	0.73	0.08	2.11
Average delay	25	3.9	2.01
Multiple ring-based nested routing (MRNR-AHP)			
Packet delivery ratio	0.85	0.05	2.08
Average delay	22	3.6	1.03
Virtual grid-based dynamic routing (VGDRA)			
Packet delivery ratio	0.33	0.08	2.11
Average delay	34	4.2	2.98
The virtual grid-based technique (VGB)			
Packet delivery ratio	0.48	0.05	1.36
Average delay	31	4.5	2.63
Nested routing			
Packet delivery ratio	0.64	0.04	1.06
Average delay	29	4.4	2.22

8 Conclusion

In this paper, an AHP-based mobile sink selection has been proposed to eliminate the energy-hole problem and to improve the packet delivery ratio. The proposed technique has been evaluated using virtual grid-based and ring-based network routing protocols. Simulation results prove that the proposed technique outperforms the state-of-the-art technique in terms of energy consumption, average delay, and packet delivery ratio. Future enhancement of this technique would be optimizing the control messages which could improve the network utilization.

References

1. Yi, W.Y., Lo, K.M., Mak, T., Leung, K.S., Leung, Y., Meng, M.L.: A survey of wireless sensor network based air pollution monitoring systems, vol. 15, no. 12 (2015)
2. Rashid, B., Rehmani, M.H.: Applications of wireless sensor networks for urban areas: a survey. J. Netw. Comput. Appl. **60**, 192–219 (2016). https://doi.org/10.1016/j.jnca.2015.09.008
3. Noel, A.B., et al.: Networks: a comprehensive survey. IEEE Commun. Surv. y Tutorials **19**(3), 1403–1423 (2017)
4. Ojha, T., Misra, S., Raghuwanshi, N.S.: Wireless sensor networks for agriculture: the state-of-the-art in practice and future challenges. Comput. Electron. Agric. **118**, 66–84 (2015). https://doi.org/10.1016/j.compag.2015.08.011
5. Xia, W.W.X., Marcin, W., Fan, X., Damaševičius, R., Li, Y.: Multi-sink distributed power control algorithm for Cyber-physical-systems in coal mine tunnels. Comput. Netw. (2019). https://doi.org/10.1016/j.comnet.2019.04.017
6. Gu, Y., Ren, F., Ji, Y., Li, J.: The evolution of sink mobility management in wireless sensor networks: a survey. IEEE Commun. Surv. Tutorials **18**(1), 507–524 (2016). https://doi.org/10.1109/COMST.2015.2388779
7. Yang, S., Adeel, U., Tahir, Y., McCann, J.A.: Practical opportunistic data collection in wireless sensor networks with mobile sinks. IEEE Trans. Mob. Comput. **16**(5), 1420–1433 (2017). https://doi.org/10.1109/TMC.2016.2595574
8. Yarinezhad, R.: Reducing delay and prolonging the lifetime of wireless sensor network using efficient routing protocol based on mobile sink and virtual infrastructure. Ad Hoc Netw. **84**, 42–55 (2019). https://doi.org/10.1016/j.adhoc.2018.09.016
9. Yarinezhad, R., Sarabi, A.: Reducing delay and energy consumption in wireless sensor networks by making virtual grid infrastructure and using mobile sink. AEU Int. J. Electron. Commun. **84**(April 2017), 144–152 (2018). https://doi.org/10.1016/j.aeue.2017.11.026
10. Tang, J., Guo, S., Yang, Y.: Delivery latency minimization in wireless sensor networks with mobile sink. In: IEEE International Conference on Communications, vol. 2015-Sept, pp. 6481–6486 (2015). https://doi.org/10.1109/ICC.2015.7249357
11. Zhang, Y., He, S., Chen, J.: Near-optimal data gathering in rechargeable sensor networks with a mobile sink. IEEE Trans. Mob. Comput. **16**(6), 1718–1729 (2017). https://doi.org/10.1109/TMC.2016.2603152
12. Wu, C., Liu, Y., Wu, F., Fan, W., Tang, B.: Graph-based data gathering scheme in WSNs with a mobility-constrained mobile sink. IEEE Access **5**, 19463–19477 (2017). https://doi.org/10.1109/ACCESS.2017.2742138
13. Miao, Y., Sun, Z., Wang, N., Cao, Y., Cruickshank, H.: Time efficient data collection with mobile sink and vMIMO technique in wireless sensor networks. IEEE Syst. J. **12**(1), 639–647 (2018). https://doi.org/10.1109/JSYST.2016.2597166

14. Zhu, C., Quan, K., Han, G., Rodrigues, J.J.P.C.: A high-available and location predictive data gathering scheme with mobile sinks for wireless sensor networks. Comput. Netw. **145**, 156–164 (2018). https://doi.org/10.1016/j.comnet.2018.08.022
15. Donta, P.K., Rao, B.S.P., Amgoth, T., Annavarapu, C.S.R., Swain, S.: Data collection and path determination strategies for mobile sink in 3D WSNs. IEEE Sens. J. **20**(4), 2224–2233 (2020). https://doi.org/10.1109/JSEN.2019.2949146
16. Mehto, A., Tapaswi, S., Pattanaik, K.K.: Virtual grid-based rendezvous point and sojourn location selection for energy and delay efficient data acquisition in wireless sensor networks with mobile sink. Wirel. Netw. **26**(5), 3763–3779 (2020). https://doi.org/10.1007/s11276-020-02293-4
17. Podvezko, V.: Application of AHP technique. J. Bus. Econ. Manag. **10**(2), 181–189 (2009). https://doi.org/10.3846/1611-1699.2009.10.181-189
18. Sallouha, H., Azari, M.M., Pollin, S.: Energy-constrained UAV trajectory design for ground node localization. In: 2018 IEEE Global Communication Conference GLOBECOM 2018—Proceedings 2019. https://doi.org/10.1109/GLOCOM.2018.8647530
19. Khan, A.W., Abdullah, A.H., Razzaque, M.A., Bangash, J.I.: VGDRA: a virtual grid-based dynamic routes adjustment scheme for mobile sink-based wireless sensor networks. IEEE Sens. J. **15**(1), 526–534 (2015). https://doi.org/10.1109/JSEN.2014.2347137

On the Characteristic Polynomial of the Subdivision-Vertex Join of Graphs

R. Pavithra and R. Rajkumar

Abstract In this paper, we consider the subdivision-vertex join $G_1 \dot{\vee} G_2$ of two graphs G_1 and G_2, where G_1 is a join of two graphs H_1 and H_2. We determine the characteristic polynomial of the adjacency, the Laplacian and the signless Laplacian matrix of this graph in terms of H_1, H_2 and G_2. This enables us to construct infinitely many pairs of adjacency, Laplacian and signless Laplacian cospectral graphs.

Keywords Subdivision-vertex join · Adjacency spectra · Laplacian spectra · Signless Laplacian spectra · Cospectral graphs

1 Introduction

Graphs considered in this paper are finite and simple. Let G be a graph with vertices $v_1, v_2, \dots, v_n$. The *adjacency matrix* $A(G)$ of G is a $n \times n$ matrix such that its (i, j)th-entry is 1, when v_i and v_j are adjacent in G and 0, otherwise. The *degree matrix* $D(G)$ of G is the diagonal matrix in which all its diagonal entries are the degrees of vertices of G. $L(G) = D(G) - A(G)$ and $Q(G) = D(G) + A(G)$ are called the *Laplacian matrix* and the *signless Laplacian matrix* of G, respectively. Let $\phi_{A(G)}(x)$ $\big($resp. $\phi_{L(G)}(x)$, $\phi_{Q(G)}(x)\big)$ denotes the *adjacency* (resp. *Laplacian, signless Laplacian*) characteristic polynomial of G. The eigenvalues of $A(G)$ (resp. $L(G)$, $Q(G)$) are known as the *adjacency* (resp. *Laplacian, signless Laplacian*) *spectrum* of G. Throughout this paper, the *adjacency* (resp. *Laplacian, signless Laplacian*) *spectrum* of G is denoted by $\lambda_1(G), \lambda_2(G), \dots, \lambda_n(G)$ (resp. $\mu_1(G) = 0, \mu_2(G), \dots, \mu_n(G)$; $\gamma_1(G), \gamma_2(G), \dots, \gamma_n(G)$). Two graphs are called *adjacency (resp. Laplacian, signless Laplacian) cospectral* if they have the same adjacency (resp. Laplacian, signless Laplacian) spectrum. The incidence matrix of a graph G on n vertices and m edges is denoted by $B(G)$ and is defined as the 0–1 matrix of order $n \times m$ whose (i, j)th entry is 1 if and only if ith vertex is incident with jth edge. It is well known that

R. Pavithra · R. Rajkumar (✉)
Department of Mathematics, The Gandhigram Rural Institute (Deemed to be University), Gandhigram 624302, Tamil Nadu, India

P. Balasubramaniam et al. (eds.), *Mathematical Modelling and Computational Intelligence Techniques*, Springer Proceedings in Mathematics & Statistics 376,
https://doi.org/10.1007/978-981-16-6018-4_18

$B(G)B(G)^T = A(G) + D(G)$. Spectra of graph matrices play an important role in several fields of science such as chemistry, physics, computer science, etc., (see [2] and its references). Finding the spectra of a graph is crucial, since a lot of information about the structural properties of a graph can be revealed from its spectra. In literature, researchers use graph operations to determine the spectrum of graphs derived from these, in terms of the spectrum of constituting graphs. Some of such well known graph operations are disjoint union, join, corona, rooted product (see, for example [1–9] and the references there in). The join $G \vee H$ of two graphs G and H is the graph obtained from G and H by joining each vertex of G to every vertex of H. $S(G)$ denote the subdivision graph of G which is the graph obtained by inserting a new vertex into every edge of G.

The subdivision-vertex join $G_1 \dot{\vee} G_2$ of G_1 and G_2 [4] is the graph constructed from $S(G_1)$ and G_2 by joining each vertex of G_1 in $S(G_1)$ with every vertex of G_2. In [4], the adjacency spectra of $G_1 \dot{\vee} G_2$ was determined when G_1 and G_2 are both regular. In [10], the adjacency (resp. Laplacian, signless Laplacian) spectra of $G_1 \dot{\vee} G_2$ were determined when G_1 is regular and G_2 is an arbitrary graph.

In this paper, we compute the adjacency (resp. Laplacian, signless Laplacian) characteristic polynomial of $G_1 \dot{\vee} G_2$ when G_1 is a join of two graphs and G_2 is an arbitrary graph. Also, we obtain the adjacency (resp. Laplacian and signless Laplacian) spectra of $G_1 \dot{\vee} G_2$ when the constituting graphs are regular. These results enable us to construct infinitely many pairs of adjacency (resp. Laplacian and signless Laplacian) cospectral graphs.

Now we furnish some notions and results which are used in the subsequent sections. Throughout this paper, $J_{n\times m}$ and $\mathbf{0}_{n\times m}$ denotes the matrix with all entries are one and the matrix with all entries are zero, respectively. The M-coronal $\Gamma_M(x)$ of a square matrix M is defined as the sum of the entries of the matrix $(xI_n - M)^{-1}$ [11, 12] that is

$$\Gamma_M(x) = J_{1\times n}(xI_n - M)^{-1}J_{n\times 1}.$$

If each row sum of M equals a constant t, then

$$\Gamma_M(x) = \frac{n}{x-t}. \tag{1}$$

In particular, for the complete bipartite graph K_{q_1,q_2} on $q_1 + q_2$ vertices,

$$\Gamma_{A(K_{q_1,q_2})}(x) = \frac{(q_1+q_2)x + 2q_1q_2}{x^2 - q_1q_2}. \tag{2}$$

Proposition 1 [2] *Let*

$$M = \begin{bmatrix} M_{11} & M_{12} \\ M_{21} & M_{22} \end{bmatrix}$$

be an $n \times n$ block matrix, where M_{11}, M_{22} are square matrices. If M_{11}, M_{22} are non-singular, then

$$\begin{aligned}\det M &= \det(M_{22})\det(M_{11} - M_{12}M_{22}^{-1}M_{21}) \\ &= \det(M_{11})\det(M_{22} - M_{21}M_{11}^{-1}M_{12}).\end{aligned}$$

Proposition 2 [10] *Let M be an $n \times n$ matrix and α be a real number. Then*

$$\det(xI_n - M - \alpha J_{n\times n}) = (1 - \alpha\Gamma_M(x))\det(xI_n - M).$$

2 Adjacency Characteristic Polynomial

In this section, we compute the adjacency characteristic polynomial of $G_1 \dot{\vee} G_2$, where G_1 is a join of two graphs and G_2 is an arbitrary graph. By using these, we determine infinitely many adjacency cospectral graphs.

Theorem 1 *Let H_i be a graph on p_i vertices for $i = 1, 2$. Let $G_1 = H_1 \vee H_2$ be a graph on n_1 vertices m_1 edges and let G_2 be a graph on n_2 vertices. Then*

$$\begin{aligned}\phi_{A(G_1\dot{\vee}G_2)}(x) &= \phi_{A(G_2)}(x)\phi_{Q(H_2)}(x^2 - p_1)\phi_{Q(H_1)}(x^2 - p_2)x^{m_1-n_1} \\ &\times \left(1 - \Gamma_{A(G_2)}(x)\Gamma_{\frac{1}{x}Q(H_2)}\left(x - \frac{p_1}{x}\right)\right)\left(1 - p(x)\Gamma_{\frac{1}{x}Q(H_1)}\left(x - \frac{p_2}{x}\right)\right),\end{aligned}$$

where

$$p(x) = \Gamma_{A(G_2)}(x) + \left(\Gamma_{A(G_2)}(x) + \frac{1}{x}\right)^2 \Gamma_{\Gamma_{A(G_2)}(x)J_{p_2\times p_2} + \frac{1}{x}Q(H_2)}\left(x - \frac{p_1}{x}\right)$$

Proof With a suitable ordering of vertices of $G_1 \dot{\vee} G_2$, we get

$$A(G_1 \dot{\vee} G_2) = \begin{bmatrix} \mathbf{0}_{n_1\times n_1} & B(G_1) & J_{n_1\times n_2} \\ B(G_1)^T & \mathbf{0}_{m_1\times m_1} & \mathbf{0}_{m_1\times n_2} \\ J_{n_2\times n_1} & \mathbf{0}_{n_2\times m_1} & A(G_2) \end{bmatrix},$$

Then

$$\phi_{A(G_1\dot{\vee}G_2)}(x) = det(xI - A(G_1 \dot{\vee} G_2)).$$

Now by using Proposition 1, we have

$$\phi_{A(G_1\dot{\vee}G_2)}(x) = det(xI - A(G_2))det(S), \tag{3}$$

where

$$S = \begin{bmatrix} xI_{n_1} - \Gamma_{A(G_2)}(x)J_{n_1\times n_1} & -B(G_1) \\ -B(G_1)^T & xI_{m_1} \end{bmatrix}.$$

Again by using Proposition 1, we have

$$det(S) = x^{m_1} det\left(xI_{n_1} - \Gamma_{A(G_2)}(x)J_{n_1\times n_1} - \frac{1}{x}B(G_1)B(G_1)^T\right). \quad (4)$$

Next,

$$xI_{n_1} - \Gamma_{A(G_2)}(x)J_{n_1\times n_1} - \frac{1}{x}B(G_1)B(G_1)^T =$$

$$\begin{bmatrix} \left(x-\frac{p_2}{x}\right)I_{p_1} - \Gamma_{A(G_2)}(x)J_{p_1\times p_1} - \frac{1}{x}Q(H_1) & -\left(\Gamma_{A(G_2)}(x)+\frac{1}{x}\right)J_{p_1\times p_2} \\ -\left(\Gamma_{A(G_2)}(x)+\frac{1}{x}\right)J_{p_2\times p_1} & \left(x-\frac{p_1}{x}\right)I_{p_2} - \Gamma_{A(G_2)}(x)J_{p_2\times p_2} - \frac{1}{x}Q(H_2) \end{bmatrix}.$$

By using Proposition 1, we have

$$det\left(xI_{n_1} - \Gamma_{A(G_2)}(x)J_{n_1\times n_1} - \frac{1}{x}B(G_1)B(G_1)^T\right)$$

$$= det\left(\left(x-\frac{p_1}{x}\right)I_{p_2} - \Gamma_{A(G_2)}(x)J_{p_2\times p_2} - \frac{1}{x}Q(H_2)\right) det(S_1), \quad (5)$$

where

$$\begin{aligned} S_1 &= \left(x-\frac{p_2}{x}\right)I_{p_1} - \Gamma_{A(G_2)}(x)J_{p_1\times p_1} - \frac{1}{x}Q(H_1) - \left(\Gamma_{A(G_2)}(x)+\frac{1}{x}\right)^2 \\ &\quad \times \Gamma_{\Gamma_{A(G_2)}(x)J_{p_2\times p_2}+\frac{1}{x}Q(H_2)}\left(x-\frac{p_1}{x}\right)J_{p_1\times p_1} \\ &= \left(x-\frac{p_2}{x}\right)I_{p_1} - p(x)J_{p_1\times p_1} - \frac{1}{x}Q(H_1), \end{aligned}$$

where

$$p(x) = \Gamma_{A(G_2)}(x) + \left(\Gamma_{A(G_2)}(x)+\frac{1}{x}\right)^2 \Gamma_{\Gamma_{A(G_2)}(x)J_{p_2\times p_2}+\frac{1}{x}Q(H_2)}\left(x-\frac{p_1}{x}\right).$$

By using Proposition 2, we have

$$det(xI_{n_1} - \Gamma_{A(G_2)}(x)J_{n_1\times n_1} - \frac{1}{x}B(G_1)B(G_1)^T)$$

$$= g(x)det\left(\left(x-\frac{p_1}{x}\right)I_{p_2} - \frac{1}{x}Q(H_2)\right) \times det\left(\left(x-\frac{p_2}{x}\right)I_{p_1} - \frac{1}{x}Q(H_1)\right), \quad (6)$$

where $g(x) = \left(1 - p(x)\Gamma_{\frac{1}{x}Q(H_1)}\left(x-\frac{p_2}{x}\right)\right)\left(1 - \Gamma_{A(G_2)}(x)\Gamma_{\frac{1}{x}Q(H_2)}\left(x-\frac{p_1}{x}\right)\right)$. Substituting (6) in (4), we have

$$det(S) = \phi_{Q(H_2)}(x^2 - p_1)\phi_{Q(H_1)}(x^2 - p_2)\left(1 - \Gamma_{A(G_2)}(x)\Gamma_{\frac{1}{x}Q(H_2)}\left(x - \frac{p_1}{x}\right)\right)$$
$$\times x^{m_1 - n_1}\left(1 - p(x)\Gamma_{\frac{1}{x}Q(H_1)}\left(x - \frac{p_2}{x}\right)\right). \quad (7)$$

By substituting (7) in (3), we get the result.

Corollary 1 *Let H_i be a r_i-regular graph on p_i vertices for $i = 1, 2$. Let $G_1 = H_1 \vee H_2$ be a graph on n_1 vertices m_1 edges and let G_2 be a graph on n_2 vertices. Then,*

$$\phi_{A(G_1 \dot{\vee} G_2)}(x) = \left(\prod_{i=2}^{p_2}(x^2 - p_1 - r_2 - \lambda_i(H_2))\right)\left(\prod_{i=2}^{p_1}(x^2 - p_2 - r_1 - \lambda_i(H_1))\right)$$
$$\times \phi_{A(G_2)}(x)\left((x^2 - p_2 - 2r_1)(x^2 - p_1 - 2r_2) - p_1p_2 - x\Gamma_{A(G_2)}(x)\right.$$
$$\left.\times (p_2(x^2 - p_2 - 2r_1) + p_1(x^2 - p_1 - 2r_2) + 2p_1p_2)\right) x^{m_1 - n_1}.$$

Proof By using (1), we have

$$\Gamma_{\Gamma_{A(G_2)}(x)J_{p_2 \times p_2} + \frac{1}{x}Q(H_2)}\left(x - \frac{p_1}{x}\right) = \frac{p_2 x}{x^2 - \Gamma_{A(G_2)}(x)p_2 x - p_1 - 2r_2}, \quad (8)$$

$$\Gamma_{\frac{1}{x}Q(H_1)}\left(x - \frac{p_2}{x}\right) = \frac{p_1 x}{x^2 - p_2 - 2r_1}, \quad (9)$$

$$\Gamma_{\frac{1}{x}Q(H_2)}\left(x - \frac{p_1}{x}\right) = \frac{p_2 x}{x^2 - p_1 - 2r_2}. \quad (10)$$

Substituting (8)–(10), $\lambda_1(H_1) = r_1, \lambda_1(H_2) = r_2$ in Theorem 1, we can get the result.

The following results can be deduced from Corollary 1 by substituting the coronal values of $A(G_2)$ from (1) and (2).

Corollary 2 *Let H_i be a r_i-regular graph on p_i vertices for $i = 1, 2$. Let $G_1 = H_1 \vee H_2$ be a graph on n_1 vertices m_1 edges and G_2 be a r-regular graph on n_2 vertices. Then,*

$$\phi_{A(G_1 \dot{\vee} G_2)}(x) = \left(\prod_{i=2}^{p_2}(x^2 - p_1 - r_2 - \lambda_i(H_2))\right)\left(\prod_{i=2}^{p_1}(x^2 - p_2 - r_1 - \lambda_i(H_1))\right)$$
$$\times \left(\prod_{i=2}^{n_2}(x - \lambda_i(G_2))\right)\left(((x^2 - p_2 - 2r_1)(x^2 - p_1 - 2r_2) - p_1p_2)\right.$$
$$\left.\times (x - r) - xn_2\left(p_2(x^2 - p_2 - 2r_1) + p_1(x^2 - p_1 - 2r_2) + 2p_1p_2\right)\right)$$
$$\times x^{m_1 - n_1}.$$

Corollary 3 *Let H_i be a r_i-regular graph on p_i vertices for $i = 1, 2$. Let $G_1 = H_1 \vee H_2$ be a graph on n_1 vertices m_1 edges. Then,*

$$\phi_{A(G_1 \dot{\vee} K_{q_1,q_2})}(x) = \left(\prod_{i=2}^{p_2}(x^2 - p_1 - r_2 - \lambda_i(H_2))\right)\left(\prod_{i=2}^{p_1}(x^2 - p_2 - r_1 - \lambda_i(H_1))\right)$$
$$\times x^{m_1-n_1+q_1+q_2-2}\left(((x^2 - p_2 - 2r_1)(x^2 - p_1 - 2r_2) - p_1p_2)\right.$$
$$\times\ (x^2 - q_1q_2) - x(n_2x + 2q_1q_2)\left(p_2(x^2 - p_2 - 2r_1) + p_1\right.$$
$$\left.\left.\times\ (x^2 - p_1 - 2r_2) + 2p_1p_2\right)\right).$$

As a consequence of Theorem 1, we construct infinitely many pairs of adjacency cospectral graphs in the following result.

Corollary 4 *Let H_1, H_2, H_3 be graphs. Let $G_1 = H_1 \vee H_2$ and $G_2 = H_3 \vee H_2$. If H_1 and H_3 are r-regular signless Laplacian cospectral graphs on p_1 vertices with $\Gamma_{\frac{1}{x}Q(H_1)}(x) = \Gamma_{\frac{1}{x}Q(H_3)}(x)$, G_3 and G_4 are adjacency cospectral graphs on n_2 vertices with $\Gamma_{A(G_3)}(x) = \Gamma_{A(G_4)}(x)$, then $G_1 \dot{\vee} G_3$ and $G_2 \dot{\vee} G_4$ are adjacency cospectral.*

3 Laplacian Characteristic Polynomial

In this section, we derive the Laplacian characteristic polynomial of $G_1 \dot{\vee} G_2$ when G_1 is the join of two graphs and G_2 is an arbitrary graph. Using these, we determine infinitely many Laplacian cospectral graphs.

Theorem 2 *Let H_i be a graph on p_i vertices for $i = 1, 2$. Let $G_1 = H_1 \vee H_2$ be a graph on n_1 vertices m_1 edges and G_2 be a graph on n_2 vertices. Then*

$$\phi_{L(G_1 \dot{\vee} G_2)}(x) = \phi_{L(G_2)}(x - n_1)det(g_1(x)I_{p_2} - g(x)J_{p_2 \times p_2} - \frac{1}{x-2}Q(H_2) - D(H_2))$$
$$\times\ det(g_2(x)I_{p_1} - p(x)J_{p_1 \times p_1} - \frac{1}{x-2}Q(H_1) - D(H_1))(x-2)^{m_1},$$

where $g(x) = \Gamma_{L(G_2)}(x - n_1)$, $g_1(x) = x - n_2 - p_1 - \frac{p_1}{x-2}$, $g_2(x) = x - n_2 - p_2 - \frac{p_2}{x-2}$ *and*

$$p(x) = g(x) + \left(g(x) + \frac{1}{x-2}\right)^2 \Gamma_{g(x)J_{p_2\times p_2} + \frac{1}{x-2}Q(H_2) + D(H_2)}g_1(x).$$

Proof With a suitable ordering of vertices of $G_1 \dot{\vee} G_2$, we get

$$L(G_1 \dot{\vee} G_2) = \begin{bmatrix} D(G_1) + n_2I_{n_1} & -B(G_1) & -J_{n_1 \times n_2} \\ -B(G_1)^T & 2I_{m_1} & \mathbf{0}_{m_1 \times n_2} \\ -J_{n_2 \times n_1} & \mathbf{0}_{n_2 \times m_1} & n_1I_{n_2} + L(G_2) \end{bmatrix}.$$

Then

$$\phi_{L(G_1\dot{\vee}G_2)}(x) = \det(xI - L(G_1 \dot{\vee} G_2)).$$

Now by using Proposition 1, we have

$$\phi_{L(G_1\dot{\vee}G_2)}(x) = \det((x - n_1)I_{n_2} - L(G_2))\det(S), \tag{11}$$

where

$$S = \begin{bmatrix} (x - n_2)I_{n_1} - \Gamma_{L(G_2)}(x - n_1)J_{n_1\times n_1} - D(G_1) & B(G_1) \\ B(G_1)^T & (x - 2)I_{m_1} \end{bmatrix}.$$

By using Proposition 1, we have

$$\begin{aligned}&\det(S)\\ &= \det\left((x - n_2)I_{n_1} - D(G_1) - \Gamma_{L(G_2)}(x - n_1)J_{n_1\times n_1} - \frac{1}{x-2}B(G_1)B^T(G_1)\right)\\ &\quad\times (x-2)^{m_1}.\end{aligned}$$

Next,

$$(x - n_2)I_{n_1} - D(G_1) - g(x)J_{n_1\times n_1} - \frac{1}{x-2}B(G_1)B^T(G_1)$$

$$= \begin{bmatrix} g_2(x)I_{p_1} - g(x)J_{p_1\times p_1} - \frac{1}{x-2}Q(H_1) - D(H_1) & -(g(x) + \frac{1}{x-2})J_{p_1\times p_2} \\ -(g(x) + \frac{1}{x-2})J_{p_2\times p_1} & g_1(x)I_{p_2} - g(x)J_{p_2\times p_2} - \frac{1}{x-2}Q(H_2) - D(H_2) \end{bmatrix}.$$

where $g(x) = \Gamma_{L(G_2)}(x - n_1)$, $g_1(x) = x - n_2 - p_1 - \frac{p_1}{x-2}$, $g_2(x) = x - n_2 - p_2 - \frac{p_2}{x-2}$.

By using Proposition 1, we have

$$\begin{aligned}\det(S) = (x-2)^{m_1}&\det\left(g_1(x)I_{p_2} - g(x)J_{p_2\times p_2} - \frac{1}{x-2}Q(H_2) - D(H_2)\right)\\ &\times\det\left(g_2(x)I_{p_1} - p(x)J_{p_1\times p_1} - \frac{1}{x-2}Q(H_1) - D(H_1)\right),\end{aligned} \tag{12}$$

where

$$p(x) = g(x) + \left(g(x) + \frac{1}{x-2}\right)^2 \Gamma_{g(x)J_{p_2\times p_2}+\frac{1}{x-2}Q(H_2)+D(H_2)}g_1(x).$$

Substituting (12) in (11), we get the result.

Corollary 5 *Let H_i be a r_i-regular graph on p_i vertices for $i = 1, 2$. Let $G_1 = H_1 \vee H_2$ and G_2 be a graph on n_2 vertices. Then,*

$$\phi_{L(G_1 \dot{\vee} G_2)}(x) - \left(\prod_{i=2}^{p_1} ((x - n_2 - r_1)(x-2) - p_2(x-1) - r_1 - \lambda_i(H_1))\right)$$
$$\times \left(\prod_{i=2}^{p_2} ((x - n_2 - r_2)(x-2) - p_1(x-1) - r_2 - \lambda_i(H_2))\right)$$
$$\times \left(\prod_{i=2}^{n_2} (x - n_1 - \mu_i(G_2))\right) (x-2)^{m_1-n_1} (((x - n_2 - r_1)(x-2)$$
$$-p_2(x-1) - 2r_1)(((x - n_2 - r_2)(x - n_1) - n_2 p_2)(x-2)$$
$$-(p_1(x-1) + 2r_2)(x - n_1)) - p_1 p_2 (x - n_1) - p_1(x-2)n_2$$
$$\times((x - n_2 - r_2)(x-2) - p_1(x-1) + 2(p_2 - r_2)))\,.$$

Proof Since H_i is r_i-regular for $i = 1, 2$, and from (12), we have

$$\det(S) = \det(((g_1(x) - r_2)(x-2) - r_2)I_{p_2} - g(x)(x-2)J_{p_2 \times p_2} - A(H_2))$$
$$\times \det(((g_2(x) - r_1)(x-2) - r_1)I_{p_1} - p(x)(x-2)J_{p_1 \times p_1} - A(H_1))$$
$$\times (x-2)^{m_1-n_1}, \tag{13}$$

where

$$p(x) = g(x) + \left(g(x) + \frac{1}{x-2}\right)^2 \Gamma_{g(x)J_{p_2 \times p_2} + \frac{1}{x-2}A(H_2) + (1+\frac{1}{x-2})r_2 I_{p_2}}\, g_1(x). \tag{14}$$

By using (1) in (14), we have

$$p(x) = g(x) + \left(g(x) + \frac{1}{x-2}\right)^2 \frac{p_2(x-2)}{(x-2)(g_1(x) - g(x)p_2 - r_2) - 2r_2}$$
$$= \frac{g(x)(x-2)^2(g_1(x) - r_2) + p_2 + 2g(x)(x-2)(p_2 - r_2)}{(x-2)^2(g_1(x) - g(x)p_2 - r_2) - 2r_2(x-2)}. \tag{15}$$

By using Proposition 2 in (13) and taking $\lambda_1(H_i) = r_i$ for $i = 1, 2$, we have

$$\det(S) = \left(\prod_{i=2}^{p_1} ((g_2(x) - r_1)(x-2) - r_1 - \lambda_i(H_1))\right) ((g_2(x) - r_1 - p(x)p_1)$$
$$\times(x-2) - 2r_1) \left(\prod_{i=2}^{p_2} ((g_1(x) - r_2)(x-2) - r_2 - \lambda_i(H_2))\right)$$
$$\times ((g_1(x) - r_2 - g(x)p_2)(x-2) - 2r_2)\,(x-2)^{m_1-n_1}. \tag{16}$$

Substituting (15) in (16), we have

$$\det(S) = (((x-2)(g_2(x)-r_1)-2r_1)\left((g_1(x)-r_2-g(x)p_2)(x-2)-2r_2\right)$$
$$-p_1g(x)(x-2)^2(g_1(x)-r_2)-p_1p_2-2p_1g(x)(x-2)(p_2-r_2))$$
$$\times\left(\prod_{i=2}^{p_1}((g_2(x)-r_1)(x-2)-r_1-\lambda_i(H_1))\right)(x-2)^{m_1-n_1}$$
$$\times\left(\prod_{i=2}^{p_2}((g_1(x)-r_2)(x-2)-r_2-\lambda_i(H_2))\right). \tag{17}$$

Substituting $g_1(x) = x - n_2 - p_1 - \frac{p_1}{x-2}$, $g_2(x) = x - n_2 - p_2 - \frac{p_2}{x-2}$ in (17) and then substituting the resulting equation in (11), we get the result.

In the following result, we obtain infinitely many pairs of Laplacian cospectral graphs by using Theorem 2 and Corollary 5.

Corollary 6

(i) *Let G_1 be the join of two graphs H_1 and H_2. Let G_2 and G_3 be Laplacian cospectral graphs on n_2 vertices. Then, the graphs $G_1 \dot{\vee} G_2$ and $G_1 \dot{\vee} G_3$ are Laplacian cospectral.*

(ii) *Let H_1, H_2, H_3, H_4 be graphs. Let $G_1 = H_1 \vee H_2$ and $G_2 = H_3 \vee H_4$. If H_1 and H_3 are r_1-regular adjacency cospectral graphs, H_2 and H_4 are r_2-regular adjacency cospectral graphs, G_3 and G_4 are Laplacian cospectral graphs, then $G_1 \dot{\vee} G_3$ and $G_2 \dot{\vee} G_4$ are Laplacian cospectral.*

4 Signless Laplacian Characteristic Polynomial

In this section, we determine the signless Laplacian characteristic polynomial of $G_1 \dot{\vee} G_2$ when G_1 is the join of two graphs and G_2 is an arbitrary graph. These results enable us to construct infinitely many signless Laplacian cospectral graphs.

Theorem 3 *Let H_i be a graph on p_i vertices. Let $G_1 = H_1 \vee H_2$ and G_2 be a graph on n_2 vertices. Then*

$$\phi_{Q(G_1\dot{\vee}G_2)}(x) = (x-2)^{m_1}\det\left(g_1(x)I_{p_2} - g(x)J_{p_2\times p_2} - \frac{1}{x-2}Q(H_2) - D(H_2)\right)$$
$$\times\det\left(g_2(x)I_{p_1} - p(x)J_{p_1\times p_1} - \frac{1}{x-2}Q(H_1) - D(H_1)\right)$$
$$\times\,\phi_{Q(G_2)}(x-n_1),$$

where $n_1 = p_1 + p_2$, $g(x) = \Gamma_{Q(G_2)}(x - n_1)$, $g_1(x) = x - n_2 - p_1 - \dfrac{p_1}{x-2}$, $g_2(x) = x - n_2 - p_2 - \dfrac{p_2}{x-2}$ *and*

$$p(x) = g(x) + \left(g(x) + \frac{1}{x-2}\right)^2 \Gamma_{g(x)J_{p_2 \times p_2} + \frac{1}{x-2}Q(H_2) + D(H_2)} g_1(x).$$

Proof Proof is similar to the proof of Theorem 2.

The following result can be deduced from Theorem 3 and by substituting the coronal value given in (1).

Corollary 7 *Let H_i be a r_i-regular graph for $i = 1, 2$. Let $G_1 = H_1 \vee H_2$ and G_2 be an arbitrary graph on n_2 vertices. Then,*

$$\begin{aligned}
\phi_{Q(G_1 \dot{\vee} G_2)}(x) = &\left(\prod_{i=2}^{p_1} ((x - n_2 - r_1)(x-2) - p_2(x-1) - r_1 - \lambda_i(H_1))\right) \\
&\times \left(\prod_{i=2}^{p_2} ((x - n_2 - r_2)(x-2) - p_1(x-1) - r_2 - \lambda_i(H_2))\right) \\
&\times \left(\prod_{i=2}^{n_2} (x - n_1 - \gamma_i(G_2))\right)(x-2)^{m_1 - n_1} (((x - n_2 - r_1)(x-2) \\
&- p_2(x-1) - 2r_1)\left((x - n_2 - \Gamma_{Q(G_2)}(x - n_1)p_2 - r_2)(x-2)\right. \\
&- p_1(x-1) - 2r_2) - p_1\Gamma_{Q(G_2)}(x - n_1)(x-2)((x - n_2 - r_2) \\
&\times (x-2) - p_1(x-1)) - p_1p_2 - 2p_1\Gamma_{Q(G_2)}(x - n_1)(x-2)(p_2 - r_2)\big).
\end{aligned}$$

The following result can be deduced from Corollary 7 by substituting the coronal value of $Q(G_2)$ by using (1).

Corollary 8 *Let H_i be a r_i-regular graph for $i = 1, 2$. Let $G_1 = H_1 \vee H_2$ and G_2 be a r-regular graph on n_2 vertices. Then,*

$$\begin{aligned}
\phi_{Q(G_1 \dot{\vee} G_2)}(x) = &\left(\prod_{i=2}^{p_1} ((x - n_2 - r_1)(x-2) - p_2(x-1) - r_1 - \lambda_i(H_1))\right) \\
&\times \left(\prod_{i=2}^{p_2} ((x - n_2 - r_2)(x-2) - p_1(x-1) - r_2 - \lambda_i(H_2))\right) \\
&\times \left(\prod_{i=2}^{n_2} (x - n_1 - \gamma_i(G_2))\right)(x-2)^{m_1 - n_1} (((x - n_2 - r_1)(x-2) \\
&- p_2(x-1) - 2r_1)(((x - n_2 - r_2)(x - n_1 - 2r) - n_2p_2)(x-2) \\
&- (p_1(x-1) + 2r_2)(x - n_1 - 2r)) - p_1p_2(x - n_1 - 2r) - p_1 \\
&\times (x-2)n_2((x - n_2 - r_2)(x-2) - p_1(x-1) + 2(p_2 - r_2)))\,.
\end{aligned}$$

In the following result, we construct infinitely many pair of signless Laplacian cospectral graphs using Corollary 7.

Corollary 9

(i) *Let G_1 be the join of two graphs H_1 and H_2. Let G_2 and G_3 be signless Laplacian cospectral graphs on n_2 vertices with $\Gamma_{Q(G_2)}(x) = \Gamma_{Q(G_3)}(x)$. Then, $G_1 \dot{\vee} G_2$ and $G_1 \dot{\vee} G_3$ are signless Laplacian cospectral.*

(ii) *Let G_1 be the join of two graphs H_1 and H_2 and let G_3 be the join of two graphs H_3 and H_4. If H_1 and H_3 are r_1-regular adjacency cospectral graphs on p_1 vertices, H_2 and H_4 are r_2-regular adjacency cospectral graphs on p_2 vertices, G_2 and G_4 are arbitary signless Laplacian cospectral graphs on n_2 vertices with $\Gamma_{Q(G_2)}(x) = \Gamma_{Q(G_4)}(x)$, then $G_1 \dot{\vee} G_2$ and $G_3 \dot{\vee} G_4$ are signless Laplacian cospectral.*

5 Conclusion

Constructing graphs by using the graph operations like subdivision-edge join, R-vertex join, R-edge join, etc., analogous to the graphs constructed in this paper, instead of using subdivision-vertex join and determining their various spectra are problems for further research.

Acknowledgements The first author is supported by INSPIRE Fellowship, Department of Science and Technology, Government of India under the grant no. DST/INSPIRE Fellowship/[IF160383] 2017.

References

1. Barik, S., Kalita, D., Pati, S., Sahoo, G.: Spectra of graphs resulting from various graph operations and products: a survey. Spec. Matrices **6**, 323–342 (2018)
2. Cvetković, D., Rowlinson, P., Simić, S.: An introduction to the theory of graph spectra. Cambridge University Press, Cambridge (2010)
3. Das, A., Panigrahi, P.: Spectra of R-vertex join and R-edge join of two graphs. Discuss. Math. Gen. Algebra Appl. **38**, 19–32 (2018)
4. Indulal, G.: Spectrum of two new joins of graphs and infinite families of integral graphs. Kragujevac J. Math. **36**, 133–139 (2018)
5. Rajkumar, R., Gayathri, M.: Spectra of (H_1, H_2)-merged subdivision graph of a graph. Indag. Math. **30**, 1061–1076 (2019)
6. Gayathri, M., Rajkumar, R.: Adjacency and Laplacian spectra of variants of neighborhood corona of graphs constrained by vertex subsets. Discrete Math. Algorithms Appl. **11**, Article No. 1950073 (2019)

7. Rajkumar, R., Pavithra, R.: Spectra of M-rooted product of graphs. Linear Multilinear Algebra (2020). https://doi.org/10.1080/03081087.2019.1709407
8. Varghese, R.P., Reji Kumar, K.: Spectra of new join of two graphs. Adv. Theor. Appl. Math. **11**, 459–470 (2016)
9. Tian, G.X., He, J.X., Cui, S.Y.: On the Laplacian spectra of some double join operations of graphs. Bull. Malays. Math. Sci. Soc. **42**, 1555–1566 (2019)
10. Liu, X., Zhang, Z.: Spectra of subdivision-vertex join and subdivision-edge join of two graphs. Bull. Malays. Math. Sci. Soc. **42**, 15–31 (2019)
11. McLeman, C., McNicholas, E.: Spectra of coronae. Linear Algebra Appl. **435**, 998–1007 (2011)
12. Cui, S.Y., Tian, G.X.: The spectrum and the signless Laplacian spectrum of coronae. Linear Algebra Appl. **437**, 1692–1703 (2012)

Genus and Book Thickness of the Inclusion Ideal Graph of a Ring

G. Gold Belsi and S. Kavitha

Abstract For a commutative ring R with unity, the *inclusion ideal graph* $\text{In}(R)$ is a graph whose vertices are all non-trivial ideals of R and two distinct ideals I and J are adjacent if and only if either $I \subset J$ or $J \subset I$. In this paper, we provide a formula for the size of $\text{In}(R)$, when R is a finite product of fields. Also we examine the genus and crosscap of $\text{In}(R)$ deliberately. Added with that, we analyse the book thickness of $\text{In}(R)$.

Keywords Inclusion ideal graph · Genus · Crosscap · Book thickness

1 Introduction

While looking back the previous decades, we can see the engrossing detections in algebraic graph theory, which helps us to solve algebraic problems using graph theory and vice versa. More specifically, a lot of discussions have been done in composing graphs from algebraic structures and studying their algebraic and graph theoretic properties. For example, one can view the graphs related with zero divisor, nilpotent, subgroup in [1, 5, 8].

Since ideals possess a decisive role in ring theory, Akbari et al. [2] initiated the study of inclusion ideal graph in 2014, which is defined as follows: For a ring R with unity, the inclusion ideal graph, denoted by $\text{In}(R)$, is a graph, whose vertex set consists of all non-trivial ideals of R and two distinct ideals I and J are adjacent in $\text{In}(R)$ if and only if either $I \subset J$ or $J \subset I$. They have studied the connectedness, clique number and chromatic number of $\text{In}(R)$. Also they are interested in learning the properties of $\text{In}(R)$ for non-commutative rings.

G. Gold Belsi (✉)
Department of Mathematics, Manonmaniam Sundaranar University, Tirunelveli 627012, Tamil Nadu, India

S. Kavitha
Department of Mathematics, Gobi Arts and Science College, Gobichettipalayam 638476, Tamil Nadu, India

P. Balasubramaniam et al. (eds.), *Mathematical Modelling and Computational Intelligence Techniques*, Springer Proceedings in Mathematics & Statistics 376,
https://doi.org/10.1007/978-981-16-6018-4_19

Subsequently, some of the basic and topological properties of In(R) was investigated by S. Kavitha and G. Gold Belsi, for a commutative ring [6]. Instigated by the above works, we are interested in investigating some other topological properties of In(R) for a commutative ring R with the finite inclusion ideal graph. Throughout this paper, we notate k_i as the number of distinct non-trivial ideals of a local ring R_i.

By a graph $G = (V, E)$, we intend an undirected simple graph with vertex set V and edge set E. A graph in which each pair of distinct vertices is joined by an edge is called a complete graph. We use K_n to denote the complete graph with n vertices. The complete bipartite graph with part sizes m and n is denoted by $K_{m,n}$. Also, for $B \subset V(G)$, $\langle B \rangle$ notates the subgraph of G induced by the vertices in B. For other definitions in graph theory, one can approach [4].

An orientable surface S_g is said to be of genus g if it is topologically homeomorphic to a sphere with g handles. A non-orientable surface N_k is said to be of crosscap k if it is topologically homeomorphic to a connected sum of k projective planes. The smallest positive integer g, for which, a graph G can be drawn without crossing on S_g, is called the genus of a graph. It is notated as, $g(G)$. Analogously, the crosscap of a graph G is the minimum positive integer k for which, G can be drawn without crossing on N_k. We designate it as $\overline{g}(G)$. A toroidal graph is a graph with genus one.

A standard n-book is established by combining n half-planes, called pages, at a common line, called spine. When embedding a graph in a book, the vertices are placed on the spine and each edge should be embedded on a single page of the book, so that no two edges intersect on a page. The *book thickness* of a graph G, notated by bt(G) is the least integer n, for which G has a n-book embedding. The pages in the book embedding are represented by the solid and dashed lines above and below the spine. To gain more knowledge about genus and book embedding, one can refer, [3, 7].

2 Preliminaries

In this section, we want to recall some theorems which are needed for the posterior sections.

Lemma 1 [7, Theorem 4.4.5] *If $n \geq 3$, then $g(K_n) = \left\lceil \frac{(n-3)(n-4)}{12} \right\rceil$*
If $n \geq 3$ and $n \neq 7$, then $\overline{g}(K_n) = \left\lceil \frac{(n-3)(n-4)}{6} \right\rceil$.

Lemma 2 [7, Theorem 4.4.7] *If $m, n \geq 2$, then $g(K_{m,n}) = \left\lceil \frac{(m-2)(n-2)}{4} \right\rceil$*
If $m, n \geq 3$, then $\overline{g}(K_{m,n}) = \left\lceil \frac{(m-2)(n-2)}{2} \right\rceil$.

Lemma 3 [7, Theorem 4.4.4] *Let G be a connected graph with $n \geq 3$ vertices and q edges. Then $g(G) \geq \left\lceil \frac{q}{6} - \frac{n}{2} + 1 \right\rceil$ and $\overline{g}(G) \geq \left\lceil \frac{q}{3} - n + 2 \right\rceil$.*

Theorem 1 [7, Theorem 4.4.2] *The genus of a connected graph is the sum of the genera of its blocks.*

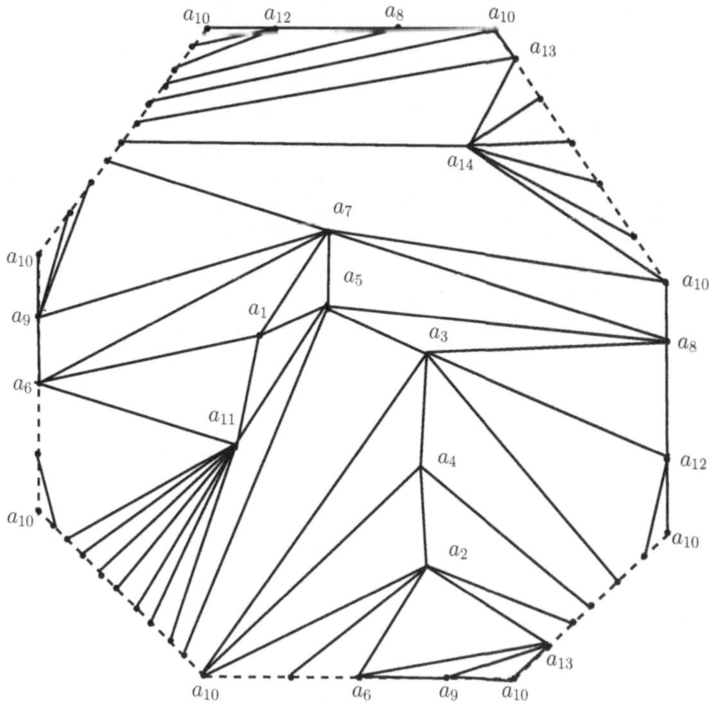

Fig. 1 An embedding of $\text{In}(R_1 \times F_1 \times F_2)$ with $k_1 = 2$ in S_2

In Fig. 1, a_i for $1 \leq i \leq 14$ denote $R_1 \times 0 \times 0, 0 \times F_1 \times 0, 0 \times 0 \times F_2, 0 \times F_1 \times F_2, R_1 \times 0 \times F_2, R_1 \times F_1 \times 0, \mathfrak{m}_1 \times 0 \times 0, \mathfrak{m}_1 \times 0 \times F_2, \mathfrak{m}_1 \times F_1 \times 0, \mathfrak{m}_1 \times F_1 \times F_1, I_1 \times 0 \times 0, a_{11}, I_1 \times 0 \times F_2, I_1 \times F_1 \times 0, I_1 \times F_1 \times F_2$, respectively.

Theorem 2 [3, Theorem 2.5] *A graph has book thickness one if and only if it is outer planar.*

Theorem 3 [3, Theorem 2.5] *The book thickness of a graph G is at most two if and only if G is a subgraph of a planar graph that has a Hamiltonian cycle.*

Theorem 4 [3, Theorem 3.4] *The book thickness of the complete graph K_n is equal to $\lceil n/2 \rceil$ when $n \geq 4$.*

3 Genus of In(R)

In this section, we give a formula for the size of In(R), when R is a product of fields. Also we find the genus of In(R).

Remark 1 Suppose R is a local ring with n distinct non-trivial ideals, then $\text{In}(R) \cong K_n$.

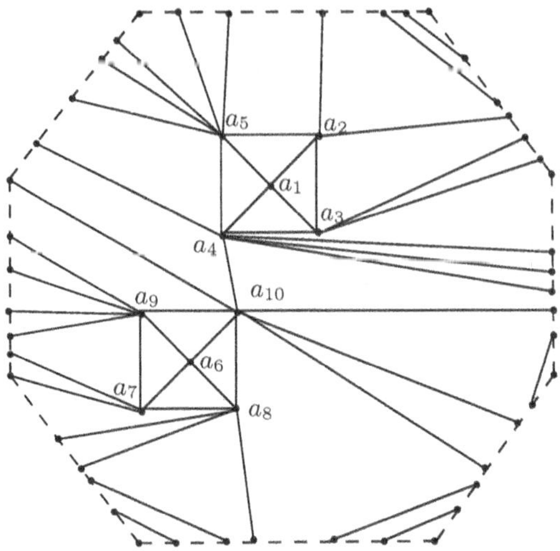

Fig. 2 An embedding of $\text{In}(R_1 \times F_1)$ with $k_1 = 4$ in S_2

In Fig. 2, a_i for $1 \leq i \leq 10$ denote $0 \times F_1, I_3 \times F_1, I_2 \times F_1, I_1 \times F_1, \mathfrak{m}_1 \times F_1, R_1 \times 0, \mathfrak{m}_1 \times 0, I_1 \times 0, I_2 \times 0, I_3 \times 0$, respectively.

Lemma 4 *For a positive integer $n \geq 2$, let $R = \prod_{i=1}^{n} F_i$ be a ring, where each F_i is a field. Then, $|E(In(R))| = \frac{1}{2}\sum_{r=1}^{n-1}\binom{n}{r}(2^{n-r} + 2^r - 2^2)$, where $r < n$.*

Proof In $\text{In}(R)$, we have $2^n - 2$ vertices. Let T_r denote the set of all ideals $I = \prod_{i=1}^{n} I_i$ in R, where $I_i = F_i$ in r places and $I_i = 0$ elsewhere. Hence, we have $\binom{n}{r}$ ideals in this collection. The degree of a vertex in T_r is the sum of the number of ways of fixing 0's in $1, 2, \ldots, (n - (r + 1))$th places and the sum of the number of ways of fixing F_i's in $1, 2, \ldots, r - 1$ places.
Therefore, the degree sum $= \sum_{s=1}^{r-1}\binom{r}{s} + \sum_{t=1}^{n-(r+1)}\binom{n-r}{t} = 2^{n-r} + 2^r - 2^2$.
The degree sum of the vertices in each T_r is $\binom{n}{r}(2^{n-r} + 2^r - 2^2)$, $1 \leq r \leq n - 1$. Since, the degree sum of the vertices in $\text{In}(R)$ is, the degree sum of the vertics in all T_r, $1 \leq r \leq n - 1$, we get that $2|E| = \sum_{r=1}^{n-1}\binom{n}{r}(2^{n-r} + 2^r - 2^2)$. Hence, the lemma is proved.

Theorem 5 *If $R = \prod_{i=1}^{n} R_i \times \prod_{j=1}^{m} F_j$, where each R_i is local with maximal ideal $\mathfrak{m}_i$ and F_j is a field and $n, m \geq 1$. Then $g(In(R)) = 2$ if and only if $n = 1, m = 2 = k_1$ or $m = 1 = n, k_1 = 4$.*

Proof Suppose $n \geq 2$ with $k_i = 1$ for all i.
Let $A = \{x_1, x_2, \ldots, x_{16}\}$ where $x_1 = R_1 \times 0 \times \ldots \times 0, x_2 = 0 \times R_2 \times 0 \times \ldots \times 0, x_3 = 0 \times 0 \times F_1 \times 0 \ldots \times 0, x_4 = 0 \times R_2 \times F_1 \times 0 \times \ldots \times 0, x_5 = R_1 \times 0 \times F_1 \times 0 \ldots \times 0, x_6 = R_1 \times R_2 \times 0 \times \ldots \times 0, x_7 = \mathfrak{m}_1 \times 0 \times \ldots \times 0, x_8 = 0 \times \mathfrak{m}_2 \times 0 \times \ldots \times 0, x_9 = \mathfrak{m}_1 \times 0 \times F_1 \times 0 \times \ldots \times 0, x_{10} = \mathfrak{m}_1 \times m_2 \times 0 \times \ldots \times 0,$

$x_{11} = 0 \times \mathfrak{m}_2 \times F_1 \times 0 \times \ldots \times 0, x_{12} = \mathfrak{m}_1 \times R_2 \times F_1 \times 0 \times \ldots \times 0, x_{13} = R_1 \times \mathfrak{m}_2 \times F_1 \times 0 \times \ldots \times 0, x_{14} = \mathfrak{m}_1 \times \mathfrak{m}_2 \times F_1 \times 0 \times \ldots \times 0, x_{15} = R_1 \times \mathfrak{m}_2 \times 0 \times \ldots \times 0, x_{16} = \mathfrak{m}_1 \times R_2 \times 0 \times \ldots \times 0$. Then, a subgraph of $\langle A \rangle$ in $\text{In}(R)$ has 16 vertices and 57 edges. Then, by Lemma 3, $g(\text{In}(R)) \geq 3$.

Next we consider the case $n = 1$.

Suppose $m \geq 3$ and $\mathfrak{m}_1$ is the only non-trivial ideal in R_1.

Let $B = \{x_1, x_2, \ldots, x_{22}\}$ where $x_1 = 0 \times F_1 \times F_2 \times F_3 \times 0 \times \ldots \times 0, x_2 = \mathfrak{m}_1 \times F_1 \times F_2 \times F_3 \times 0 \times \ldots \times 0, x_3 = R_1 \times 0 \times F_2 \times F_3 \times 0 \ldots \times 0, x_4 = R_1 \times F_1 \times 0 \times F_3 \times 0 \times \ldots \times 0, x_5 = R_1 \times F_1 \times F_2 \times 0 \ldots \times 0, x_6 = 0 \times F_1 \times 0 \times \ldots \times 0, x_7 = 0 \times 0 \times F_2 \times \ldots \times 0, x_8 = 0 \times 0 \times F_3 \times 0 \times \ldots \times 0, x_9 = R_1 \times 0 \times 0 \times \ldots \times 0, x_{10} = R_1 \times F_1 \times 0 \times \ldots \times 0, x_{11} = R_1 \times 0 \times 0 \times F_3 \times 0 \times \ldots \times 0, x_{12} = 0 \times 0 \times F_2 \times F_3 \times 0 \times \ldots \times 0, x_{13} = 0 \times F_1 \times 0 \times F_3 \times 0 \times \ldots \times 0, x_{14} = 0 \times F_1 \times F_2 \times 0 \times \ldots \times 0, x_{15} = R_1 \times 0 \times F_2 \times 0 \times \ldots \times 0, x_{16} = \mathfrak{m}_1 \times 0 \times 0 \times \ldots \times 0, x_{17} = \mathfrak{m}_1 \times 0 \times F_2 \times F_3 \times 0 \times \ldots \times 0, x_{18} = \mathfrak{m}_1 \times F_1 \times 0 \times F_3 \times 0 \times \ldots \times 0, x_{19} = \mathfrak{m}_1 \times F_1 \times F_2 \times 0 \times \ldots \times 0, x_{20} = \mathfrak{m}_1 \times 0 \times F_1 \times 0 \times \ldots \times 0, x_{21} = \mathfrak{m}_1 \times 0 \times 0 \times F_3 \times 0 \times \ldots \times 0, x_{22} = \mathfrak{m}_1 \times 0 \times F_2 \times 0 \times \ldots \times 0$. Then, a subgraph of $\langle B \rangle$ in $\text{In}(R)$ has 22 vertices and 93 edges. Then, by Lemma 3, $g(\text{In}(R)) \geq 6$.

Now assume that $m \leq 2$.

Case 1. Suppose $m = 2$. Then $R = R_1 \times F_1 \times F_2$.

Suppose R_1 has at least 3 distinct non-trivial ideals such that, $I_2 \subset I_1 \subset \mathfrak{m}_1$. Let $C = \{x_1, x_2, \ldots, x_{18}\}$ where $x_1 = R_1 \times 0 \times \ldots \times 0, x_2 = 0 \times F_1 \times 0 \times \ldots \times 0, x_3 = 0 \times 0 \times F_2 \times 0 \ldots \times 0, x_4 = 0 \times F_1 \times F_2 \times 0 \times \ldots \times 0, x_5 = R_1 \times 0 \times F_2 \times 0 \ldots \times 0, x_6 = R_1 \times F_1 \times 0 \times \ldots \times 0, x_7 = \mathfrak{m}_1 \times 0 \times \ldots \times 0, x_8 = \mathfrak{m}_1 \times 0 \times F_2 \times 0 \times \ldots \times 0, x_9 = \mathfrak{m}_1 \times F_1 \times 0 \times \ldots \times 0, x_{10} = \mathfrak{m}_1 \times F_1 \times F_2 \times 0 \times \ldots \times 0, x_{11} = I_1 \times 0 \times 0 \times \ldots \times 0, x_{12} = I_1 \times 0 \times F_2 \times 0 \times \ldots \times 0, x_{13} = I_1 \times F_1 \times 0 \times \ldots \times 0, x_{14} = I_1 \times F_1 \times F_2 \times 0 \times \ldots \times 0, x_{15} = I_2 \times 0 \times \ldots \times 0, x_{16} = I_2 \times 0 \times F_2 \times 0 \times \ldots \times 0, x_{17} = I_2 \times F_1 \times 0 \times \ldots \times 0, x_{18} = I_2 \times F_1 \times F_2 \times 0 \times \ldots \times 0$. Then $\langle C \rangle$ in $\text{In}(R)$ contains a subdivision of $K_{5,5}$. Then, by Lemma 2, $g(\text{In}(R)) \geq 3$.

Suppose R_1 has exactly two distinct non-trivial ideals such that, $I_1 \subset \mathfrak{m}_1$. Then, the vertex set of R is $D = \{x_1, x_2, \ldots, x_{14}\}$ where $x_1 = R_1 \times 0 \times 0, x_2 = 0 \times F_1 \times 0, x_3 = 0 \times 0 \times F_2, x_4 = 0 \times F_1 \times F_2, x_5 = R_1 \times 0 \times F_2, x_6 = R_1 \times F_1 \times 0, x_7 = \mathfrak{m}_1 \times 0 \times 0, x_8 = \mathfrak{m}_1 \times 0 \times F_2, x_9 = \mathfrak{m}_1 \times F_1 \times 0, x_{10} = \mathfrak{m}_1 \times F_1 \times F_2, x_{11} = I_1 \times 0 \times 0, x_{12} = I_1 \times 0 \times F_2, x_{13} = I_1 \times F_1 \times 0, x_{14} = I_1 \times F_1 \times F_2$. Then, $\langle D \rangle$ in $\text{In}(R)$ has 14 vertices and 45 edges and by Lemma 3, $g(\text{In}(R)) \geq 2$. By Fig. 1, $g(\text{In}(R)) = 2$.

Case 2. Suppose $m = 1$. Then, $R = R_1 \times F_1$

Suppose $k_1 \geq 5$. Let the chain of ideals be $I_n \subset \ldots \subset I_4 \subset I_3 \subset I_2 \subset I_1 \subset \mathfrak{m}_1$. Let $G = \{x_1, x_2, \ldots, x_{12}\}$ where $x_1 = R_1 \times 0 \times \ldots \times 0, x_2 = \mathfrak{m}_1 \times 0 \times \ldots \times 0, x_3 = 0 \times F_1 \times 0 \times \ldots \times 0, x_4 = \mathfrak{m}_1 \times F_1 \times 0 \times \ldots \times 0, x_j = I_i \times 0 \times 0 \ldots \times 0, x_k = I_i \times F_1 \times 0 \times \ldots \times 0, 5 \leq j \leq 8, 9 \leq k \leq 12, 1 \leq i \leq 4$.

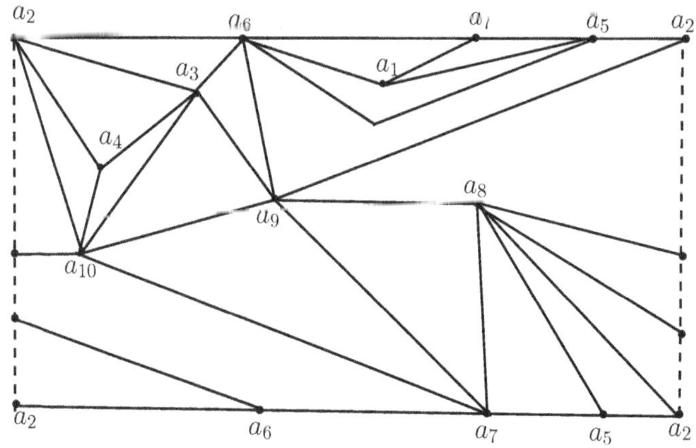

Fig. 3 An embedding of $\text{In}(R_1 \times R_2)$ with $k_1 = 2, k_2 = 1$ in S_1

In Fig. 3, a_i for $1 \leq i \leq 10$ denote $0 \times R_2$, $I_1 \times 0$, $\mathfrak{m}_1 \times 0$, $R_1 \times 0$, $I_1 \times R_2$, $\mathfrak{m}_1 \times R_2$, $0 \times \mathfrak{m}_2$, $I_1 \times \mathfrak{m}_2$, $\mathfrak{m}_1 \times \mathfrak{m}_2$, $R_1 \times \mathfrak{m}_2$, respectively.
Then, a subgraph of $\langle G \rangle$ in $\text{In}(R)$ has 12 vertices and 45 edges and by Lemma 3, $g(\text{In}(R)) \geq 3$.
Suppose $k_1 = 4$.
Let $I_3 \subset I_2 \subset I_1 \subset \mathfrak{m}_1$ be the distinct non-trivial ideals of R_1. Let $H = \{x_1, x_2, \ldots, x_{10}\}$ where $x_1 = R_1 \times 0, x_2 = \mathfrak{m}_1 \times 0, x_3 = 0 \times F_1, x_4 = \mathfrak{m}_1 \times F_1$, $x_j = I_i \times 0, x_k = I_i \times F_1, 5 \leq j \leq 7, 8 \leq k \leq 10, 1 \leq i \leq 3$. Then, $\langle H \rangle$ in $\text{In}(R)$ has two blocks of K_5. Hence, by Lemma 1 and Theorem 1, $g(\text{In}(R)) \geq 2$. By Fig. 2, $g(\text{In}(R)) = 2$.

Theorem 6 *If R is a product of n local rings (but nor a field), then*

(i) $g(In(R)) = 1$ *if and only if* $n = k_1 = 2, k_2 = 1$ *or* $n = 1, k_1 = 5, 6$ *or* 7.
(ii) $g(In(R)) = 2$ *if and only if* $n = 1, k_1 = 8$.

Proof Assume $n \geq 3$ and $\mathfrak{m}_i$'s are the only non-trivial ideals in R_i.
Let $J = \{x_1, x_2, \ldots, x_{25}\}$ where $x_1 = 0 \times R_2 \times R_3 \times 0 \times \ldots \times 0, x_2 = 0 \times \mathfrak{m}_2 \times R_3 \times 0 \times \ldots \times 0, x_3 = 0 \times R_2 \times \mathfrak{m}_3 \times 0 \ldots \times 0, x_4 = 0 \times \mathfrak{m}_2 \times \mathfrak{m}_3 \times 0 \times \ldots \times 0, x_5 = R_1 \times 0 \times R_3 \times 0 \ldots \times 0, x_6 = \mathfrak{m}_1 \times 0 \times R_3 \times 0 \times \ldots \times 0, x_7 = R_1 \times 0 \times \mathfrak{m}_3 \times \ldots \times 0, x_8 = \mathfrak{m}_1 \times 0 \times \mathfrak{m}_3 \times 0 \times \ldots \times 0, x_9 = R_1 \times R_2 \times 0 \times \ldots \times 0, x_{10} = \mathfrak{m}_1 \times R_2 \times 0 \times \ldots \times 0, x_{11} = R_1 \times \mathfrak{m}_2 \times 0 \times \ldots \times 0, x_{12}=\mathfrak{m}_1 \times \mathfrak{m}_2 \times 0 \times \ldots \times 0, x_{13} = \mathfrak{m}_1 \times \mathfrak{m}_2 \times \mathfrak{m}_3 \times 0 \times \ldots \times 0, x_{14} = \mathfrak{m}_1 \times R_2 \times R_3 \times 0 \times \ldots \times 0, x_{15} = R_1 \times \mathfrak{m}_2 \times R_3 \times 0 \times \ldots \times 0, x_{16} = R_1 \times R_2 \times \mathfrak{m}_3 \times 0 \times \ldots \times 0, x_{17} = \mathfrak{m}_1 \times 0 \times 0 \times \ldots \times 0, x_{18} = 0 \times \mathfrak{m}_2 \times 0 \times \ldots \times 0, x_{19} = 0 \times 0 \times \mathfrak{m}_3 \times 0 \times \ldots \times 0, x_{20} = 0 \times R_2 \times 0 \times \ldots \times 0, x_{21} = R_1 \times 0 \times 0 \times \ldots \times 0, x_{22} = 0 \times 0 \times R_3 \times 0 \times \ldots \times 0, x_{23} = \mathfrak{m}_1 \times \mathfrak{m}_2 \times R_3 \times 0 \times \ldots \times 0, x_{24} = R_1 \times \mathfrak{m}_2 \times \mathfrak{m}_3 \times 0 \times \ldots \times 0, x_{25} = \mathfrak{m}_1 \times R_2 \times \mathfrak{m}_3 \times 0 \times \ldots \times 0$. Then, $\langle J \rangle$ in $\text{In}(R)$ contains $K_{6,7}$. Then, by Lemma 2, $g(\text{In}(R)) \geq 5$.

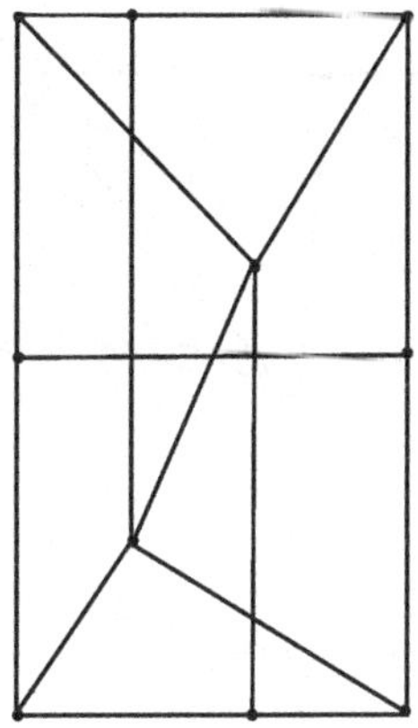

Fig. 4 Minimal forbidden subgraph for projective plane

Next we consider the case $n = 2$.
Case 1. Suppose R_1 has at least 3 distinct non-trivial ideals such that, $I_2 \subset I_1 \subset \mathfrak{m}_1$, then let $M = \{A \times B\} \setminus \{0 \times 0, R_1 \times R_2\}$, where $A = \{0, I_1, I_2, \mathfrak{m}_1, R_1\}$ and $B = \{0, \mathfrak{m}_2, R_2\}$. Then, $\langle M \rangle$ in $\text{In}(R)$ has 13 vertices and 53 edges. Then, by Lemma 3, $g(\text{In}(R)) \geq 4$.
Case 2. Suppose R_1 and R_2 have exactly two distinct non-trivial ideals such that, $I_1 \subset \mathfrak{m}_1$ and $I_2 \subset \mathfrak{m}_2$. Then, let $N = \{C \times D\} \setminus \{0 \times 0, R_1 \times R_2\}$, where $C = \{0, I_1, \mathfrak{m}_1, R_1\}$ and $D = \{0, I_2, \mathfrak{m}_2, R_2\}$. Then, $\langle N \rangle$ in $\text{In}(R)$ has 14 vertices and 55 edges. Then, by Lemma 3, $g(\text{In}(R)) \geq 4$.
Case 3. Suppose R_1 has exactly two distinct non-trivial ideal such that, $I_1 \subset \mathfrak{m}_1$. Then, let $O = \{G \times H\} \setminus \{0 \times 0, R_1 \times R_2\}$, where $G = \{0, I_1, \mathfrak{m}_1, R_1\}$ and $H = \{0, \mathfrak{m}_2, R_2\}$. Then, $\langle O \rangle$ in $\text{In}(R)$ has 10 vertices and 27 edges. Then, by Lemma 3, $g(\text{In}(R)) \geq 1$. By Fig. 3, $g(\text{In}(R)) = 1$. When $n = 1$, the proof is clear by Remark 1 and Theorem 1.

Theorem 7 *Let $R = \prod_{i=1}^{n} F_i$, where each F_i is a field. If $g(In(R)) \leq 2$, then $n \leq 4$.*

Proof By Lemmas 3 and 4, we get that $g(\text{In}(R)) \geq 9$ for $n \geq 5$.

4 Crosscap of In(*R*)

In this section, we investigate the crosscap of $\text{In}(R)$ and prove that only one type of ring can be embeddable in N_2.

Theorem 8 *Let R be a commutative ring with identity with finite inclusion ideal graph. Then*

(i) $\overline{g}(In(R)) = 1$ *if and only if R is local with five or six distinct non-trivial ideals.*
(ii) $\overline{g}(In(R)) = 2$ *if and only if* $R = R_1 \times R_2$ *with* $k_1 = 2$ *and* $k_2 = 1$.

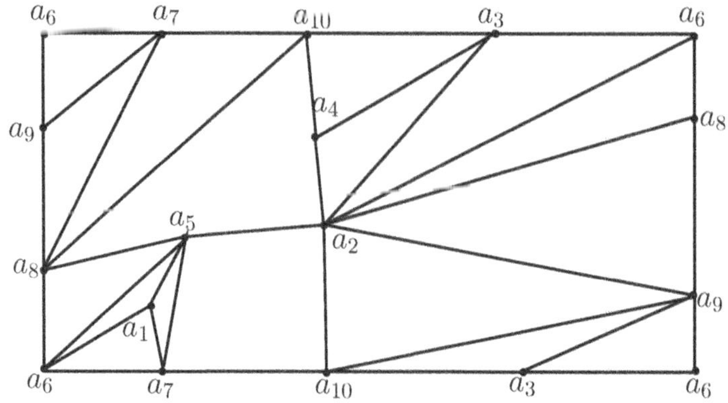

Fig. 5 An embedding of $\text{In}(R_1 \times R_2)$ with $k_1 = 2$ and $k_2 = 1$ in N_2. The $a_i's$ are the same as in Fig. 3

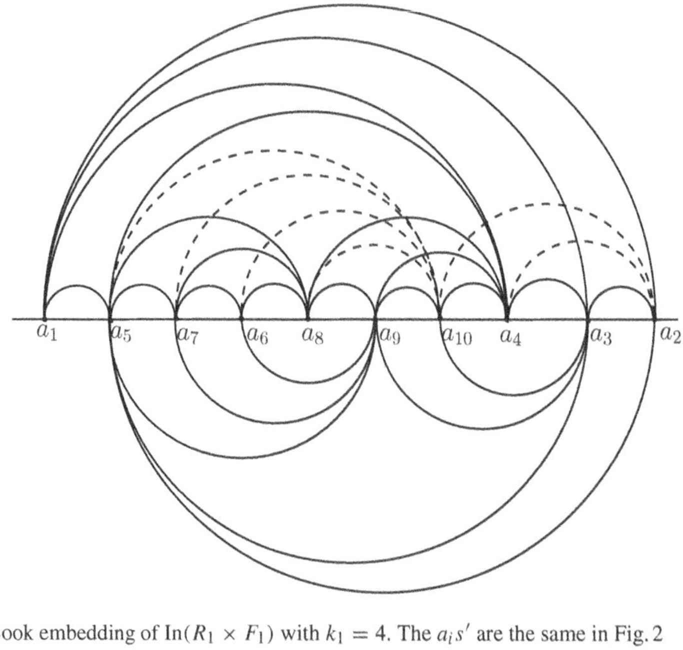

Fig. 6 Book embedding of $\text{In}(R_1 \times F_1)$ with $k_1 = 4$. The a_is' are the same in Fig. 2

Proof Using Remark 1 and Theorem 1, (*i*) is clear. Applying Lemmas 2 and 3 in Theorem 5, we get that $\overline{g}(\text{In}(R)) \geq 3$ in all cases. Suppose R is a product of fields, then, Lemmas 3 and 4 yield, $\overline{g}(\text{In}(R)) \geq 3$. Also applying Lemmas 2 and 3 in cases 1 and 2 of Theorem 6, we get that $\overline{g}(\text{In}(R)) \geq 3$. Now examining case 3 ($n = 2$ $k_1 = 2, k_2 = 1$), we obtain a minimal forbidden subgraph of $\text{In}(R)$ as in Fig. 4. Hence, $\overline{g}(\text{In}(R)) \geq 2$. Now by Fig. 5, $\overline{g}(\text{In}(R)) = 2$.

5 Book Thickness of In(R)

In this section we determine the book thickness of the inclusion ideal graph that has genus atmost two, except for the ring $R = R_1 \times F_1 \times F_2$ with $k_1 = 1, k_2 = 2$. For that, first we provide the list of rings that has planar inclusion ideal graph.

Theorem 9 [6, Theorem 3.2, 3.3, Remark 1] *Let R be a commutative ring. Then, In(R) is outerplanar if and only if R is any one of the following:*

(i) $R = F_1 \times F_2$ *or* $R = F_1 \times F_2 \times F_3$.
(ii) $R = R_1 \times F_1, k_1 \leq 2$ *where each* $(R_i, \mathfrak{m}_i)$ *is local and each* F_i *is a field.*
(iii) R *is local with atmost* 3 *distinct non-trivial ideals.*

Theorem 10 [6, Theorems 4.1–4.3] *Let R be a commutative ring. Then, In(R) is planar if and only if R is any one of the following:*

(i) $R = F_1 \times F_2 \times F_3 \times F_4$.
(ii) $R = R_1 \times F_1 \times F_2$ *with* $k_1 = 1$ *or* $R = R_1 \times F_1$ *with* $k_1 = 3$.
(iii) $R = R_1 \times R_2$ *with* $k_1 = k_2 = 1$ *or R is local with* 4 *distinct non-trivial ideals, where each* $(R_i, \mathfrak{m}_i)$ *is a local ring,* F_j *is a field.*

Theorem 11 *For a commutative ring R,*

(i) $bt(In(R)) = 1$ *if and only if R is any one of the following:*
$R = F_1 \times F_2$, $F_1 \times F_2 \times F_3$ *or* $R_1 \times F_1, k_1 \leq 2$ *or R is local with atmost* 3 *distinct non-trivial ideals.*
(ii) $bt(In(R)) = 2$ *if and only if R is any one of the following:*
$R = F_1 \times F_2 \times F_3 \times F_4$, $R_1 \times F_1 \times F_2$ *with* $k_1 = 1$, $R_1 \times F_1$ *with* $k_1 = 3$, $R = R_1 \times R_2$ *with* $k_1 = k_2 = 1$ *or R is local with* 4 *distinct non-trivial ideals.*
(iii) $bt(In(R)) = 3$ *if and only if R is any one of the following:*
$R = R_1 \times R_2$ *with* $k_1 = k_2 = 1$, $R_1 \times F_1$ *with* $k_1 = 4$ *or R is local with* 5 *or* 6 *distinct non-trivial ideals, where each* $(R_i, \mathfrak{m}_i)$ *is a local ring,* F_j *is a field.*
(iv) $bt(In(R)) = 4$ *if and only if R is local with* 7 *or* 8 *distinct non-trivial ideals.*

Proof Since the inclusion ideal graph of the rings in (i) and (ii) produce outer planar and planar graphs, by Theorems 2 and 3, they receive book thickness 1 and 2, respectively. We know that, a two page book embedding corresponds to a planar embedding. Hence, the inclusion ideal graph of the rings in (iii) have book thickness at least three. By Figs. 6, 7 and Theorem 4, we obtain that the book thickness of the respective graphs is 3. Suppose R is local with 7 or 8 distinct non-trivial ideals, using Theorem 4 and Remark 1, their book thickness is 4.

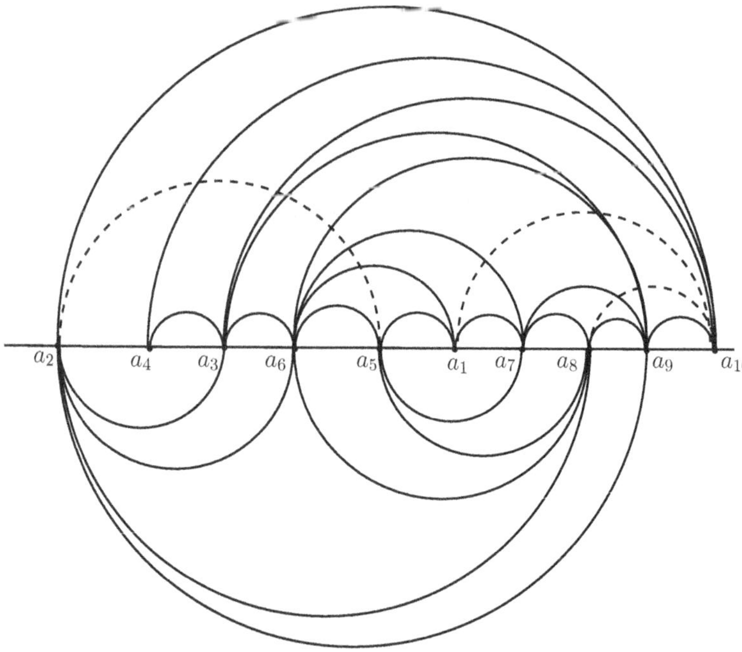

Fig. 7 Book embedding of $\text{In}(R_1 \times R_2)$ with $k_1 = 1, k_2 = 1$. The $a_i s'$ are the same in Fig. 3

6 Conclusion

Topological graph theory is a captivating area which deals with embedding of graphs in topological structures such as sphere and book. Impressed by the researches in this area, in this paper, we have investigated the genus, crosscap and book thickness of the inclusion ideal graph of a commutative ring.

Acknowledgements The support for this research work is provided by MANF program (201718-MANF- 2017 -18- TAM- 82372) of University Grants Commission, Government of India for the first author. Also the authors are grateful to the referees for their valuable suggestions.

References

1. Ahmadi, H., Taeri, B.: A graph related to join of subgroups of a finite group. Rend. Sem. Mat. Univ. Padov. **131**, 281–292 (2014). https://doi.org/10.4171/RSMUP/131-17
2. Akbari, S., Habibi, M., Majidinya, A., Manaviyat, R.: The inclusion ideal graph of rings. Commun. Algebra **43**, 2457–2465 (2015). https://doi.org/10.1080/00927872.2014.894051
3. Bernhart, F.R., Kainen, P.C.: The book thickness of a graph. J. Comb. Theory Ser. B **27**, 320–331 (1979). https://doi.org/10.1016/0095-8956(79)90021-2

4. Bondy, J.A., Murty, U.S.R.: Graph Theory and Its Applications. American Elsevier, New York (1976)
5. Kala, R., Kavitha, S.: On nilpotent graph of genus one. Discrete Math. Algorithms Appl. **63**, 1–10 (2014). https://doi.org/10.1142/S1793830914500372
6. Kavitha, S., Gold Belsi, G.: On the planarity of the inclusion ideal graph of a commutative ring. Malay. J. Math. S, 22–26 (2020). https://doi.org/10.26637/MJM0S20/005
7. Mohar, B., Thomassen, C.: Graphs on Surfaces. The Johns Hopkins University Press, Baltimore and London (1956)
8. Redmond, S.P.: On zero-divisor graphs of small finite commutative rings. Discrete Math. **307**, 1155–1166 (2007). https://doi.org/10.1016/j.disc.2006.07.025

Inventory Control

An EOQ Inventory Model with Shortage Backorders and Incorporating a Learning Function in Fuzzy Parameters

S. Ganesan and R. Uthayakumar

Abstract This paper analyzes the impact of Wright's learning function in an economic order quantity (EOQ) model with backorders and fuzzy parameters. All the input parameters are fuzzified. Wright's learning function is incorporated in all fuzzy parameters to provide a generalized treatment. All the previous similar research works used iterative one-dimensional search algorithms to find near-optimal solutions. We use an approximate value of generalized harmonic numbers and the calculus method as an alternate for deriving optimal inventory policies. We provide numerical examples for crisp, fuzzy, and fuzzy learning models. The models are compared based on the learning rates. An exponential regression equation is fitted to determine the relationship between learning rates and total cost. The results show that the fuzzification of input parameters has impact on the optimal values, whereas the fuzzification of the decision variables does not affect the optimal values. We also find that the fuzzy learning model converges with the crisp model when the learning rate approaches 1.

Keywords EOQ inventory · Shortage · Backorder · Learning · Trapezoidal fuzzy numbers · Generalized harmonic numbers

1 Introduction

An inventory control problem aims to design, operate, and control an inventory system. The problem is studied extensively ever since from the origin of Harris [16] model. There are three broad categories of inventory control problems based on the pattern of parameters involved in the system. They are deterministic, probabilistic, and fuzzy inventory models. Researchers made many alterations in the parameters that suit their requirements to create mathematical models that conform to real-world

S. Ganesan (✉)
Department of Mathematics, Government Polytechnic College, Usilampatti, Madurai District, Tamil Nadu, India

R. Uthayakumar
Department of Mathematics, The Gandhigram Rural Institute (Deemed to be University), Gandhigram, Tamil Nadu, India

P. Balasubramaniam et al. (eds.), *Mathematical Modelling and Computational Intelligence Techniques*, Springer Proceedings in Mathematics & Statistics 376,
https://doi.org/10.1007/978-981-16-6018-4_20

inventories more closely. Traditional methods for handling uncertainties in inventory control mainly depended on probability models. The other way is the application of fuzzy set theory. We find one of the early traces of an application of fuzzy logic in inventory decision in Kakprzyk and Staniewski [19]. Park [31] devised a fuzzy version of the inventory model of Harris [16] with fuzzy cost parameters. An inventory item's order quantity depends on the demand, inventory level, warehouse space, and various related costs. Petrovic and Sweeney [33] mentioned the following characterizations of most of the parameters, which encourages the use of fuzzy number representation to them. (1) Values of parameters and their relations are uncertain and imprecise. Parameter estimation may depend on the subjective beliefs of the inventory operator. (2) It may not be possible to measure the parameters either because there is no measurement unit or no quantitative criterion for representing their values. (3) The knowledge available about their values and relations is incomplete. (4) Some of them are vaguely and not clearly defined. Learning is a process that results in change or improvement in a person's knowledge or behavior due to experience. The personal experiences of the inventory manager and repetitive decision-making process have an impact on the inventory policies. Learning functions are the best way to represent performance improvement. The study of learning functions was first initiated by Wright [40] on studying costs in the aircraft industry. A learning function is a relationship between production costs and the collective production output over the whole period of production activity [34]. Learning functions are developed on the principle that learning by doing encourages improvement in the process. Learning functions contribute a prime role in management policies because the growing knowledge of costs and demands in a business firm ultimately reduces unexpected loss and increases profit due to forecasting [8]. Frequency plays a significant role in the learning or performance of a task, as a decision-maker's experience inculcates masterly in the way of implementation [1]. Deciding the value of an imprecise parameter depends on the experience and knowledge of the inventory manager. An inventory manager's learning capabilities reduce the impreciseness and help predict a better estimate of parameters. Thus, learning functions have a significant role to play in fuzzy inventory models.

A considerable amount of research has been carried out in EOQ, EPQ, and supply chain (SC) models recently. Ganesan and Uthayakumar [9] developed two EPQ models for an imperfect manufacturing system with warm-up production run and hybrid maintenance periods. Malleeswaran and Uthayakumar [29] developed an SC model with price discount for backorders and price-sensitive demand. Latha et al. [27] presented a two-level production–distribution EOQ model with lead time reduction and discount on backorder price. Sundararajan et al. [38] constructed and analyzed an EOQ inventory model for non-instantaneous deteriorating items by considering the impacts of inflation on a multivariate demand function. Tharani and Uthayakumar [39] constructed a production inventory model with a remanufacturing setup and stock-dependent production and demand rates. Karthick and Uthayakumar [20] developed a sustainable EPQ model with trapezoidal fuzzy demand. Ganesan and Uthayakumar [10] developed an EPQ model with learning-dependent demand and production rates. Several inventory models considered the representation of demand

and cost components using fuzzy numbers. Bjork [3] and Kazemi et al. [21] are two different fuzzy versions of the same EOQ model with backorders which is the first extension of the model of Harris [16]. Li et al. [28] recently examined supply contract problem using triangular and trapezoidal random fuzzy variables to model the uncertain parameters. De and Mahata [7] used cloudy fuzzy numbers to represent imprecise demand rate and proved the total cost of the fuzzy inventory system merge with crisp inventory system over a long run time. Hemalatha and Annadurai [17] developed a fuzzy inventory model with advance payments. Shekarian et al. [36] provide a comprehensive literature survey on fuzzy inventory models. We can see in an organization that learning may happen at every level with different learning rates at different levels. Kim et al. [26] approximated multiple-level learning and forgetting in an organization by an aggregated single-level model. Gosling et al. [14] conducted a study on the managerial activities related to encouragement of learning in an organization and analyzed the empirical data using the time constant learning function. Goyal et al. [15] developed a fuzzy model for an imperfect lot with partial backordering, where the demand is assumed as fuzzy and logistic learning function is used to reduce the number of defective items. Jaber [18] provides a category of learning functions in an industrial setup. Glock et al. [11] are a systematic literature review of applications of learning functions in production and operations management.

Bera et al. [2], and Pal et al. [30] initiated the study of learning effects in fuzzy parameters in inventory models. They studied fuzzy EPQ models. They incorporated learning in setup cost and production cost, which are fuzzy parameters. Glock et al. [13] is one of the earliest works in which a classical EOQ model is modified in terms of the number of orders in a planning period to apply learning effect in the fuzzy parameter. Glock and Jaber [12] investigated the impact of learning in a multistage production inventory model with rework and scrap. Kazemi et al. [22] developed a fuzzy EOQ model for imperfect quality items with fuzzy parameters and studied the effect of learning impacts. Kazemi et al. [25] incorporated human learning in the model of Bjork [3]. Kazemi et al. [23] carried out an empirical study to support the learning effects. Kazemi et al. [24] constructed a model with two-stage composite learning consisting of cognitive and motor capabilities. Sonia et al. [37] discussed demand uncertainty and learning in fuzziness in a continuous review inventory model. All the ten models mentioned above used Wright's learning function models. Shekerian et al. [35] developed a fuzzy EOQ model with two different holding costs, viz. holding cost for perfect items and holding cost for defective items. Patro et al. [32] considered a fuzzy EOQ model for deteriorating items with imperfect quality using proportionate discount under learning effects. The latter two models studied the impact of learning effects using the S-shaped logistic learning function. All the works mentioned above used fuzzy triangular numbers to represent fuzzy parameters.

In all the models of Kazemi et al. [23–25], there are two decision variables, namely the number of orders and maximum inventory level. They could not calculate the optimal number of orders analytically due to the cost function's complexity and adopted one-dimensional search algorithms. In the present paper, we fuzzify only the input parameters and keep the decision variables as crisp values. We substitute an approximate value of generalized harmonic numbers to some parts of the annual total cost function and enable it differentiable with respect to the decision variables. Then, the simultaneous equations obtained from the first-order partial derivatives are solved for numerical values. This approach facilitates more accurate optimal values for the decision variables. The remaining parts of this paper are structured in the following way: In Sect. 2, we provide some basic concepts about fuzzy numbers and learning functions. The fuzzy version of a classical inventory model with backorders is formulated in Sect. 3. In Sect. 4, we present the fuzzy learning model and derive optimal solutions. The model is complemented with some numerical examples, which are presented in Sect. 5. The observations of sensitivity analysis are presented in Sect. 6. Some conclusions are presented in Sect. 7.

2 Review of Basic Concepts

2.1 Fuzzy Numbers

Let X denotes the universe of discourse and denotes its elements by x. A fuzzy set $\breve{A}$ in X is an ordered pair $(X, \mu_{\breve{A}})$ where $\mu_{\breve{A}}$ is a function defined from X to the unit interval $[0, 1]$. Hence, $\breve{A} = \{(x, \mu_{\breve{A}}(x)) : x \in X\}$, and we call $\mu_{\breve{A}}$ as the membership function of the fuzzy set $\breve{A}$ and $\mu_{\breve{A}}(x)$ is known as the membership grade of x in $\breve{A}$. The maximum value of $\mu_{\breve{A}}(x)$ is called the height of the fuzzy set. If the height of a fuzzy set $\breve{A}$ is 1, then the fuzzy set is called a normalized fuzzy set. Assume that the universal set X is defined in k-dimensional Euclidean vector space $\mathbb{R}^k$, $k \geq 1$. If the relation $\mu_{\breve{A}}(x) \geq \min\{\mu_{\breve{A}}(y), \mu_{\breve{A}}(z)\}$ where $x = \lambda y + (1 - \lambda)z;\ y, z \in \mathbb{R}^k;\ \lambda \in [0, 1]$ holds then $\breve{A}$ is called a convex fuzzy set. A convex and normalized fuzzy set $\breve{A}$ whose membership function is defined in $\mathbb{R}$ and piecewise continuous is called a fuzzy number. The fuzzy number $\breve{A}$ is said to be a trapezoidal fuzzy number if it is fully determined by (a, b, c, d) of crisp numbers where $a < b < c < d$ and its membership function is given by

$$\mu_{\breve{A}}(x) = \begin{cases} l(x) = \frac{x-a}{b-a} & \text{when} \quad a \leq x \leq b \\ 1 & \text{when} \quad b \leq x \leq c \\ r(x) = \frac{x-d}{c-d} & \text{when} \quad c \leq x \leq d \\ 0 & \text{otherwise} \end{cases} \tag{1}$$

2.2 Arithmetic Operations on Fuzzy Numbers

Arithmetic operations on trapezoidal fuzzy numbers were developed by Chen [4] using function principle. Let $\breve{A} = (x_1, x_2, x_3, x_4)$ and $\breve{B} = (y_1, y_2, y_3, y_4)$ be two trapezoidal fuzzy numbers. The arithmetic operations on trapezoidal fuzzy numbers are defined by the following rules.

1. Addition $\oplus$:

$$\breve{A} \oplus \breve{B} = (x_1, x_2, x_3, x_4) \oplus (y_1, y_2, y_3, y_4) = (x_1 + y_1, x_2 + y_2, x_3 + y_3, x_4 + y_4)$$

   where $+$ represents usual real number addition.
2. Subtraction $\ominus$:

$$\breve{A} \ominus \breve{B} = (x_1, x_2, x_3, x_4) \ominus (y_1, y_2, y_3, y_4) = (x_1 - y_4, x_2 - y_3, x_3 - y_2, x_4 - y_1)$$

   where $-$ represents usual real number subtraction.
3. Multiplication by a scalar:
   If $k \geq 0$, then $k\breve{A} = (kx_1, kx_2, kx_3, kx_4)$, and if $k < 0$, then $k\breve{A} = (kx_4, kx_3, kx_2, kx_1)$.
4. Multiplication $\otimes$:
   $\breve{A} \otimes \breve{B} = (x_1, x_2, x_3, x_4) \otimes (y_1, y_2, y_3, y_4) = (z_1, z_2, z_3, z_4)$ where $z_1 = \min\{x_1 \times y_1, x_1 \times y_4, x_4 \times y_1, x_4 \times y_4\}$, $z_2 = \min\{x_2 \times y_2, x_2 \times y_3, x_3 \times y_2, x_3 \times y_3\}$, $z_3 = \max\{x_2 \times y_2, x_2 \times y_3, x_3 \times y_2, x_3 \times y_3\}$, $z_4 = \max\{x_1 \times y_1, x_1 \times y_4, x_4 \times y_1, x_4 \times y_4\}$ and $\times$ represents usual real number multiplication. We note that if $x_1, x_2, x_3, x_4, y_1, y_2, y_3, y_4$ are positive real numbers then $\breve{A} \otimes \breve{B} = (x_1 \times y_1, x_2 \times y_2, x_3 \times y_3, x_4 \times y_4)$, $(-\breve{A}) \otimes \breve{B} = \breve{A} \otimes (-\breve{B}) = -(\breve{A} \otimes \breve{B})$ and $(-\breve{A}) \otimes (-\breve{B}) = \breve{A} \otimes \breve{B}$.
5. Multiplicative inverse:
   If y_1, y_2, y_3, y_4 are positive real numbers, the inverse of the fuzzy numbers $\breve{B} = (y_1, y_2, y_3, y_4)$ is given by $\breve{B}^{-1} = \left(\frac{1}{y_4}, \frac{1}{y_3}, \frac{1}{y_2}, \frac{1}{y_1}\right)$.
6. Division $\oslash$:
   If $x_1, x_2, x_3, x_4, y_1, y_2, y_3, y_4$ are positive real numbers then $\breve{A} \oslash \breve{B} = \breve{A} \otimes \breve{B}^{-1} = \left(\frac{x_1}{y_4}, \frac{x_2}{y_3}, \frac{x_3}{y_2}, \frac{x_4}{y_1}\right)$ where / represents usual real number division.

2.3 *Defuzzification of Fuzzy Numbers*

The need for defuzzification arises in every system that uses fuzzy control and producing a crisp deterministic interpretation. Here, we employ the graded mean integration representation (GMI) method developed by Chen and Hseih [5] for the defuzzification of the trapezoidal fuzzy numbers. Let l^{-1} and r^{-1} be the inverse function of l and r respectively (refer the definition of a trapezoidal fuzzy number in Sect. 2.1). The h-level graded mean value of a trapezoidal fuzzy number $\breve{A} = (a, b, c, d)$ is defined as $\frac{h}{2}[l^{-1}(h) + r^{-1}(h)]$. The GMI representation GMI($\breve{A}$) of $\breve{A}$ is calculated by the following formula.

$$\mathrm{GMI}(\breve{A}) = \frac{\int_0^1 \frac{h}{2}[l^{-1}(h) + r^{-1}(h)]\, dh}{\int_0^1 h\, dh} \tag{2}$$

The explicit $\mathrm{GMI}(\breve{A})$ formula is given below.

$$\mathrm{GMI}(\breve{A}) = \frac{1}{6}(a + 2b + 2c + d) \tag{3}$$

2.4 Wright's Learning Function

The learning function model of Wright [40] is described by the equation

$$y_x = y_0 x^{-m} \tag{4}$$

where m is the learning rate, x is the cumulative number of times a task is performed, y_x is the value of the parameter y at the xth cumulative performance, and y_0 is the value of the parameter y at the first performance. The learning rate assumes values between 0 and 1, where 0 corresponds to the lowest learning, and 1 corresponds to the highest learning. In the present paper, trapezoidal fuzzy numbers' fuzziness limits (refer to Sect. 4) are taken as parameters. An increment in the cumulative number of performances and learning rate reduces the parameter values and reduces fuzziness in the system.

3 Formulation of Fuzzy Model and Its Solution Procedure

The model we consider in this work is one of the earliest extensions of the basic EOQ model of Harris [16] to include backorders. We take this model in the form studied by Kazemi et al. [25]. The following notations are used in the development of the crisp model.

Notation	Description
D	The demand for the product per year
C_o	The ordering cost per order
C_h	The holding cost per unit item per year
C_p	The shortage penalty cost per unit item per year
Q	The order quantity per order (decision variable)
I_m	The maximum inventory level (decision variable)
T	The annual total cost

The inventory behavior of the model is shown in Fig. 1. The total cost function is given by

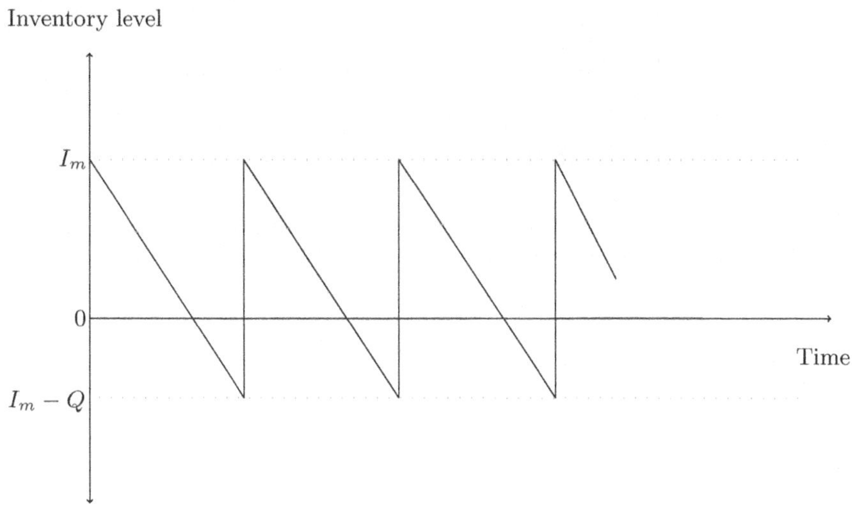

Fig. 1 Inventory behavior of the model

$$T = \frac{C_o D}{Q} + \frac{I_m^2 C_h}{2Q} + \frac{(Q - I_m)^2 C_p}{2Q} \tag{5}$$

The optimal values of Q and I_m are calculated using partial derivatives as follows.

$$Q^* = \sqrt{\frac{2DC_o(C_p + C_h)}{C_p C_h}} \tag{6}$$

$$I_m^* = \sqrt{\frac{2C_o C_p D}{C_h(C_h + C_p)}} \tag{7}$$

Bjork [3] has developed a fuzzy version of the classical EOQ model presented in Eq. (5), where an input parameter (demand) and a decision variable (maximum inventory level) were fuzzified using fuzzy triangular numbers. Kazemi et al. [21] converted all the input parameters and decision variables of the same model into trapezoidal fuzzy numbers. The results were deduced for fuzzy triangular numbers as a particular case. They applied Kuhn–Tucker conditions for the formula derivation of optimal values because a conventional partial derivative approach leads to a complex function that is hard to solve. In this section, we develop a similar model but not as a full fuzzy model. We consider all the input parameters as trapezoidal fuzzy numbers but keep the decision variables to be crisp numbers. This kind of approach is intended to show that both the full fuzzy model of Kazemi et al. [21], and our present fuzzy model gives the same optimal policies. Hence, the fuzzification of decision variables does not make any difference in the optimal values. We have the following notations.

$$\breve{C}_o = \text{Ordering cost} = (C_o - a_1, C_o - a_2, C_o + a_3, C_o + a_4)$$
$$\breve{C}_h = \text{Holding cost} = (C_h - a_5, C_h - a_6, C_h + a_7, C_h + a_8)$$
$$\breve{C}_p = \text{Penalty cost} = (C_p - a_9, C_p - a_{10}, C_p + a_{11}, C_p + a_{12})$$
$$\breve{D} = \text{Demand} = (D - a_{13}, D - a_{14}, D + a_{15}, D + a_{16})$$

where $a_1 \geq a_2$, $a_4 \geq a_3$, $a_5 \geq a_6$, $a_8 \geq a_7$, $a_9 \geq a_{10}$, $a_{12} \geq a_{11}$, $a_{13} \geq a_{14}$, $a_{16} \geq a_{15}$. We retain the decision variables Q and I_m as crisp numbers. Thus, the fuzzy total cost function of this model becomes as below.

$$\breve{T}(Q, I_m) = \frac{\breve{C}_o \breve{D}}{Q} + \frac{\breve{C}_h I_m^2}{2Q} + \frac{(Q - I_m)^2 \breve{C}_p}{2Q} \tag{8}$$

After substituting the fuzzy numbers and employing the fuzzy arithmetic operations, the total cost function gets transformed into

$$\breve{T}(Q, I_m) = (\Gamma_1, \Gamma_2, \Gamma_3, \Gamma_4) \tag{9}$$

where

$$\Gamma_1 = \frac{(C_o - a_1)(D - a_{13})}{Q} + \frac{(C_h - a_5)I_m^2}{2Q} + \frac{(Q - I_m)^2(C_p - a_9)}{2Q}$$
$$\Gamma_2 = \frac{(C_o - a_2)(D - a_{14})}{Q} + \frac{(C_h - a_6)I_m^2}{2Q} + \frac{(Q - I_m)^2(C_p - a_{10})}{2Q}$$
$$\Gamma_3 = \frac{(C_o + a_3)(D + a_{15})}{Q} + \frac{(C_h + a_7)I_m^2}{2Q} + \frac{(Q - I_m)^2(C_p + a_{11})}{2Q}$$
$$\Gamma_4 = \frac{(C_o + a_4)(D + a_{16})}{Q} + \frac{(C_h + a_8)I_m^2}{2Q} + \frac{(Q - I_m)^2(C_p + a_{12})}{2Q}$$

We use the graded mean integration method $y(\breve{T}(Q, I_m)) = \frac{1}{6}(\Gamma_1 + 2\Gamma_2 + 2\Gamma_3 + \Gamma_4)$ for defuzzification. Hence, the defuzzified total cost function is given by

$$\begin{aligned} y(\breve{T}(Q, I_m)) = \tfrac{1}{6}&\left[\tfrac{(C_o-a_1)(D-a_{13})}{Q} + \tfrac{(C_h-a_5)I_m^2}{2Q} + \tfrac{(Q-I_m)^2(C_p-a_9)}{2Q}\right] \\ +\tfrac{2}{6}&\left[\tfrac{(C_o-a_2)(D-a_{14})}{Q} + \tfrac{(C_h-a_6)I_m^2}{2Q} + \tfrac{(Q-I_m)^2(C_p-a_{10})}{2Q}\right] \\ +\tfrac{2}{6}&\left[\tfrac{(C_o+a_3)(D+a_{15})}{Q} + \tfrac{(C_h+a_7)I_m^2}{2Q} + \tfrac{(Q-I_m)^2(C_p+a_{11})}{2Q}\right] \\ +\tfrac{1}{6}&\left[\tfrac{(C_o+a_4)(D+a_{16})}{Q} + \tfrac{(C_h+a_8)I_m^2}{2Q} + \tfrac{(Q-I_m)^2(C_p+a_{12})}{2Q}\right] \end{aligned} \tag{10}$$

The first- and second-order partial derivative expressions of $y(\breve{T}(Q, I_m))$ with respect to Q and I_m are presented below.

$$\begin{aligned}\frac{\partial}{\partial Q}y(\check{T}(Q,I_m)) = -\frac{1}{12Q^2}\{&[2(C_o-a_1)(D-a_{13})+4(C_o-a_2)(D-a_{14})\\&+4(C_o+a_3)(D+a_{15})+2(C_o+a_4)(D+a_{16})]\\&+[(C_h-a_5)+2(C_h-a_6)+2(C_h+a_7)+(C_h+a_8)\\&+(C_p-a_9)+2(C_p-a_{10})+2(C_p+a_{11})+(C_p+a_{12})]I_m^2\\&-[(C_p-a_9)+2(C_p-a_{10})+2(C_p+a_{11})+(C_p+a_{12})]Q^2\}\end{aligned} \quad (11)$$

$$\begin{aligned}\frac{\partial}{\partial I_m}y(\check{T}(Q,I_m)) = \frac{1}{6Q}\{&[(C_h-a_5)+2(C_h-a_6)+2(C_h+a_7)+(C_h+a_8)\\&+(C_p-a_9)+2(C_p-a_{10})+2(C_p+a_{11})+(C_p+a_{12})]I_m\\&-[(C_p-a_9)+2(C_p-a_{10})+2(C_p+a_{11})+(C_p+a_{12})]Q\}\end{aligned} \quad (12)$$

$$\begin{aligned}\frac{\partial^2}{\partial Q^2}y(\check{T}(Q,I_m)) = -\frac{1}{144Q^4}\{&-24Q^3[(C_p-a_9)+2(C_p-a_{10})+2(C_p+a_{11})\\&+(C_p+a_{12})]-24Q[[2(C_o-a_1)(D-a_{13})\\&+4(C_o-a_2)(D-a_{14})+4(c_o+a_3)(D+a_{15})\\&+2(C_o+a_4)(D+a_{16})]+[(C_h-a_5)+2(C_h-a_6)\\&+2(C_h+a_7)+(C_h+a_8)+(C_p-a_9)+2(C_p-a_{10})\\&+2(C_p+a_{11})+(C_p+a_{12})]I_m^2-[(C_p-a_9)\\&+2(C_p-a_{10})+2(C_p+a_{11})+(C_p+a_{12})]Q^2]\}\end{aligned} \quad (13)$$

$$\begin{aligned}\frac{\partial^2}{\partial I_m^2}y(\check{T}(Q,I_m)) = \frac{1}{6Q}&[(C_h-a_5)+2(C_h-a_6)+2(C_h+a_7)+(C_h+a_8)\\&+(C_p-a_9)+2(C_p-a_{10})+2(C_p+a_{11})+(C_p+a_{12})]\end{aligned} \quad (14)$$

$$\begin{aligned}\frac{\partial^2}{\partial Q\partial I_m}y(\check{T}(Q,I_m)) = \frac{\partial^2}{\partial I_m\partial Q}y(\check{T}(Q,I_m)) = -\frac{1}{6Q^2}&[(C_h-a_5)+2(C_h-a_6)\\&+2(C_h+a_7)+(C_h+a_8)+(C_p-a_9)+2(C_p-a_{10})\\&+2(C_p+a_{11})+(C_p+a_{12})]I_m\end{aligned} \quad (15)$$

Solving the equations $\frac{\partial}{\partial Q}y(\check{T}(Q,I_m))=0$ and $\frac{\partial}{\partial I_m}y(\check{T}(Q,I_m))=0$, we obtain the optimal values of Q and I_m as follows.

$$Q^* = \sqrt{\frac{\begin{array}{l}[(C_h - a_5) + 2(C_h - a_6) + 2(C_h + a_7) + (C_h + a_8) + (C_p - a_9)\\ +2(C_p - a_{10}) + 2(C_p + a_{11}) + (C_p + a_{12})] \times [2(C_o - a_1)(D - a_{13})\\ +4(C_o - a_2)(D - a_{14}) + 4(c_o + a_3)(D + a_{15}) + 2(C_o + a_4)(D + a_{16})]\end{array}}{\begin{array}{c}[(C_p - a_9) + 2(C_p - a_{10}) + 2(C_p + a_{11}) + (C_p + a_{12})]\\ \times[(C_h - a_5) + 2(C_h - a_6) + 2(C_h + a_7) + (C_h + a_8)]\end{array}}} \tag{16}$$

$$I_m^* = \sqrt{\frac{\begin{array}{l}[(C_p - a_9) + 2(C_p - a_{10}) + 2(C_p + a_{11}) + (C_p + a_{12})]\\ \times[2(C_o - a_1)(D - a_{13}) + 4(C_o - a_2)(D - a_{14})\\ +4(c_o + a_3)(D + a_{15}) + 2(C_o + a_4)(D + a_{16})]\end{array}}{\begin{array}{l}[(C_h - a_5) + 2(C_h - a_6) + 2(C_h + a_7) + (C_h + a_8)\\ +(C_p - a_9) + 2(C_p - a_{10}) + 2(C_p + a_{11}) + (C_p + a_{12})]\\ \times[(C_h - a_5) + 2(C_h - a_6) + 2(C_h + a_7) + (C_h + a_8)]\end{array}}} \tag{17}$$

At (Q^*, I_m^*) the determinant value of the Hessian matrix

$$H = \begin{bmatrix} \frac{\partial^2}{\partial Q^2} y(\breve{T}(Q, I_m)) & \frac{\partial^2}{\partial Q \partial I_m} y(\breve{T}(Q, I_m)) \\ \frac{\partial^2}{\partial Q \partial I_m} y(\breve{T}(Q, I_m)) & \frac{\partial^2}{\partial I_m^2} y(\breve{T}(Q, I_m)) \end{bmatrix}$$

is calculated as

$$\frac{1}{36}\left\{\frac{\begin{array}{c}[(C_p - a_9) + 2(C_p - a_{10}) + 2(C_p + a_{11}) + (C_p + a_{12})]^2\\ \times[(C_h - a_5) + 2(C_h - a_6) + 2(C_h + a_7) + (C_h + a_8)]^2\end{array}}{\begin{array}{l}[(C_h - a_5) + 2(C_h - a_6) + 2(C_h + a_7) + (C_h + a_8) + (C_p - a_9)\\ +2(C_p - a_{10}) + 2(C_p + a_{11}) + (C_p + a_{12})] \times [2(C_o - a_1)(D - a_{13})\\ +4(C_o - a_2)(D - a_{14}) + 4(c_o + a_3)(D + a_{15}) + 2(C_o + a_4)(D + a_{16})]\end{array}}\right\} \tag{18}$$

which is found to be a positive value by simple observation. We also have $\frac{\partial^2}{\partial Q^2} y(\breve{T}(Q, I_m)) > 0$ at (Q^*, I_m^*). Thus, we proved the defuzzified total cost function $y(\breve{T}(Q, I_m))$ is convex and attains its minimum at (Q^*, I_m^*). The minimum total cost is given by

$$y^*(\breve{T}(Q, I_m)) = \sqrt{\frac{\begin{array}{l}[(C_h - a_5) + 2(C_h - a_6) + 2(C_h + a_7) + (C_h + a_8)]\\ \times[(C_p - a_9) + 2(C_p - a_{10}) + 2(C_p + a_{11}) + (C_p + a_{12})]\\ \times[(C_o - a_1)(D - a_{13}) + 2(C_o - a_2)(D - a_{14})\\ +2(c_o + a_3)(D + a_{15}) + (C_o + a_4)(D + a_{16})]\end{array}}{\begin{array}{l}18[(C_h - a_5) + 2(C_h - a_6) + 2(C_h + a_7) + (C_h + a_8)\\ +(C_p - a_9) + 2(C_p - a_{10}) + 2(C_p + a_{11}) + (C_p + a_{12})]\end{array}}} \tag{19}$$

4 Impact of Wright's Learning Function on Fuzzy Input Parameters

We face many real-life situations where demand and costs are not precisely predictable and do not follow any probability distribution. In such situations, we can predict a rough estimate of expenses to be met out due to keeping inventory by representing the input parameters using fuzzy numbers. This approach is a reliable prediction using a mathematical model to determine an approximation of the inventory level and order quantity, maintaining an equilibrium between reducing inventory cost and shortages. The learning function works depending on the number of orders in one year, which is denoted by n. We treat n as a continuous variable to differentiate the cost function with respect to n. To apply the learning function, we eliminate Q replacing it with $\frac{D}{n}$. Now the defuzzified total cost function for one year takes the form

$$\begin{aligned} y_{\text{ann}} = &\frac{1}{6}\left[\frac{(C_o-a_1)(D-a_{13})n}{D} + \frac{(C_h-a_5)I_m^2 n}{2D} + \frac{(C_p-a_9)(D-I_m n)^2}{2Dn}\right] \\ &+\frac{2}{6}\left[\frac{(C_o-a_2)(D-a_{14})n}{D} + \frac{(C_h-a_6)I_m^2 n}{2D} + \frac{(C_p-a_{10})(D-I_m n)^2}{2Dn}\right] \\ &+\frac{2}{6}\left[\frac{(C_o+a_3)(D+a_{15})n}{D} + \frac{(C_h+a_7)I_m^2 n}{2D} + \frac{(C_p+a_{11})(D-I_m n)^2}{2Dn}\right] \\ &+\frac{1}{6}\left[\frac{(C_o+a_4)(D+a_{16})n}{D} + \frac{(C_h+a_8)I_m^2 n}{2D} + \frac{(C_p+a_{12})(D-I_m n)^2}{2Dn}\right] \end{aligned} \quad (20)$$

The average defuzzified total cost for each of the n orders is

$$\begin{aligned} y_{\text{avg}} = &\frac{1}{6}\left[\frac{(C_o-a_1)(D-a_{13})}{D} + \frac{(C_h-a_5)I_m^2}{2D} + \frac{(C_p-a_9)(D-I_m n)^2}{2Dn^2}\right] \\ &+\frac{2}{6}\left[\frac{(C_o-a_2)(D-a_{14})}{D} + \frac{(C_h-a_6)I_m^2}{2D} + \frac{(C_p-a_{10})(D-I_m n)^2}{2Dn^2}\right] \\ &+\frac{2}{6}\left[\frac{(C_o+a_3)(D+a_{15})}{D} + \frac{(C_h+a_7)I_m^2}{2D} + \frac{(C_p+a_{11})(D-I_m n)^2}{2Dn^2}\right] \\ &+\frac{1}{6}\left[\frac{(C_o+a_4)(D+a_{16})}{D} + \frac{(C_h+a_8)I_m^2}{2D} + \frac{(C_p+a_{12})(D-I_m n)^2}{2Dn^2}\right] \end{aligned} \quad (21)$$

Learning functions have a vital role in reducing the fuzziness of the input parameters. In Sect. 3, we used the notations a_i: $i = 1, 2, 5, 6, 9, 10, 13, 14$ to denote the lower fuzziness limits and a_j: $j = 3, 4, 7, 8, 11, 12, 15, 16$ to denote the upper fuzziness limits. The duration between two consecutive orders is $\frac{365}{n}$ days and the time of xth order is $(x-1)\frac{365}{n}$. If a_i is the value of the fuzziness limit at the beginning of the planning period, and $0 < m < 1$ is an appropriate learning exponent which suits the business firm, then we let

$$a_{ix} = \begin{cases} a_i & \text{if } i = 1, 2, \ldots, 16 \;\&\; x = 1 \\ a_i\left(\frac{365(x-1)}{n}\right)^{-m} & \text{if } i = 1, 2, \ldots, 16 \;\&\; x = 2, 3, \ldots, n \end{cases}$$

to denote the fuzziness limit corresponding to a_i in the xth order. The defuzzified learning total cost function at the xth order $(x > 1)$ is given below.

$$
\begin{aligned}
y_x = \frac{1}{6}\Bigg[&\frac{(C_o-a_1\left(\frac{n}{365}\right)^m(x-1)^{-m})(D-a_{13}\left(\frac{n}{365}\right)^m(x-1)^{-m})}{D} \\
&+\frac{(C_h-a_5\left(\frac{n}{365}\right)^m(x-1)^{-m})I_m^2}{2D}+\frac{(C_p-a_9\left(\frac{n}{365}\right)^m(x-1)^{-m})(D-I_mn)^2}{2Dn^2}\Bigg] \\
+\frac{2}{6}\Bigg[&\frac{(C_o-a_2\left(\frac{n}{365}\right)^m(x-1)^{-m})(D-a_{14}\left(\frac{n}{365}\right)^m(x-1)^{-m})}{D} \\
&+\frac{(C_h-a_6\left(\frac{n}{365}\right)^m(x-1)^{-m})I_m^2}{2D}+\frac{(C_p-a_{10}\left(\frac{n}{365}\right)^m(x-1)^{-m})(D-I_mn)^2}{2Dn^2}\Bigg] \\
+\frac{2}{6}\Bigg[&\frac{(C_o+a_3\left(\frac{n}{365}\right)^m(x-1)^{-m})(D+a_{15}\left(\frac{n}{365}\right)^m(x-1)^{-m})}{D} \\
&+\frac{(C_h+a_7\left(\frac{n}{365}\right)^m(x-1)^{-m})I_m^2}{2D}+\frac{(C_p+a_{11}\left(\frac{n}{365}\right)^m(x-1)^{-m})(D-I_mn)^2}{2Dn^2}\Bigg] \\
+\frac{1}{6}\Bigg[&\frac{(C_o+a_4\left(\frac{n}{365}\right)^m(x-1)^{-m})(D+a_{16}\left(\frac{n}{365}\right)^m(x-1)^{-m})}{D} \\
&+\frac{(C_h+a_8\left(\frac{n}{365}\right)^m(x-1)^{-m})I_m^2}{2D}+\frac{(C_p+a_{12}\left(\frac{n}{365}\right)^m(x-1)^{-m})(D-I_mn)^2}{2Dn^2}\Bigg]
\end{aligned}
\tag{22}
$$

For $x = 1$, the cost function is given by Eq. (21). The defuzzified learning total cost function for the whole year is given below.

$$
\begin{aligned}
y = &\frac{1}{6}\left[\frac{(C_o-a_1)(D-a_{13})}{D}+\frac{(C_h-a_5)I_m^2}{2D}+\frac{(C_p-a_9)(D-I_mn)^2}{2Dn^2}\right] \\
&+\frac{2}{6}\left[\frac{(C_o-a_2)(D-a_{14})}{D}+\frac{(C_h-a_6)I_m^2}{2D}+\frac{(C_p-a_{10})(D-I_mn)^2}{2Dn^2}\right] \\
&+\frac{2}{6}\left[\frac{(C_o+a_3)(D+a_{15})}{D}+\frac{(C_h+a_7)I_m^2}{2D}+\frac{(C_p+a_{11})(D-I_mn)^2}{2Dn^2}\right] \\
&+\frac{1}{6}\left[\frac{(C_o+a_4)(D+a_{16})}{D}+\frac{(C_h+a_8)I_m^2}{2D}+\frac{(C_p+a_{12})(D-I_mn)^2}{2Dn^2}\right] \\
&+\frac{1}{6}\sum_{x=2}^{n}\Bigg[\frac{(C_o-a_1\left(\frac{n}{365}\right)^m(x-1)^{-m})(D-a_{13}\left(\frac{n}{365}\right)^m(x-1)^{-m})}{D} \\
&+\frac{(C_h-a_5\left(\frac{n}{365}\right)^m(x-1)^{-m})I_m^2}{2D}+\frac{(C_p-a_9\left(\frac{n}{365}\right)^m(x-1)^{-m})(D-I_mn)^2}{2Dn^2}\Bigg] \\
&+\frac{2}{6}\sum_{x=2}^{n}\Bigg[\frac{(C_o-a_2\left(\frac{n}{365}\right)^m(x-1)^{-m})(D-a_{14}\left(\frac{n}{365}\right)^m(x-1)^{-m})}{D} \\
&+\frac{(C_h-a_6\left(\frac{n}{365}\right)^m(x-1)^{-m})I_m^2}{2D}+\frac{(C_p-a_{10}\left(\frac{n}{365}\right)^m(x-1)^{-m})(D-I_mn)^2}{2Dn^2}\Bigg] \\
&+\frac{2}{6}\sum_{x=2}^{n}\Bigg[\frac{(C_o+a_3\left(\frac{n}{365}\right)^m(x-1)^{-m})(D+a_{15}\left(\frac{n}{365}\right)^m(x-1)^{-m})}{D} \\
&+\frac{(C_h+a_7\left(\frac{n}{365}\right)^m(x-1)^{-m})I_m^2}{2D}+\frac{(C_p+a_{11}\left(\frac{n}{365}\right)^m(x-1)^{-m})(D-I_mn)^2}{2Dn^2}\Bigg]
\end{aligned}
$$

$$+\frac{1}{6}\sum_{x=2}^{n}\left[\frac{(C_o+a_4\left(\frac{n}{365}\right)^m(x-1)^{-m})(D+a_{16}\left(\frac{n}{365}\right)^m(x-1)^{-m})}{D}\right.$$
$$\left.+\frac{(C_h+a_8\left(\frac{n}{365}\right)^m(x-1)^{-m})I_m^2}{2D}+\frac{(C_p+a_{12}\left(\frac{n}{365}\right)^m(x-1)^{-m})(D-I_mn)^2}{2Dn^2}\right] \quad (23)$$

After some algebraic manipulations and rearrangement of terms, the cost function takes the form

$$y=\frac{1}{12D}\left[12nC_oD+6nC_hI_m^2+6C_p\frac{(D-I_mn)^2}{n}\right]$$
$$+\frac{1}{12D}\left[2U_1+U_3I_m^2+U_4\frac{(D-I_mn)^2}{n^2}\right]\left[1+\left(\frac{n}{365}\right)^m\sum_{x=2}^{n}\frac{1}{(x-1)^m}\right]$$
$$+\frac{U_2}{6D}\left[1+\left(\frac{n}{365}\right)^{2m}\sum_{x=2}^{n}\frac{1}{(x-1)^{2m}}\right] \quad (24)$$

where

$$U_1=-(C_oa_{13}+Da_1)-2(C_oa_{14}+Da_2)+2(C_oa_{15}+Da_3)+(C_oa_{16}+Da_4)$$
$$U_2=a_1a_{13}+2a_2a_{14}+2a_3a_{15}+a_4a_{16}$$
$$U_3=-a_5-2a_6+2a_7+a_8$$
$$U_4=-a_9-2a_{10}+2a_{11}+a_{12}$$

Here, the terms of y involve a partial sum of the series $s=\sum_{h=1}^{\infty}\frac{1}{h^p}$. We know that this series is divergent for $p\leq 1$, and there does not exist any general closed-form expression for its partial sum (known as generalized harmonic number)

$$s_n=\sum_{h=1}^{n}\frac{1}{h^p}=1+\frac{1}{2^p}+\frac{1}{3^p}+\cdots+\frac{1}{n^p} \quad (25)$$

Hence, it is challenging to differentiate y with respect to n, and proving its convexity both in n and I_m analytically is not possible. Therefore, Kazemi et al. [23–25] proved the convexity of y with respect to I_m and reduced the problem into finding a local near-optimum value for n by an iterative procedure. In this section, we try to formulate simultaneous equations in the variables I_m and n by inserting an approximate but most close acceptable expression of the partial sum $\sum_{x=2}^{n}\frac{1}{(x-1)^m}$ into y and differentiate it with respect to I_m and n. Then, we solve the simultaneous equations for numerical values. Chlebus [6] proposed the following formula as an approximate sum estimate of the series given in Eq. (25).

$$s_n = \begin{cases} \frac{(n+1)^{1-p}+n^{1-p}-2}{2(1-p)} + \frac{1}{2} & \text{if } p < 1 \\ \frac{1+\log n+\log(n+1)}{2} & \text{if } p = 1 \end{cases} \tag{26}$$

We can verify that the sum's relative approximate error resulting from Eq. (26) from the actual sum decreases for the higher values of n and increases for increasing values of p. For example, for the values $n = 100$, $p = 0.9$ the relative approximate error is -1.0878% and for the values $n = 400$, $p = 0.9$, the relative approximate error is -0.7962% (Table 1, Chlebus [6]). It shows that the relative approximate error is a negligible quantity. Now, from Eq. (26), we deduce that

$$\sum_{x=2}^{n} \frac{1}{(x-1)^m} = \begin{cases} \frac{n^{1-m}+(n-1)^{1-m}-2}{2(1-m)} + \frac{1}{2} & \text{if } m < 1 \\ \frac{1+\log(n-1)+\log n}{2} & \text{if } m = 1 \end{cases} \tag{27}$$

4.1 Case-1: $0 \le m < 0.5$

In this case, we have $m < 0.5$ & $2m < 1$, and hence, the cost function becomes

$$\begin{aligned} y = &\frac{1}{12D}\left[12nC_oD + 6nC_hI_m^2 + 6C_p\frac{(D-I_mn)^2}{n}\right] \\ &+ \frac{1}{12D}\left[2U_1 + U_3I_m^2 + U_4\frac{(D-I_mn)^2}{n^2}\right] \\ &\times \left[1 + \left(\frac{n}{365}\right)^m\left(\frac{n^{1-m}+(n-1)^{1-m}-2}{2(1-m)} + \frac{1}{2}\right)\right] \\ &+ \frac{U_2}{6D}\left[(1 + \left(\frac{n}{365}\right)^{2m}\left(\frac{n^{1-2m}+(n-1)^{1-2m}-2}{2(1-2m)} + \frac{1}{2}\right)\right] \end{aligned} \tag{28}$$

The partial derivatives of y with respect to I_m and n are given below.

$$\begin{aligned} \frac{\partial y}{\partial I_m} = &\frac{1}{D}\left[(C_hI_mn - C_p(D-I_mn)\right] + \frac{1}{6D}\left[U_3I_m - U_4\frac{(D-I_mn)}{n}\right] \\ &\times \left[\left(\frac{n}{365}\right)^m\left(\frac{(n-1)^{1-m}+n^{1-m}-2}{2m-2} - \frac{1}{2}\right) - 1\right] \end{aligned} \tag{29}$$

$$\begin{aligned} \frac{\partial y}{\partial n} = &\frac{1}{2D}\left[C_hI_m^2 + 2C_oD - C_p\frac{(D-I_mn)^2}{n^2} - 2C_p\frac{(D-I_mn)I_m}{n}\right] \\ &- \frac{1}{12D}\left[\frac{m}{365}\left(\frac{n}{365}\right)^{m-1}\left(\frac{n^{1-m}+(n-1)^{1-m}-2}{2m-2} - \frac{1}{2}\right)\right. \\ &\left.- \frac{1}{2m-1}\left(\frac{n}{365}\right)^m\left(\frac{m-1}{n^m} + \frac{m-1}{(n-1)^m}\right)\right] \end{aligned}$$

$$\times \left[2U_1 + U_3 I_m^2 + U_4 \frac{(D - I_m n)^2}{n^2}\right]$$
$$- \frac{U_2}{6D}\left[\frac{2m}{365}\left(\frac{n}{365}\right)^{2m-1}\left(\frac{n^{1-2m} + (n-1)^{1-2m} - 2}{4m-2} - \frac{1}{2}\right)\right.$$
$$\left. - \frac{1}{4m-2}\left(\frac{n}{365}\right)^{2m}\left(\frac{2m-1}{n^{2m}} + \frac{2m-1}{(n-1)^{2m}}\right)\right]$$
$$+ \frac{U_4}{6D}\left[\left(\frac{n}{365}\right)^m\left(\frac{n^{1-m} + (n-1)^{1-m} - 2}{2m-2} - \frac{1}{2}\right) - 1\right]$$
$$\times \left[\frac{(D - I_m n)^2}{n^3} + \frac{(D - I_m n) I_m}{n^2}\right] \tag{30}$$

4.2 Case-2: $m = 0.5$

In this case, we have $m = 0.5$ and $2m = 1$, and hence, the cost function becomes

$$y = \frac{1}{12D}\left[12nC_o D + 6nC_h I_m^2 + 6C_p \frac{(D - I_m n)^2}{n}\right]$$
$$+ \frac{1}{12D}\left[2U_1 + U_3 I_m^2 + U_4 \frac{(D - I_m n)^2}{n^2}\right]$$
$$\times \left[1 + \left(\frac{n}{365}\right)^m\left(\frac{n^{1-m} + (n-1)^{1-m} - 2}{2(1-m)} + \frac{1}{2}\right)\right]$$
$$+ \frac{U_2}{6D}\left[1 + \left(\frac{n}{365}\right)^m\left(\frac{1 + \log(n-1) + \log n}{2}\right)\right] \tag{31}$$

The partial derivatives of y with respect to I_m and n are given below.

$$\frac{\partial y}{\partial I_m} = \frac{1}{D}\left[C_h I_m n - C_p (D - I_m n)\right]$$
$$- \frac{1}{6D}\left[\left(\frac{n}{365}\right)^m\left(\frac{n^{1-m} + (n-1)^{1-m} - 2}{2m-2} - \frac{1}{2}\right) - 1\right]$$
$$\times \left[U_3 I_m - U_4 \frac{(D - I_m n)}{n}\right] \tag{32}$$

$$\frac{\partial y}{\partial n} = \frac{1}{2D}\left[C_h I_m^2 + 2C_o D - \frac{C_p (D - I_m n)^2}{n^2} - 2C_p I_m \frac{(D - I_m n)}{n}\right]$$
$$- \frac{1}{12D}\left[\frac{m}{365}\left(\frac{n}{365}\right)^{m-1}\left(\frac{n^{1-m} + (n-1)^{1-m} - 2}{2m-2} - \frac{1}{2}\right)\right.$$

$$
\begin{aligned}
& -\frac{1}{2m-2}\left(\frac{n}{365}\right)^m\left(\frac{m-1}{n^m}+\frac{m-1}{(n-1)^m}\right)\Bigg] \\
& \times\left[2U_1+U_3I_m^2+U_4\frac{(D-I_mn)^2}{n^2}\right] \\
& +\frac{U_2}{6D}\left[\left(\frac{n}{365}\right)^m\left(\frac{1}{2n}+\frac{1}{2n-2}\right)\right. \\
& \left.+\frac{m}{365}\left(\frac{n}{365}\right)^{m-1}\left(\frac{\log n}{2}+\frac{\log(n-1)}{2}+\frac{1}{2}\right)\right] \\
& +\frac{U_4}{6D}\left[\left(\frac{n}{365}\right)^m\left(\frac{n^{1-m}+(n-1)^{1-m}-2}{2m-2}-\frac{1}{2}\right)-1\right] \\
& \times\left[\frac{(D-I_mn)^2}{n^3}+\frac{(D-I_m)I_m}{n^2}\right]
\end{aligned} \tag{33}
$$

In both cases, the optimal values of n and I_m are found by solving the equations $\frac{\partial y}{\partial I_m}=0$ and $\frac{\partial y}{\partial n}=0$. Since the Hessian is lengthy, we do not mention it here. The graph of the cost function, shown in Fig. 2, shows that the cost function is convex. Therefore, cost function attains its minimum value at the optimum point. It is challenging to find closed-form solutions for n and I_m due to the equations' complexity. Hence, MATLAB 2014 is used to solve numerical problems. Since we treated n as a continuous variable, we arrive at decimal n values in most numerical examples. We take $\lfloor n \rfloor$ and $\lceil n \rceil$ values and calculate the corresponding maximum inventory level, order quantity, and total cost and choose the appropriate inventory policies (refer to Sect. 5).

5 Numerical Examples

This section presents numerical examples to demonstrate the working of the fuzzy learning model developed in Sect. 4 and how the learning function affects the inventory decisions. The crisp input values of this example are acquired from Bjork [3] which are as follows: Demand, $D=50{,}000$ kg/year, purchase price= 1 Euro/kg, holding cost, $C_h=25\%$ of purchase price $=0.25$ Euro/kg in a year, ordering cost, $C_o=200$ Euros/order, penalty cost, $C_p=5$ Euros/kg in a year. The optimal values of the crisp model are $Q^*=9165.15$, $I_m^*=8728.72$ and $T^*=2182.18$ according to Bjork [3]. Next, let us acquire two sets of fuzzy input parameter values from Kazemi et al. [21], all of whose defuzzified values of the first set of values follow a 10% and the second set of values follow a -10% deviation from the corresponding crisp values. The fuzzy inventory policies are presented in Table 1.
Comparing the optimal solutions of Table 1 of this paper and Table 3 of Kazemi et al. [21], we observe that the optimal values Q^*, I_m^* and y^* of the present model are the same as that of the optimal values of Kazemi et al. [21], in which the decision variables are also fuzzified in addition to the fuzzification of the input parameters.

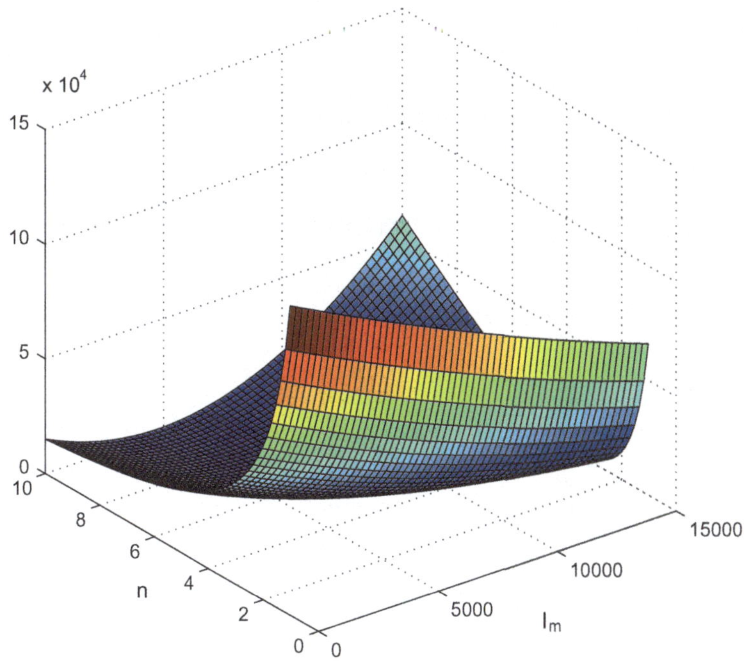

Fig. 2 Three dimensional cost curve for the fuzzy learning case

Table 1 Fuzzy inventory policy

Set	Fuzzy input value	Lower and upper fuzziness limits	Optimal solution
1	$\check{C}_o = (150, 180, 255, 300)$	$a_1 = 50, a_2 = 20, a_3 = 55, a_4 = 100$	$Q^* = 10{,}036$
	$\check{C}_h = (0.13, 0.17, 0.33, 0.52)$	$a_5 = 0.12, a_6 = 0.08, a_7 = 0.08, a_8 = 0.27$	$I_m^* = 9558.2$
	$\check{C}_p = (3.6, 4.2, 6.3, 8.4)$	$a_9 = 1.4, a_{10} = 0.8, a_{11} = 1.3, a_{12} = 3.4$	$y^* = 2628.5$
	$\check{D} = (29{,}000, 41{,}000, 63{,}000, 93{,}000)$	$a_{13} = 21{,}000, a_{14} = 9000, a_{15} = 13{,}000, a_{16} = 43{,}000$	
2	$\check{C}_o = (120, 130, 230, 260)$	$a_1 = 80, a_2 = 70, a_3 = 30, a_4 = 60$	$Q^* = 9249.6$
	$\check{C}_h = (0.11, 0.13, 0.3, 0.38)$	$a_5 = 0.14, a_6 = 0.12, a_7 = 0.05, a_8 = 0.13$	$I_m^* = 8809.2$
	$\check{C}_p = (1.5, 3.0, 6.0, 7.5)$	$a_9 = 3.5, a_{10} = 2.0, a_{11} = 1.0, a_{12} = 2.5$	$y^* = 1982.1$
	$\check{D} = (20{,}000, 35{,}000, 55{,}000, 70{,}000)$	$a_{13} = 30{,}000, a_{14} = 15{,}000, a_{15} = 5000, a_{16} = 20{,}000$	

Table 2 Fuzzy inventory policies for parameters with same defuzzified values

Set	$\breve{C}_o$	$\breve{C}_h$	$\breve{C}_p$	$\breve{D}$	Q^*	I_m^*	$y^*(\breve{T}(Q, I_m))$
1	(198, 199, 201, 202)	(0.23, 0.24, 0.26, 0.27)	(4.5, 4.8, 5.2, 5.5)	(45,000, 47,000, 53,000, 55,000)	9167.6	8731	2182.8
2	(150, 180, 220, 250)	(0.05, 0.15, 0.35, 0.45)	(1.5, 3, 7, 8.5)	(25,000, 35,000, 65,000, 75,000)	9443.5	8993.8	2248.5
3	(145, 175, 225, 255)	(0.1, 0.13, 0.37, 0.4)	(2, 2.5, 7.5, 8)	(20,000, 30,000, 70,000, 80,000)	9561.4	9106.1	2276.5
4	(138, 185, 215,262)	(0.02, 0.11, 0.39, 0.48)	(1.2, 2, 8, 8.8)	(15,000, 25,000, 75,000, 85,000)	9600.8	9143.7	2285.9

Therefore, we conclude that the fuzzification of the decision variables has no impact on the optimal values. The fuzzy inventory policy depends on the input parameters' defuzzified values and the fuzziness level of the input parameters. It depends on the defuzzified value of the resultant fuzzy total cost function $\breve{T}(Q, I_m)$ after the fuzzy arithmetic operations are performed. Table 2 enumerates some more fuzzy inventory policies. In each of them, the defuzzified values of each of the fuzzy input parameters are the crisp value of the respective parameter. The optimal values are different in each case. The set with the smallest interval length (minimum fuzziness) for all input parameters matches the crisp optimal values.

Next, we determine the optimum inventory policy for the fuzzy learning case. Let us take the first set of fuzzy input values from Table 1 and fix learning rates at different levels. The suitable optimal inventory policies for the nearest integer number of orders are shown in bold letters. The results are shown in Table 3.

6 Sensitivity Analysis

To understand more about the impact of fuzziness in the input parameters on the model constructed in Sect. 4, we perform a sensitivity analysis. The results are summarized in Table 4. We take the fourth set of values from Table 2 as base values, whose defuzzified values are the respective crisp values and the analysis is conducted for the learning rate $m = 0.4$. We assign values to $\breve{C}_o$, $\breve{C}_h$, $\breve{C}_p$ and $\breve{D}$ at -20%, -10%, 10%, 20% difference levels of the defuzzified values from their respective defuzzified base values (i.e., crisp values). The fuzzy learning

Table 3 Fuzzy learning inventory policy for different learning rates

LR	Optimal policy for continuous n				Optimal policy for integer less than n				Optimal policy for integer greater than n			
m	n	I_m^*	Q^*	y^*	$\lfloor n \rfloor$	I_m^*	Q^*	y^*	$\lceil n \rceil$	I_m^*	Q^*	y^*
0.0	4.9820	9558.20	10,036.00	2628.50	4	11,905.00	12,500.00	2692.10	**5**	**9523.80**	**10,000.00**	**2628.50**
0.1	5.2368	9093.07	9547.72	2463.09	**5**	**9523.80**	**10,000.00**	**2465.70**	6	7936.50	8333.33	2485.90
0.2	5.3737	8861.57	9304.65	2376.71	**5**	**9523.80**	**10,000.00**	**2382.90**	6	7936.50	8333.33	2391.10
0.3	5.4480	8740.59	9177.62	2329.81	**5**	**9523.80**	**10,000.00**	**2338.40**	6	7936.50	8333.33	2340.60
0.4	5.4898	8674.10	9107.81	2303.43	5	9523.80	10,000.00	2313.50	**6**	**7936.50**	**8333.33**	**2312.50**
0.5	5.5016	8655.50	9088.27	2293.60	5	9523.80	10,000.00	2304.10	**6**	**7936.50**	**8333.33**	**2302.20**

Table 4 Sensitivity analysis for fuzzy learning model with $m = 0.4$

Parameters	% change	Parameter values	n^*	I_m^*	Q^*	y^*	% change in I_m^*	% change in y^*
$\breve{C}_o$	−20	(98, 145, 175, 222)	6	7936.5	8333.3	1973.4	−16.67	−10.76
	−10	(118, 165, 195, 242)	6	7936.5	8333.3	2093.4	−16.67	−5.34
	10	(158, 205, 235, 282)	5	9523.8	10,000.0	2311.4	0.00	4.52
	20	(178, 225, 255, 302)	5	9523.8	10,000.0	2411.4	0.00	9.04
$\breve{C}_h$	−20	(0.01, 0.06, 0.34, 0.39)	5	9523.8	10,000.0	1984.6	0.00	−10.26
	−10	(0.035, 0.085, 0.365, 0.415)	5	9523.8	10,000.0	2098.0	0.00	−5.13
	10	(0.045, 0.135, 0.415, 0.505)	6	7936.5	8333.3	2307.9	−16.67	4.36
	20	(0.07, 0.16, 0.44, 0.53)	6	7936.5	8333.3	2402.3	−16.67	8.63
$\breve{C}_p$	−20	(0.2, 1, 7, 7.8)	5	9523.8	10,000.0	2200.0	0.00	−0.52
	−10	(0.7, 1.5, 7.5, 8.3)	5	9523.8	10,000.0	2205.7	0.00	−0.26
	10	(1.7, 2.5, 8.5, 9.3)	5	9523.8	10,000.0	2217.0	0.00	0.25
	20	(2.2, 3, 9, 9.8)	5	9523.8	10,000.0	2222.8	0.00	0.52
$\breve{D}$	−20	(5000, 15,000, 65,000, 75,000)	5	7619.0	8000.0	1978.5	−20.00	−10.53
	−10	(10,000, 20,000, 70,000, 80,000)	5	8571.4	9000.0	2094.6	−10.00	−5.28
	10	(20,000, 30,000, 80,000, 90,000)	6	8730.2	9166.7	2310.6	−8.33	4.49
	20	(25,000, 35,000, 85,000, 95,000)	6	9523.8	10,000.0	2408.2	0.00	8.90

optimal policy for the base values for $m = 0.4$ is $n^* = 5.45,\ I_m^* = 8737.9,\ Q^* = 9174.8.2,\ y^* = 2203.2$ and for integer value of n the optimal values are $n^* = 5,\ I_m^* = 9523.8,\ Q^* = 10{,}000,\ y^* = 2211.4$. The % difference in the decision variables and total cost are calculated from $n^* = 5,\ I_m^* = 9523.8,\ Q^* = 10{,}000,\ y^* = 2211.4$.

We draw the following observations from the above sensitivity analysis. The total cost is moderately sensitive to the parameters ordering cost ($\check{C}_o$), holding cost ($\check{C}_h$), and demand ($\check{D}$) but less sensitive to penalty cost ($\check{C}_p$). Changes in each of the moderate sensitive parameters have the same amount of impact on the system. The cost increases concerning an increase in each of the input parameter values. % change in the total cost for the moderate sensitive parameters ordering cost ($\check{C}_o$), holding cost ($\check{C}_h$), and demand ($\check{D}$), is slightly greater for negative changes comparing to positive changes. Positive changes in holding cost ($\check{C}_h$) and demand ($\check{D}$) increase the optimum number of orders by one, whereas in the case of ordering cost ($\check{C}_o$) negative changes increase the optimum number of orders by one. Changes in penalty cost ($\check{C}_p$) do not impact the optimum number of orders.

7 Conclusion

In this paper, we studied a fuzzy learning lot-sizing problem with backorders. The inventory operator assumes learning from past experiences over the planning period. This model treats all the input parameters as fuzzy, and learning is incorporated in all the fuzzy parameters. In the fuzzy learning model, the number of orders is used as a decision variable instead of order quantity per order. The amount of learning and reduction in fuzziness is directly proportional to the planning period's number of orders. From Table 3, we note that when there is no learning (i.e., $m = 0$), the total costs are the same in the fuzzy learning model and fuzzy model. Since the fuzzy learning model and fuzzy model are equivalent when $m = 0$, the fuzzy learning model can be considered a generalization of the fuzzy model. The total cost decreases as the learning rate increases. A simple exponential regression analysis between learning rate (m) and the total cost (y) gives the result $y = 2567.5 \times 0.77^m$ with a strong negative correlation coefficient -0.93 between m and y. When $m = 1$, we get $y = 1977.0$, which shows the fuzzy learning model's convergence tendency to crisp model when the learning rate approaches 1.

Inventory operators could effectively use learning functions for their decision-making but should keep some aspects in mind. The prior knowledge one should possess for the successful handling of learning functions is the appraisal of learning rate. The mean value learning rate of a particular industrial sector may not work for a particular company. The work style varies from place to place, and learning is purely individual. A learning function depends on the initial value of a parameter estimated carefully from past experiences. Finally, we emphasize that when demand and other parameters are imprecise, the present fuzzy learning model with an appropriate learning rate helps build a better inventory policy close to the crisp inventory policy.

Acknowledgements The authors would like to thank the Editor(s) and the Reviewers for their valuable and constructive comments that have led to a significant improvement in the manuscript.
Funding This work was supported by the University Grants Commission—Special Assistance Program (DSA-I), Government of India, New Delhi.
Conflict of Interest The authors declare that they have no conflict of interest.

References

1. Ausubel, D.P.: The role of frequency in learning and retention: a cognitive structure interpretation. J. Gen. Psychol. **72**, 359–368 (1965)
2. Bera, U.K., Mahapatra, N.K., Maiti, M.: An imperfect fuzzy production-inventory model over a finite time horizon under the effect of learning. Int. J. Math. Oper. Res. **1**(3), 351–371 (2009)
3. Bjork, K.M.: An analytic solution to a fuzzy economic order quantity problem. Int. J. Approx. Reason. **50**, 485–493 (2009)
4. Chen, S.H.: Operations on fuzzy numbers with function principle. Tamkang J. Manag. Sci. **6**(1), 13–26 (1985)
5. Chen, S.H., Hseih, C.H.: Graded mean integration representation of generalized fuzzy number. J. Chin. Fuzzy Syst. **5**, 1–7 (1999)
6. Chlebus, E.: An approximate formula for a partial sum of the divergent p-series. Appl. Math. Lett. **22**, 732–737 (2009)
7. De, S.K., Mahata, G.C.: Decision of a fuzzy inventory with fuzzy backorder model under cloudy fuzzy demand rate. Int. J. Appl. Comput. Math. **3**(3), 2593–2609 (2017)
8. Dutton, J.M., Thomas, A., Butler, J.E.: The history of progress functions as a managerial technology. Bus. Hist. Rev. **58**, 204–233 (1984)
9. Ganesan, S., Uthayakumar, R.: EPQ models for an imperfect manufacturing system considering warm-up production run, shortages during hybrid maintenance period and partial backordering. Adv. Ind. Manuf. Eng. **1**, 100005 (2020)
10. Ganesan, S., Uthayakumar, R.: EPQ models with bivariate random imperfect proportions and learning-dependent production and demand rates. J. Manag. Anal. **8**(1), 134–170 (2021)
11. Glock, C.H., Grosse, E.H., Jaber, M.Y., Smunt, T.L.: Applications of learning functions in production and operations management: a systematic literature review. Comput. Ind. Eng. **131**, 422–441 (2019)
12. Glock, C.H., Jaber, M.Y.: A multi-stage production-inventory model with learning and forgetting effects, rework and scrap. Comput. Ind. Eng. **64**, 708–720 (2013)
13. Glock, C.H., Schwindl, K., Jaber, M.Y.: An EOQ model with fuzzy demand and learning in fuzziness. Int. J. Serv. Oper. Manag. **12**, 90–100 (2012)
14. Gosling, J., Abouarghoub, W., Naim, M., Moone, B.: Constructing supplier learning functions to evaluate relational gain in engineering projects. Comput. Ind. Eng. **131**, 502–514 (2019)
15. Goyal, S.K., Singh, S.R., Yadav, D.: Economic order quantity model for imperfect lot with partial backordering under the effect of learning and advertisement dependent imprecise demand. Int. J. Oper. Res. **29**(2), 197–218 (2017)
16. Harris, F.W.: How many parts to make at once. Mag. Manag. **10**, 135–136 (1913)
17. Hemalatha, S., Annadurai, K.: A fuzzy EOQ inventory model with advance payment and various fuzzy numbers. Mater. Today: Proc. (2020). https://doi.org/10.1016/j.matpr.2020.10.769
18. Jaber, M.Y.: Learning Curves: Theory, Models, and Applications. CRC Press, Boca Raton, FL (2011)
19. Kakprzyk, J., Staniewski, P.: Long term inventory policy-making through fuzzy decision making models. Fuzzy Sets Syst. **8**, 117–132 (1982)
20. Karthick, B., Uthayakumar, R.: A multi-item sustainable manufacturing model with discrete setup cost and carbon emission reduction under deterministic and trapezoidal fuzzy demand. Process Integr. Optim. Sustain. 1–39 (2021). https://doi.org/10.1007/s41660-021-00159-6

21. Kazemi, N., Ehsani, E., Jaber, M.Y.: An inventory model with backorders with fuzzy parameters and decision variables. Int. J. Approx. Reason. **51**(8), 964–972 (2010)
22. Kazemi, N., Olugu, E.U., Rashid, S.H.A., Ghazilla, R.A.B.R.: Development of a fuzzy economic order quantity model for imperfect quality items using the learning effect on fuzzy parameters. J. Intell. Fuzzy Syst. **28**, 2377–2389 (2015)
23. Kazemi, N., Olugu, E.U., Rashid, S.H.A., Ghazilla, R.A.B.R.: A fuzzy EOQ model with backorders and forgetting effect on fuzzy parameters: an empirical study. Comput. Ind. Eng. **96**, 140–148 (2016)
24. Kazemi, N., Rashid, S.H.A., Shekarian, E., Bottani, E., Montanari, R.: A fuzzy lot-sizing problem with two-stage composite human learning. Int. J. Prod. Res. **54**(16), 5010–5025 (2016)
25. Kazemi, N., Shekarian, E., Barrón, L.E.C., Olugu, E.U.: Incorporating human learning into a fuzzy EOQ inventory model with backorders. Comput. Ind. Eng. **87**, 540–542 (2015)
26. Kim, I., Springer, M., George Zhang, Z., Park, Y.-S.: Organizational learning: approximation of multiple-level learning and forgetting by an aggregated single-level model. Comput. Ind. Eng. **131**, 442–454 (2019)
27. Latha, K.F.M., Kumar, M.G., Uthayakumar, R.: Two echelon economic lot sizing problems with geometric shipment policy backorder price discount and optimal investment to reduce ordering cost. OPSEARCH 1–31 (2021). https://doi.org/10.1007/s12597-021-00515-7
28. Li, W., Liu, Y., Chen, Y.: Modeling a two-stage supply contract problem in a hybrid uncertain environment. Comput. Ind. Eng. **123**, 289–302 (2018)
29. Malleeswaran, B., Uthayakumar, R.: An integrated vendor–buyer supply chain model for backorder price discount and price-dependent demand using service level constraints and carbon emission cost. Int. J. Syst. Sci.: Oper. Logist. 1–10 (2020). https://doi.org/10.1080/23302674.2020.1833258
30. Pal, S., Maiti, M.K., Maiti, M.: An EPQ model with price discounted promotional demand in an imprecise planning horizon via genetic algorithm. Comput. Ind. Eng. **57**(1), 181–187 (2009)
31. Park, K.: Fuzzy-set theoretic interpretation of economic order quantity. IEEE Trans. Syst. Man Cybern. **17**, 1082–1084 (1987)
32. Patro, R., Acharya, M., Nayak, M.M.: A fuzzy EOQ model for deteriorating items with imperfect quality using proportionate discount under learning effects. Int. J. Manag. Decis. Mak. **17**(2), 171–198 (2018)
33. Petrovic, D., Sweeney, E.: Fuzzy knowledge-based approach to treating uncertainty in inventory control. Comput. Integr. Manuf. Syst. **7**(3), 147–152 (1994)
34. Preston, L.E., Keachie, E.C.: Cost functions and progress functions: an integration. Am. Econ. Rev. **54**(2), 100–107 (1964)
35. Shekarian, E., Olugu, E.U., Rashid, S.H.A., Kazemi, N.: An economic order quantity model considering different holding costs for imperfect quality items subject to fuzziness and learning. J. Intell. Fuzzy Syst. **30**, 2985–2997 (2016)
36. Shekarian, E., Rashid, S.H.A., Bottani, E., De, S.K.: Fuzzy inventory models: a comprehensive review. Appl. Soft Comput. **55**, 588–621 (2017)
37. Sonia, H.N., Sarkarb, B., Joshi, M.: Demand uncertainty and learning in fuzziness in a continuous review inventory model. J. Intell. Fuzzy Syst. **33**, 2595–2608 (2017)
38. Sundararajan, R., Palanivel, M., Uthayakumar, R.: An inventory system of non-instantaneous deteriorating items with backlogging and time discounting. Int. J. Syst. Sci.: Oper. Logist. **7**(3), 233–247 (2020)
39. Tharani, S., Uthayakumar, R.: A comparative study on stock dependent production model for deteriorating items with carbon emission, rework and multiple production setups. AIP Conf. Proc. **2261**(1), 030065 (2020)
40. Wright, T.P.: Factors affecting the cost of airplanes. J. Aeronaut. Sci. **3**(4), 122–128 (1936)

A Comparison Between Fuzzy and Intuitionistic Fuzzy Optimization Technique for Profit and Production of Crops in Ariyalur District

S. Angammal and Hannah Grace. G

Abstract In agriculture, crop area planning plays an important role. A study was undertaken in Ariyalur district medium farm holder to optimally allocate land area under various crops such as paddy, groundnut, cotton, pearl millet, sweet corn, onion, brinjal, sesame in kharif, rabi and summer seasons to maximize the production and profit. In agriculture land allocation problem, the parameter such as water supply, labour requirement, fertility usage and food requirements are all uncertainty in nature. To deal with such an inexplicit information fuzzy and intuitionistic fuzzy optimization techniques has been used for crop land allocation problem and it maximize the degree of acceptance and minimize the degree of rejection. In this study, multi-objective fuzzy linear programming (MOFLP) technique provided 81,896 kg of production and Rs.16,39,300 profit with 0.712 degree of acceptance, whereas multi-objective intuitionistic fuzzy linear programming (MOIFLP) technique provided 81,989 kg of production and Rs. 16,44,100 profit with 0.7402 degree of acceptance and 0.1776 degree of rejection. By comparing the result, MOIFLP technique has given a compromised solution to this problem.

Keywords Crop planning · Fuzzy optimization technique · Intuitionistic fuzzy optimization technique · Linear programming problem

1 Introduction

Agriculture is the science and art of cultivating plants and livestock. India is an agriculture country. In India, 70–75% of population depends on agriculture. One-third of India's National income comes from agriculture. Agronomy plays an important role within the lifetime of an economy. It is the pillar of our financial system. Agriculture

---

S. Angammal · H. Grace. G (✉)
Vellore Institute of Technology, Chennai, India
e-mail: hannahgrace.g@vit.ac.in

S. Angammal
e-mail: angammal.s2019@vitstudent.ac.in

P. Balasubramaniam et al. (eds.), *Mathematical Modelling and Computational Intelligence Techniques*, Springer Proceedings in Mathematics & Statistics 376,
https://doi.org/10.1007/978-981-16-6018-4_21

not only grants food and staple but also provides the platform to employees in the large quantity of population. Farming activities are crucial from each sociable and monetary aspect. In our country, land and water are consider to be the important and essential assets but due to indiscriminate exploitation these assets are reducing gently. Geographical situation of India is specific for farming as it gives numerous favourable weather situations such as long growing season, simple area, fertile soil and extensive variant in weather situation. Apart from these situations India has been constantly making progressive efforts by using technological know—how and technology to increase production. In agriculture crop planning, the most commonly used objective functions are maximization of net revenue, profit, employment, overall contribution of agriculture sector or minimization of input cost, water usage, erosion and natural resources. From farm to regional level, stakeholders are confronted with various limited resources and multiple options to achieve the desired benefits. Thus, researchers emphasized that agricultural planning issues involve multiple conflicting goals such as overall profit maximization, crop production maximization, labour expenditure minimization, water use minimization and other input costs without compromising sustainability of natural resources [19, 20]. In agricultural production planning problem, the aim is to increase the gain with the lowest investment under the fixed conditions concerning categories of land, labour, available capital, etc. If the farmer wants to improve their profit and product within the availability of land and water and also under many interferences related to agriculture, they have to know the technique of optimum allocation of agriculture land to the different types of crops [1]. For an allotment of species and pursuit to region in farming, the land use allocations are specially used [2]. Optimization approaches can be broadly put into linear and nonlinear categories. Linear programming (LP) has been widely applied in agriculture since 1950. Generally, LPP is used for the trouble of crop making plans, but many undetermined elements occur in agriculture manufacturing plans. Because of such highlights of yield creation issue the crisp multi-objective strategies are not reasonable for building up the cropping models. Therefore, fuzzy multi-objective techniques are used for crop production problems. To determine the maximum yield of various types of crops, fuzzy optimization techniques are used [3, 4]. The gain or drop of particular crop is usually relying on the fluctuation of demand and cost. There is no certainty in agricultural process. An intuitionistic fuzzy optimization technique is the effective method to increase the profit for small farm holder [5]. Ram Karan Singh et al. [6] used Fuzzy Topsis method for sustainable cash crop harvesting and the selection of cash crops are based on different criteria. This method used to derive the crop selection and comparison with other MCDM techniques include AHP, also the derived weights are depending on the criteria comparison. If there is no satisfactory vegetable cash crop manufacturing scheme, then it will be critical to harvest the cash crop with maximum profit. Therefore, in agriculture, Topsis methodology will be a support to stakeholders. By using MCA (Multi-objective Constrained Algorithm) and NSGA-II, Sams Jain et al. [24] solve the single-cropped land optimization problem with the two objective functions, maximization of total profit and minimization of expenditure and the constraints namely land, capital, area and import bound constraints. The goal of sustainably growing crop production will no longer be executed without improving, as properly as, maintaining soil fertility too. To maximize crop yields and sustain soil fertility, Genetic algorithm is used. To determine the optimum cropping pattern with optimal water usage for small farm Dutta et al. [25] had used fuzzy stochastic simulation based genetic algorithm.

The model of fuzzy goal programming is effectively used for optimal production of numerous seasonal crops. Inequality and unpredictability in choice making problem may be largely used in fuzzy-based methods. The important goal of the production management is to maximize their gain and satisfy the customer's demand. Fuzzy optimization technique is used to satisfy the decision maker's aspiration level [7]. For the planning of land allocation Babita Mishra et al. [8] used fuzzy multi-fractional programming technique and this study state that the ratio type optimization of the set of goals is better than their individual optimization. Atanassov [15] has discussed some basic ideas of intuitionistic fuzzy set that helps to solve the optimization problem. To obtain the Pareto optimal solution of multi-objective optimization problem in an intuitionistic environment Razmi et al. [12] have proposed two-step goal programming technique. Pawar et al. [13] have used intuitionistic fuzzy optimization technique in an irrigation system and the result was compared with multi-objective fuzzy optimization technique. Sankar Kumar Roy et al. [14] have used intuitionistic fuzzy programming and goal programming to minimize the transportation cost and compare the result, and it was clear that intuitionistic programming was superior to the result of goal programming. Several authors have used Intuitionistic fuzzy optimization techniques to solve the problem which comes under the uncertainty environment, and it helps to determine the compromising optimal solution [16–18]. To maximize the production and minimize the expenditure with land, labour, water, machine hour and fertility requirement constraints. Biswas and Modak [23] had used fuzzy multi-objective chance constrained programming model with the help of fuzzy goal programming technique. The fuzzy-based multi-criteria method is the best tool for farmers and agricultural policymakers. It helps to prepare the complete coverage for reliable farming system that is an international ongoing essential need for reliable farming system.

2 Study Area and Crops

As per census in 2011, Ariyalur district has a population of 7, 54,894 out of which 3, 74,703 are male and 3, 80,191 are female. In 2019, (estimates as per aadhar uidai.gov.in Feb 2019 data) it has a population of 7,97,677 amongst this 4,80,604 are literate people, out of which 2,69,582 are male and 2,11,022 are female. There are 3, 59,851 workers are in Ariyalur district amongst this 2, 12,547 are men and 1, 47,304 are women. The total number of cultivators in this district are 94,912 amongst this the number of men and women are 63,868 and 31,044. The agriculture labour in this district is 1, 06,252 with 56,334 men and 49,918 women. The sex ratio in this district is 1015 female per 1000 male.70% of the people in Ariyalur district are engaged with agricultural activity. Agriculture plays a major role in this district economy, amongst

1933.38 km^2 area of Ariyalur district 1.118 lakhs Ha are the gross cropped area. The principal goal of the division of agribusiness in this locale is to guarantee the strength of farming creation and furthermore to improve the yield creation in a possible way to accomplish the prerequisite of nourishment for the expanding populace, likewise it assists with giving the crude material to horticulture-based businesses. With this, the division is useful in giving the working occasions to the country individuals. In spite of the fact that this area gets precipitation in all the season yet the significant precipitation got in North East Monsoon and the mean yearly precipitation in this locale is 954 mm.

The soil texture in this district is loamy and the shading contrasting from red at the outward to yellow at the base. The soils are of moderate profundity with dependable water, additionally it is liberated from the conglomeration of calcium carbonate and salt and the soil pH varying from 6.5 to 8 and it consist of small quantity of organic matter, Phosphorus and Nitrogen but normally it has sufficient quantity of lime and potash. The major sources which are used for irrigation in this district are tank, canal, tube well and open well, amongst this bore well and tube well are used to irrigate the major areas. Since a large portion of the individuals in this region do farming and they produce various kinds of yields. Paddy, cotton, ground nut, maize and sugarcane are the significant yields developed in this region. The study deals with eight crops: paddy, cotton, ground nut, pearl millet, sesame, sweet corn, onion and brinjal. The maximum harvesting time for brinjal, groundnut, sesame and onion are 3 months. Pearl millet needs 70 days, sweet corn and paddy needs 4 months and cotton needs 6 months for harvesting. In one acre, it is possible to harvest at least 12500 kg of brinjal, 15–20 bundles of groundnut each bundle contains 80 kg of groundnut, 12–15 bundles of pearl millet each bundle contains 100 kg, 3–4 bundles of sesame each bundle contains 80 kg, and 8 tonnes of onion. Figure 1, help us to analyse the land holding pattern of Ariyalur district.

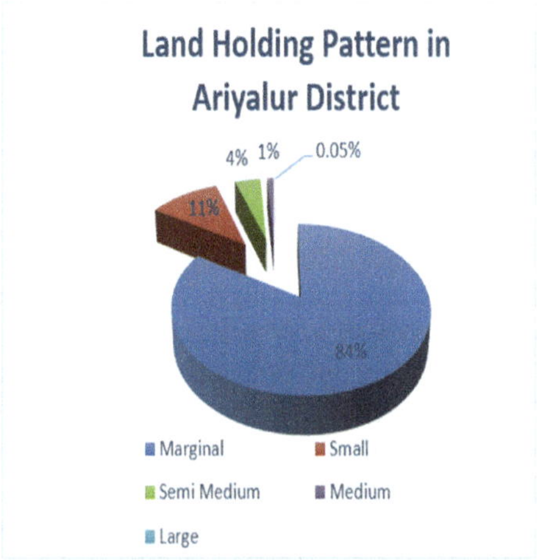

Fig. 1 Land holding pattern in Ariyalur District *Source* tn.data.gov.in

3 Fuzzy Optimization Technique

Multi-target enhancement has been applied in numerous fields of science, including designing, financial aspects and transportation where ideal choices should be taken within the sight of compromises between at least two clashing goals. If the objective and constraints of the multi-objective optimization problem are approximate then fuzzy optimization technique is considered as the best tool to find the compromise solution to the problem. Generally, LPP is used for the difficulty of yield making arrangements; however, numerous unsure components happen in agriculture manufacturing plans. Because of such highlights of yield creation issue the crisp multi-objective strategies are not reasonable for building up the cropping models. Thus, the concept of fuzzy set used by Zimmermann was given by Zadeh with fuzzy programming and LPP with multi objectives [9, 10]. The multi-objective decision-making (MODM) problem is defined as follows:

$$\text{Max} \quad Z_r(y) = \sum_{i=1}^{n} C_{ri}\, y_i$$

$$\text{Subject to} \quad By_i \leq b \;\text{ and }\; y_i \geq 0 \forall y_i \in \mathrm{Y} \tag{1}$$

where $\mathrm{Y} = y_1, y_2, \ldots, y_n$, y_i is an n-dimensional vector of decision variable and $Z_1(y), Z_2(y) \ldots Z_r(y)$ are r different linear objective function of the decision variable y. $C_{ri} = (C_{r1}, C_{r2}, \ldots C_{rn})$ is the cost factor vector, the constraint matrix is mentioned by B with the order $m \times n$ and b is m dimensional resource vector.

$$\text{To find } y$$

$$\text{Such that} \quad Z_r(y) \geq Z_r\left(y'\right) \forall r$$

Here, $Z_r\left(y'\right)$ is the analogous goal and all objective functions are to be maximized. Here, Eq. (1) objective functions are turns into fuzzy constraints. If the tolerance of fuzzy constraints is given then a set of feasible solution is characterized by its membership functions.

$$\mu_k(y) = \text{Min}\{\mu_1(y), \mu_2(y) \ldots \mu_r(y)\}$$

Now the solution can be received by the way of solving the problem of $\text{Max}\{\mu_k(y)\}$ such that $y \in Y$.

$$\text{(i.e.)} \quad \text{Max}\{\text{Min}_r \mu_r(y)\} \forall y \in Y.$$

Let $\lambda = \text{Min}_r \mu_r(y)$ be the overall satisfactory stage of compromise then the equivalent model is as follows:

$$\text{Max } \lambda$$

$$\text{Subject to } \lambda \le \mu_r(y) \forall r, y \in Y.$$

Here, membership functions of objective characteristic are predicted by obtaining pay off matrix of positive ideal solution (PIS) and anticipate that the membership functions are of the sort of non-decreasing linear or hyperbolic, etc.

3.1 Computational Algorithm

Step 1: Find the solution of each objective function one by one with the given constraints.
Step 2: From each solution of step 1 find the solution of other objective functions.
Step 3: Find the lower and upper bound of the objective function and consider the membership function between them. Let it be

$$\mu_k(\mathrm{y}) = \begin{cases} 1 & if\ Z_r(\mathrm{y}) = T_k^B \\ \frac{Z_r(\mathrm{y}) - T_k^A}{T_k^B - T_k^A} & if\ T_k^A \le T_k(\mathrm{y}) \le T_k^B \\ 0 & if\ T_k(\mathrm{y}) < T_k^A \end{cases}$$

where $T_k^A \& T_k^B$ are lower and upper bound of the PIS.
Step 4: Convert the MOLPP into LPP.

$$\text{Max } \lambda$$

$$\text{Such that } \mu_k(\mathrm{y}) = \frac{Z_r(\mathrm{y}) - T_k^A}{T_k^B - T_k^A} \ge \lambda$$

$$By_i \le b \text{ and } y_i \ge 0\ \forall y_i \in Y.$$

3.2 Notations

$Z_r(y)$	Objective function.
C_r	Benefit of ith crop.
B	Constraint matrix.
b	Resource vector.
y_i	ith crop area in acre.
y	Decision variable.
α, β	Scaling factor.

T_k^A	Lower bound of PIS.
T_k^B	Upper bound of PIS.
$T_k^{B\mu}$	Upper bound of objective's membership function.
$T_k^{A\mu}$	Lower bound of objective's membership function.
$T_k^{B\gamma}$	Upper bound of objective's non-membership function.
$T_k^{A\gamma}$	Lower bound of objective's non-membership function.
M_k^0	Aspiration level of the objective function.
T_i^a	Tolerance of constraints membership function.
T_i^r	Tolerance of constraints non-membership function.
G	Decision set.
μ_k (y)	Degree of acceptance of decision set.
γ_k (y)	Degree of rejection of decision set.
λ	Degree of minimum acceptance of decision set.
δ	Degree of maximum rejection of decision set.
r	Number of objectives.
n	Number of constraints.

4 Intuitionistic Fuzzy Optimization Technique

Definition 1 Let S be the universal set. An intuitionistic fuzzy set A is defined as $A = \{y, \mu_A(y), \gamma_A(y)\}$, where $\mu_A(y)$: $S \to [0, 1]$ and $\gamma_A(y)$: $S \to [0, 1]$ are the degree of membership and non-membership function and for every $y \in$ S, $0 \leq \mu_A(y) \leq \gamma_A(y)$ [26].

Definition 2 The point $x^* \in X$ is said to be Pareto optimal solution if there does not exist $x \in X$ such that $Z_i(x) \geq Z_i(x^*)$ $\forall$ i and $Z_j(x) \geq Z_j(x^*)$ for at least one j [27].

Definition 3 If P and Q are two intuitionistic fuzzy set of S then.

(i) $P \subset Q \Leftrightarrow \mu_P(y) \leq \mu_Q(y)$ and $\gamma_P(y) \geq \gamma_Q(y) \forall y \in S$.
(ii) $P \subset Q$ and $Q \subset P \Leftrightarrow P = Q$.
(iii) $P \cap Q = \{(y, \min(\mu_P(y), \mu_Q(y)), \max(\gamma_P(y), \gamma_Q(y))/y \in S\}$
(iv) $P \cup Q = \{(y, \max(\mu_P(y), \mu_Q(y)), \min(\gamma_P(y), \gamma_Q(y))/y \in S\}$ [26].

Intuitionistic fuzzy optimization technique is an extension of fuzzy optimization technique. In a fuzzy set, the degree of acceptance is the complement of degree of rejection, but it is not true in intuitionistic fuzzy set. In intuitionistic fuzzy set, both degree of membership and non-membership functions are defined separately as in Fig. 2. Thus, the intuitionistic model for Eq. (2) is classified as follows:

$$\max_D\{\mu_k(y)\}, k = 1, 2 \ldots r + n.$$

$$\min_D\{\gamma_k(y)\}, k = 1, 2 \ldots r + n.$$

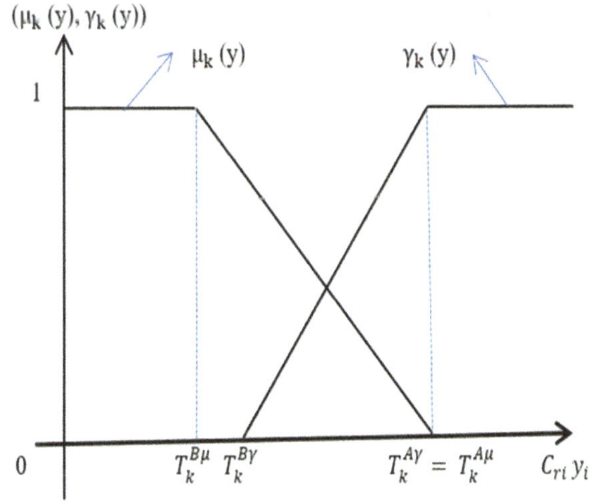

Fig. 2 Membership and non-membership function of objective function

$$\text{Such that} \quad \mu_k(y) \geq \gamma_k(y)$$

$$\gamma_k(y) \geq 0$$

$$\mu_k(y) + \gamma_k(y) \leq 1, k = 1, 2 \ldots r + n.$$

Here, $\mu_k(y)$ and $\gamma_k(y)$ are the degree of acceptance and rejection, respectively.

Now, the intersection of integrated intuitionistic fuzzy objectives and constraints will make the decision set G.

$$\begin{aligned} G &= Z \cap C \\ &= \{(y, \min(\mu_Z(y), \mu_C(y)), \max(\gamma_Z(y), \gamma_C(y)))\} \\ G &= \{y, \lambda, \delta / y \epsilon Y\} \\ \lambda &= \min(\mu_Z(y), \mu_C(y)) = \min_{k=1}^{r+n} \mu_k(y) \\ \delta &= \max(\gamma_Z(y), \gamma_C(y)) = \max_{k=1}^{r+n} \gamma_k(y) \end{aligned}$$

where λ and δ denote the degree of minimum acceptance and maximum rejection of the decision set. Also

$$\lambda \leq \mu_k(y)$$

$$\delta \geq \gamma_k(y) \forall k = 1, 2 \ldots r + n.$$

Thus, the above system can be changes as the following format

$$\mu_k(y) \geq \lambda$$
$$\gamma_k(y) \leq \delta$$
$$\lambda + \delta \leq 1$$
$$\lambda \geq \delta$$
$$\beta \geq 0 \quad \forall y \in Y, k = 1, 2 \ldots r + n.$$

By using intuitionistic fuzzy optimization technique problem (1) can be changed into the following linear programming problem.

$$\text{Max}(\lambda - \delta)$$

Subject to the constraints $\mu_k(y) \geq \lambda$

$$\gamma_k(y) \leq \delta$$
$$\lambda + \delta \leq 1$$
$$\lambda \geq \delta$$
$$\delta \geq 0 \quad \forall y \in Y, k = 1, 2 \ldots r + n.$$

This LPP problem can be easily solved by using MATLAB coding.

4.1 Algorithm

Intuitionistic fuzzy optimization problem is utilized to discover the Pareto optimal solution of the optimization problem that increases the degree of acceptance and decrease the degree of rejection simultaneously. Angelov [11] developed a programming model to find an effective solution of the multi-objective optimization problem in an intuitionistic fuzzy condition. The following algorithm is developed by using Angelov method.

Step1: Solve the multi-objective optimization problem as a single-objective optimization problem.
Step 2: By utilizing the solution of step 1 find the solution of remaining objective function.
Step 3: From step 1 and step 2 find all the Positive Ideal Solution (PIS) and from the PIS determine the lower and upper bounds for each objective function.
Step 4: Create a lower and upper bound for membership and non-membership function

$$T_k^{B\mu} = \text{Max}(Z_r(y))$$

$$T_k^{A\mu} = \text{Min}(Z_r(y))$$
$$T_k^{B\gamma} = T_k^{B\mu} - \alpha\left(T_k^{B\mu} - T_k^{A\mu}\right)$$
$$T_k^{A\gamma} = T_k^{A\mu}, 0 < \alpha < 1, \text{ in this problem consider } \alpha = 0.1.$$

Step 5: Convert the problem (1) is as follows:

$$\text{Find } y$$
$$Z_r(y) \gtrsim M_k^0 \forall k$$

where M_k^0 stands for the aspiration level for the objective function which is given by the decision maker. In this problem, the maximum value in the PIS is consider as the tolerance limit of the objective function.
Step 6: Create a membership and non-membership function of the intuitionistic fuzzy objectives & constraints.

Objective function (Maximization)

$$\mu_k(Z_r(y)) = \begin{cases} 1 & Z_r(y) \geq T_k^{B\mu} \\ \frac{Z_r(y) - T_k^{A\mu}}{T_k^{B\mu} - T_k^{A\mu}} & T_k^{A\mu} < Z_r(y) < T_k^{B\mu} \\ 0 & Z_r(y) \leq T_k^{A\mu} \end{cases}$$

$$\gamma_k(Z_r(y)) = \begin{cases} 0 & Z_r(y) \geq T_k^{B\gamma} \\ \frac{T_k^{B\gamma} - Z_r(y)}{T_k^{B\gamma} - T_k^{A\gamma}} & T_k^{A\gamma} < Z_r(y) < T_k^{B\gamma} \\ 1 & Z_r(y) \leq T_k^{A\gamma} \end{cases}$$

Objective function (Minimization)

$$\mu_k(Z_r(y)) = \begin{cases} 0 & Z_r(y) \geq T_k^{B\mu} \\ \frac{T_k^{B\mu} - Z_r(y)}{T_k^{B\mu} - T_k^{A\mu}} & T_k^{A\mu} < Z_r(y) < T_k^{B\mu} \\ 1 & Z_r(y) \leq T_k^{A\mu} \end{cases}$$

$$\gamma_k(Z_r(y)) = \begin{cases} 1 & Z_r(y) \geq T_k^{B\gamma} \\ \frac{Z_r(y) - T_k^{A\gamma}}{T_k^{B\gamma} - T_k^{A\gamma}} & T_k^{A\gamma} < Z_r(y) < T_k^{B\gamma} \\ 0 & Z_r(y) \leq T_k^{A\gamma} \end{cases}$$

Constraints ($\leq$ int)

$$\mu_i(\mathrm{B}(y_i)) = \begin{cases} 1 & \mathrm{B}(y_i) \le b_i \\ \frac{T_i^a + b_i - \mathrm{B}(y_i)}{T_i^a} & b_i < \mathrm{B}(y_i) < b_i + T_i^a \\ 0 & \mathrm{B}(y_i) \ge b_i + T_i^a \end{cases}$$

$$\gamma_i(\mathrm{B}(y_i)) = \begin{cases} 0 & \mathrm{B}(y_i) \le b_i + T_i^r \\ \frac{\mathrm{B}(y_i) - (b_i + T_i^r)}{T_i^a - T_i^r} & b_i + T_i^r < \mathrm{B}(y_i) < b_i + T_i^a \\ 1 & \mathrm{B}(y_i) \ge b_i + T_i^a \end{cases}$$

Constraints ($\ge$ int)

$$\mu_i(\mathrm{B}(y_i)) = \begin{cases} 1 & \mathrm{B}(y_i) \ge b_i \\ \frac{\mathrm{B}(y_i) - (b_i - T_i^a)}{T_i^a} & b_i - T_i^a < \mathrm{B}(y_i) < b_i \\ 0 & \mathrm{B}(y_i) \le b_i - T_i^a \end{cases}$$

$$\gamma_i(\mathrm{B}(y_i)) = \begin{cases} 0 & \mathrm{B}(y_i) \ge b_i - T_i^r \\ \frac{(b_i - T_i^r) - \mathrm{B}(y_i)}{T_i^a - T_i^r} & b_i - T_i^a < \mathrm{B}(y_i) < b_i - T_i^r \\ 1 & \mathrm{B}(y_i) \le b_i + T_i^a \end{cases}$$

where T_i^a and T_i^r are the tolerance of membership and non-membership function, the tolerance of the non-membership function is defined as $T_i^r = \beta T_i^a, 0 < \beta < 1$, also consider $\beta = 0.2$ in this problem.

Step 7: By using intuitionistic fuzzy optimization technique, the multi-objective optimization problem is converted into the following single LPP

$$\mathrm{Max}(\lambda - \delta)$$

$$\begin{aligned} \text{Such that} \quad & \mu_k(y) \ge \lambda \\ & \gamma_k(y) \le \delta \\ & \lambda + \delta \le 1 \\ & \lambda \ge \delta \\ & \delta \ge 0 \quad \forall y \in Y, k = 1, 2 \ldots r + n. \end{aligned}$$

5 Problem Illustration

Generally, the farmers are categorized as marginal (< 1 hectare), small (1–2 hectare), semi medium (2–4 hectare), medium (4–10 hectare) and large (10 and above) farm holders. In Ariyalur district, 1% of farmers comes under medium farm holder which is less than the percentage of marginal, small and semi medium farm holder. To improve the production and profit of medium farm holder, MOFLP and MOIFLP

Table 1 Crops, yield, return and labour per acre

Season	Crops	Yield (kg/ac)	Returns (1000 of Rs/ac)	Labour/ac
Kharif	Paddy (y_1)	1725	15.525	68
	Groundnut (y_2)	1324	43.692	56
	Cotton (y_3)	1084	57.225	258
Rabi	Paddy (y_4)	1548	13.158	85
	Groundnut (y_5)	1360	41.25	64
	Pearl millet (y_6)	1324	19.198	48
	Sweet corn (y_7)	2478	29.736	75
Summer	Paddy (y_8)	1742	15.678	76
	Onion (y_9)	8000	136	72
	Brinjal (y_{10})	12,500	250	48
	Sesame (y_{11})	240	12	30

techniques are used in this study. The data of this study was collected from Ariyalur district medium farm holder who has a land of 15 acres and they cultivate paddy, groundnut, cotton, pearl millet, sweet corn, onion, brinjal, sesame in kharif, rabi and summer seasons. The important goal of this problem is to increase the production and profit with land, labour, water and food requirement constraints. The farmer needs at least 1548 kg of paddy and 230 kg of pearl millet per annum and there are 260-man days available in each season. Here the constraints which are labour, water and food requirements are comes under fuzzy environment so that the tolerance level of labour, water and food requirement constraints are consider as 2, 5, 3, respectively. Table 1 indicates the type of crops, yield, return and labour per acre in each season.

By using Table 1, the problem can be illustrating mathematically as follows:

$$\begin{aligned}\text{Max} z_1 &= 15.525y_1 + 43.692y_2 + 57.225y_3 + 13.158y_4 + 41.25y_5 \\ &\quad + 19.198y_6 + 29.736y_7 + 15.678y_8 + 136y_9 + 250y_{10} + 12y_{11} \\ \text{Max} z_2 &= 1725y_1 + 1324y_2 + 1084y_3 + 1548y_4 + 1360y_5 \\ &\quad + 1324y_6 + 2478y_7 + 1742y_8 + 8000y_9 + 12500y_{10} + 240y_{11}\end{aligned}$$

Land Constraints

$$\begin{aligned}&y_1 + y_2 + y_3 \le 15 \\ &y_4 + y_5 + y_6 + y_7 \le 15 \\ &y_8 + y_9 + y_{10} + y_{11} \le 15\end{aligned}$$

Labour Constraints

$$\begin{aligned}
&68y_1 + 56y_2 + 258y_3 \le 260 \\
&85y_4 + 64y_5 + 48y_6 + 75y_7 \le 260 \\
&76y_8 + 72y_9 + 48y_{10} + 30y_{11} \le 260
\end{aligned} \quad (2)$$

Water Constraints

$$\begin{aligned}
&y_1 + y_2 + y_3 \le 258.75 \\
&y_4 + y_5 + y_6 + y_7 \le 334.75 \\
&y_8 + y_9 + y_{10} + y_{11} \le 345.5
\end{aligned}$$

Food Requirement Constraints

$$\begin{aligned}
&1725y_1 + 1548y_4 + 1742y_8 \ge 1548 \\
&1324y_6 \ge 230
\end{aligned}$$

$$y_1, y_2, y_3, y_4, y_5, y_6, y_7, y_8, y_9, y_{10}, y_{11} \ge 0$$

The PIS of problem (2) is tabulated in Table 2.

By using step 4 of fuzzy optimization algorithm, the multi-objective optimization problem (2) can be changed into the following single LPP

$$\text{Max } \lambda$$

Such that

$$\begin{aligned}
&15.525y_1 + 43.692y_2 + 57.225y_3 + 13.158y_4 + 41.25y_5 + 19.198y_6 \\
&\quad + 29.736y_7 + 15.678y_8 + 136y_9 + 250y_{10} + 12y_{11} - 172.3\lambda \ge 1516.6 \\
&1725y_1 + 1324y_2 + 1084y_3 + 1548y_4 + 1360y_5 + 1324y_6 + 2478y_7 \\
&\quad + 1742y_8 + 8000y_9 + 12500y_{10} + 240y_{11} - 3310\lambda \ge 79539 \\
&y_1 + y_2 + y_3 \le 15 \\
&y_4 + y_5 + y_6 + y_7 \le 15 \\
&y_8 + y_9 + y_{10} + y_{11} \le 15
\end{aligned}$$

Table 2 Positive ideal solution

	$MaxZ_1$	$MaxZ_2$	Max–Min
Z_1	1688.9	1516.6	172.3
Z_2	79,539	82,849	3310

Table 3 MOFLP solution

y_1	0.8974
y_2	3.5532
y_5	0.8081
y_6	0.1737
y_7	2.6659
y_{10}	5.4167
λ	0.712
$\text{Max}Z_1$	1639.3
$\text{Max}Z_2$	81,896

$$68y_1 + 56y_2 + 258y_3 \leq 26$$
$$85y_4 + 64y_5 + 48y_6 + 75y_7 \leq 260$$
$$76y_8 + 72y_9 + 48y_{10} + 30y_{11} \leq 260$$
$$y_1 + y_2 + y_3 \leq 258.75$$
$$y_4 + y_5 + y_6 + y_7 \leq 334.75$$
$$y_8 + y_9 + y_{10} + y_{11} \leq 345.5$$
$$1725y_1 + 1548y_4 + 1742y_8 \geq 1548$$
$$1324y_6 \geq 230$$
$$y_1, y_2, y_3, y_4, y_5, y_6, y_7, y_8, y_9, y_{10}, y_{11} \geq 0$$

The solution of the above problem is tabulated as in Table 3.

By using Step 5 of intuitionistic fuzzy optimization algorithm problem (2) has change as

$$\text{Find } y$$

$$\text{Such that}$$

$$15.525y_1 + 43.692y_2 + 57.225y_3 + 13.158y_4 + 41.25y_5 + 19.198y_6 + 29.736y_7 + 15.678y_8 + 136y_9 + 250y_{10} + 12y_{11} \geq 1688.9$$
$$1725y_1 + 1324y_2 + 1084y_3 + 1548y_4 + 1360y_5 + 1324y_6 + 2478y_7 + 1742y_8 + 8000y_9 + 12500y_{10} + 240y_{11} \geq 82849$$
$$y_1 + y_2 + y_3 \leq 15$$
$$y_4 + y_5 + y_6 + y_7 \leq 15$$
$$y_8 + y_9 + y_{10} + y_{11} \leq 15$$
$$68y_1 + 56y_2 + 258y_3 \leq_{\text{int}} 260$$
$$85y_4 + 64y_5 + 48y_6 + 75y_7 \leq_{\text{int}} 260$$

Table 4 MOIFLP solution

y_1	0.8969
y_2	3.563
y_5	0.9032
y_6	0.1731
y_7	2.592
y_{10}	5.4275
λ	0.7402
δ	0.1776
$\text{Max}Z_1$	1644.1
$\text{Max}Z_2$	81,989

$$76y_8 + 72y_9 + 48y_{10} + 30y_{11} \leq_{\text{int}} 260$$
$$y_1 + y_2 + y_3 \leq_{\text{int}} 258.75$$
$$y_4 + y_5 + y_6 + y_7 \leq_{\text{int}} 334.75$$
$$y_8 + y_9 + y_{10} + y_{11} \leq_{\text{int}} 345.5$$
$$1725y_1 + 1548y_4 + 1742y_8 \geq_{\text{int}} 1548$$
$$1324y_6 \geq_{\text{int}} 230$$
$$y_1, y_2, y_3, y_4, y_5, y_6, y_7, y_8, y_9, y_{10}, y_{11} \geq 0$$

By using step 6 and step 7 of intuitionistic fuzzy optimization algorithm, the above problem is changed into single-objective optimization problem, and the solution of crop planning problem obtained from this step is tabulated in Table 4.

6 Conclusion

Crop planning problem has received considerable attention from the operation management and agricultural economic literature which is associated with the issues of risk and uncertainty during any process. To overcome this unpredictability and to maximize the yield and returns of the medium farm holder in Ariyalur district, fuzzy optimization and fuzzy intuitionistic optimization techniques are used, and the solutions are tabulated in the third and fourth table. The farmer has got the return of Rs.16, 39,300 with the yield of 81, 896 kg by using fuzzy optimization technique, whilst using fuzzy intuitionistic optimization technique the returns and yield are improved to Rs. 16, 44,100 and 81, 989 kg. Further comparing both the results, it is clear from Fig. 3 that fuzzy intuitionistic optimization technique is superior to the fuzzy optimization technique. In this study, an intuitionistic fuzzy optimization technique deals with the degree of acceptance and degree of rejection only. Garai and Roy [21] had used the concept of hesitation index which is the extension of Angelov method. In a decision-making process, if the degree of hesitation is maximum then it will be difficult to find the best solution to the problem. Hence, it is important to minimize the

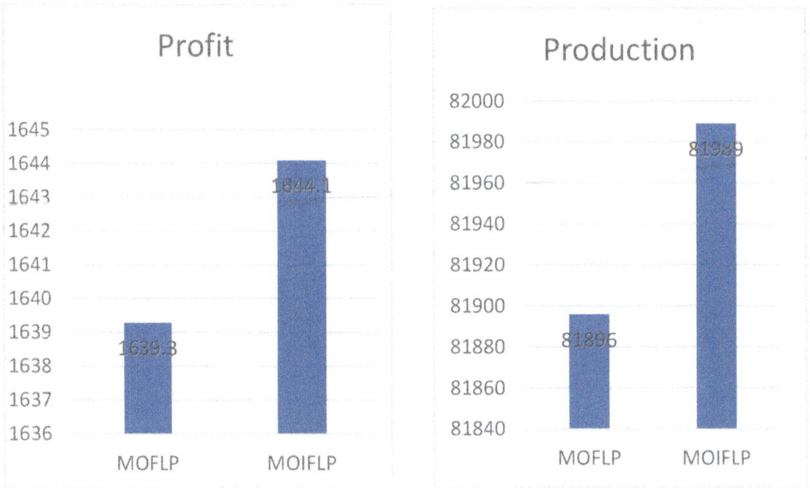

Fig. 3 Comparative analysis with MOFLP & MOIFLP

degree of hesitation, and it will be the future work. Olofintoye et al. [22] established the new evolutionary algorithm named as Combined Pareto multi-objective differential evolution (CPMDE). CPMDE is the good alternative for solving constrained and unconstrained multi-objective optimization problems. Multi-objective differential evolution algorithm, Multi-objective constrained algorithm and NSGA-II are also used in land allocation problem, and it is the key finding for forgoing study.

Acknowledgements The authors are thankful to Mr. Sivakumar and family for active support to supply the agriculture data to implement the model.

References

1. Manibhushan, Upadhya, A., Singh, A.K., Batta, R.K., Singh, A.S., Rahman, A., Kumar, S.: Decision support system for design, layout and cost estimation pressurized irrigation system. Int. J. Agric. Stat. Sci. **10**(1), 231–236 (2014)
2. Menmah, M.M., Lescourret, F., Yao, X., Lavigne, C.: Metaheuristics for agricultural land use optimization. Agron. Sustain. Dev **35**(3), 975–998 (2015)
3. Ponnalagu, K., Madhumathi, D., Pavithra, A.: Fuzzy optimization approaches to determine the maximum yield of different variety of crops. Int. J. Sci. Technol. **29**, 388–399 (2020)
4. Thomas Felix, K., Murali Gopal, S., Senthilnathan, S.: Crop production planning for different group of farmers using fuzzy multi objective linear programme. Int. Res. J. Agric. Econ. Stat. **8**, 287–292 (2017)
5. Bharati, S.K., Singh, S.R.: Intuitionistic fuzzy optimization technique in agricultural production planning: a small farm holder perspective. Int. J. Comput. Appl. **89**(6), 17–23 (2014)
6. Singh, R.K., Mallick, J.: Fuzzy based multi-criteria method for sustainable green chamber farming practices. Indian J. Agric. Sci. **89**(10), 1732–1736 (2019)

7. Kavitha, C., Vijayalakshmi, C.: Design and implementation of fuzzy multi objective optimization model for production planning. Indian J. Appl. Res. **3**(12), 372–375 (2011)
8. Mishra, B., Nishad, A.K., Singh, S.R.: Fuzzy multi-fractional programming for land use planning in agriculture production system. Fuzzy Inf. Eng. **6**, 245–262 (2014)
9. Zimmermann, H.J.: Fuzzy programming and linear programming with several objective function. Fuzzy Sets Syst. **1**, 45–55 (1978)
10. Zadeh, L.A.: Fuzzy sets. Inf. Control **8**, 338–353 (1965)
11. Angelov, P.P.: Optimization in an intuitionistic fuzzy environment. Fuzzy Sets Syst. **86**(3), 299–306 (1997)
12. Razmi, J., Jafarian, E., Amin, S.H.: An intuitionistic fuzzy goal programming approach for finding pareto-optimal solutions to multi-objective programming problems. Expert Syst. Appl. **65**,181–193 (2016)
13. Pawar, S.V., Patel, P.L., Mirajkar, A.B.: Intuitionistic fuzzy approach in multi-objective optimization for KRBMC irrigation system, India. ISH J. Hydraulic Eng., 1781700 (2020)
14. Roy, S.K., Ebrahimnejad, A., Verdegay, J.V., Das, S.: New approach for solving intuitionistic fuzzy multi-objective transportation problem. Sadhana **43**(3) (2018)
15. Atanassaov, K.T.: Remarks on the intuitionistic fuzzy set–III. Fuzzy Sets syst. **75**, 401–402 (1995)
16. Jana, B, Tapan Kumar Roy.: Multi-objective fuzzy linear programming and its application in transportation problem. Notes on Intuitionistic Fuzzy Sets **13**(1), 34–51 (2007)
17. Chakraborty, D., Jana, D.K.., Roy, T.K.: A new approach to solve intuitionistic fuzzy optimization problem using possibility, necessity and creadibility measures. Int. J. Eng. Math., Article ID 593185 (2014)
18. Bharati, S.K., Nishad, A.K., Singh, S.R.: Solution of multi objective linear programming problems in intuitionistic fuzzy environment. Adv. Intell. Syst. Comput. **236**, 161–171 (2014)
19. Soltani, J., Karbasi, A.R., Fahimifard, S.M.: Determining optimum cropping pattern using fuzzy goal programming model. Afr. J. Agric. Res. **6**(14), 3305–3310 (2011)
20. Mirkarimi, S.H., Joolaie, R., Eshraghi, F., Abadi, F.S.B.: Application of fuzzy goal programming in cropping pattern management of selected crops in Mazandaran province (Case study Amol township). Int. J. Agric. Crop Sci. **6**(15), 1062–1067 (2013)
21. Garai, A., Roy, T.K.: Intuitionistic fuzzy optimization: usage of hesitation index. Int. J. Comput. Technol. **10**(4),1489–1495 (2013)
22. Olofintoye, Adeyemo, J., Otieno, F.: A Combined Pareto Differential Evolution Approaches for Multi-objective Optimization, pp. 213–231. Springer International Publishing, Switzerland (2014)
23. Biswas, A., Modak, N.: A multiobjective fuzzy chance constrained programming model for land allocation in agricultural sector: a case study. Int. J. Comput. Intell. Syst. **10**(1), 196–211 (2017)
24. Jarin, S., Khatun, M.K., Shafie, A.A.: Multi-objective constrained algorithm (MCA) and non-dominated sorting genetic algorithm (NSGA-II) for solving multi-objective crop planning problem. ARPN J. Eng. Appl. Sci. **11**(6), 4079–4086 (2016)
25. Dutta, S., Sahoo, B.C., Mishra, R., Acharya, S.: Fuzzy stochastic genetic algorithm for obtaining optimum crops pattern and water balance in a farm. Water Resour. Manage. **30**, 4097–4123 (2016)
26. Atanassov, K.T: Intuitionistic Fuzzy Set, Theory and Application. Springer-Verlag Berlin Heidelberg GmbH, Bulgaria (1999)
27. Liu, G.P., Yang, J.B., Whidborne, J.F.: Multi Objective Optimization and Control. Prentice-Hall of India Private Limited, New Delhi (2008)

The manufacturer's authorised representative in the EU is Springer Nature Customer Service Centre GmbH, Europaplatz 3, 69115 Heidelberg, Germany. If you have any concerns regarding our products, please contact ProductSafety@springernature.com

Printed and bound by CPI Group (UK) Ltd, Croydon, CR0 4YY
30/09/2026
02225293-0011